Phonon Scattering
in Solids

Phonon Scattering in Solids

Edited by

L. J. Challis, V. W. Rampton, and A. F. G. Wyatt
University of Nottingham

Springer Science+Business Media, LLC

Library of Congress Cataloging in Publication Data

International Conference on Phonon Scattering in Solids, 2d, University of Nottingham, 1975.
Phonon scattering in solids.

Includes bibliographical references and index.
1. Phonons—Scattering—Congresses. I. Challis, Lawrence John, 1933-   II.
Rampton, V. W. III. Wyatt, Adrian Frederick George, 1938-   IV. Title.
QC176.8.P5I59   1975   539.7′217   75-40492

ISBN 978-1-4613-4273-1   ISBN 978-1-4613-4271-7 (eBook)
DOI 10.1007/978-1-4613-4271-7

Proceedings of the Second International Conference on
Photon Scattering in Solids held at the University of
Nottingham, August 27-30, 1975

INTERNATIONAL ADVISORY COMMITTEE

LOCAL ORGANIZING COMMITTEE

Preface

The Second International Conference on Phonon Scattering in Solids was held at the University of Nottingham from August 27th – 30th 1975. It was attended by 192 delegates from 24 countries who were accompanied by 43 members of their families. Eleven invited papers were read and 96 contributed papers; the contributed papers were in two parallel sessions. The Conference included the topics of the two International Conferences held in France in 1972, in Paris and at Ste Maxime.

The Conference brought together workers concerned with many aspects of phonon scattering in solids and liquid helium. Some of the work reported were studies of the intrinsic properties of dielectric materials such as the effects of anharmonicity, dispersion and anisotropy on phonon propagation and the conditions for the existence of zero sound and second sound modes. Work was also presented on various aspects of phonon interaction with free electrons in metals and semiconductors. A substantial part of the Conference was devoted to phonon spectroscopy – investigations of the energy levels of ions or neutral impurities by observing the resonant absorption or scattering of phonons. The materials being studied include paramagnetic and paraelectric solids, amorphous systems in which the 'impurities' appear to be intrinsic, and semiconductors. Work was reported on the use of phonons to observe phase transitions; in some cases the cooperative phase also arises through strong spin-phonon coupling. One of the intriguing unsolved problems discussed in detail at the Conference is the Kapitza conductance problem. The problem here is to understand the reason for the very large transmission coefficient for phonons at an interface between 'normal' solids and low sound velocity, low density liquids and solids such as helium and hydrogen. A number of new and important advances in

technique were reported. These include methods of generating monochromatic phonons at frequencies greater than 1 THz using tunnel
junctions or infra-red lasers, an X-ray technique for observing
electrically amplified sound and advances in time-reversed ultrasonic
echo techniques leading to holography and a number of other valuable
applications.

The Nottingham Conference was arranged at the suggestion of the
International Committee of the 1972 Paris Conference on Phonon Scattering in Solids under the Chairmanship of Professor H J Albany.
It was later made a satellite conference of the Fourteenth International Conference of Low Temperature Physics held in Helsinki in
August 1975 in order to reduce overlap in the subjects of the two
Conferences and we are very grateful to Professor J Huiskamp and
the Programme Committee of LT14 for their cooperation in this. It
was sponsored by the International Union of Pure and Applied Physics,
the European Physical Society, the Institute of Physics, the Royal
Society and the University of Nottingham. We are very grateful for
financial support and support in kind from the British Council,
Brookdeal Electronics Ltd, Edwards High Vacuum Ltd, A C Gill Ltd,
Hewlett-Packard Ltd, The International Union of Pure and Applied
Physics, the Oxford Instrument Co Ltd, Plessey Co Ltd, and the Royal
Society.

The Conference was formally opened by Professor J S L Leach,
Pro Vice Chancellor of the University of Nottingham. The scientific
programme included discussion groups on amorphous systems and on
Kapitza conductance and there was also a tour of the areas of the
research laboratories of the Physics Department of the University of
Nottingham of special interest to the delegates. An exhibition of
scientific equipment and books was presented by seven companies.
The social programme included a University reception, a Demonstration Lecture on Explosives by Dr B D Shaw, an Elizabethan Banquet at
the Crown Hotel, Bawtry and tours of Haddon Hall and Hardwick Hall.
We are most grateful to Mr David Phillips of the Nottingham Castle
Museum for arranging the exhibition on George Green (1793-1841) in
the Conference lounge.

We acknowledge with thanks the help of the International Comm-
ittee in selecting the invited speakers and contributed papers. We
are also most grateful to the many people who helped with the organ-
ization and running of the Conference. We mention particularly
Mr D R Ringer, Mr V M Conway, Miss S Stiegler, Dr G A Toombs, Dr
P J King, Dr R M Bowley, Dr O H Hughes, Dr M W S Parsons, Mr D R S
Cameron, Miss M R Jackson, Mr L Lee, Mr P M Mustoe, Mr C D Bull,
Mr P J Burke, Mr R J A Chettle, Mr C D Hackett, Mr C F Stapleton,
Dr F W Sheard, Dr J R Fletcher, Mrs D R Ringer and Mrs L J Challis
and add the many other members of the department of Physics who
acted as guides, drivers and cochairmen. We also acknowledge grate-
fully the help provided by the late Mr J Houldgate and the staff of

the University Printing Unit and Mr K L F Jacobs and the staff of Cripps Hall, where most of the delegates stayed.

Finally we are very grateful to the authors for their cooperation and the care with which they prepared their manuscripts for this Proceedings.

L J Challis
V W Rampton
A F G Wyatt

Contents

SECOND AND ZERO SOUND

Session Chairmen: A. Ikushima and K. Luszcynskii

AMORPHOUS SYSTEMS

Session Chairmen: H. M. Rosenberg and J. Jäckle

CO-OPERATIVE MAGNETISM AND CRITICAL PHENOMENA

Session Chairmen: J. Joffrin and J. P. Harrison

FREE CARRIERS

Session Chairmen: J. P. Maneval, J. A. Rayne,
                and P. Lindenfeld

LOCALIZED HOLES AND ELECTRONS

Session Chairmen: V. Narayanamurti and H. J. Albany

ANHARMONIC EFFECTS AND PHONON-PHONON SCATTERING

Session Chairmen: V. W. Rampton, J. E. Parrott,
                and P. G. Klemens

DEFECT SCATTERING AND PHONON FOCUSING

Session Chairman: R. Berman

NEW TECHNIQUES

Session Chairmen: C. H. Anderson, A. M. de Goër, and P. Gosar

THE THERMAL BOUNDARY RESISTANCE*

A. C. Anderson

Department of Physics and Materials Research Laboratory

University of Illinois, Urbana, Illinois 61801 U.S.A.

This paper will provide a brief introduction to the intriguing problem of the Kapitza thermal boundary resistance (R_B), and will serve as background material for the contributions to this topic being presented at this conference.

A phonon which encounters a well-defined interface between two dissimilar media is scattered; from a naive viewpoint, the phonon is either reflected or transmitted. Therefore, establishment of a time independent (dc) thermal flux $\dot{Q}$ across the interface produces a temperature difference ΔT which is distinct from any temperature gradients that may exist within the two materials [1]. By definition $R_B = (\Delta T/\dot{Q})$. We are assuming that at least one of the two materials is nonmetallic so that electronic thermal transport across the interface will not occur. The energy flux $\dot{Q}$, and thus R_B, can be calculated in a manner which is similar to that used to calculate the phonon thermal conductivity [2]:

$$\dot{Q} = (\tfrac{1}{2}) \sum_j \iint E_{1j}(\omega) w_{1j}(\omega,\theta) c_{1j} \cos\theta \, \sin\theta \, d\theta d\omega$$

$$- (\tfrac{1}{2}) \sum_j \iint E_{2j}(\omega) w_{2j}(\omega,\theta) c_{2j} \cos\theta \, \sin\theta \, d\theta d\omega \, . \tag{1}$$

Here $E_j(\omega)$ is the energy density of phonons of mode j having frequency ω, and $w(\omega,\theta)$ is the probability that a phonon (or its energy) incident on the interface at angle θ will be transmitted. For convenience, we will use subscript 1 to indicate phonons incident from the reference or "cell" side of the interface, 2 from the sample side, although R_B of course is independent of the direction of heat flow. Equation (1) is of general validity; its limitations are determined by the model adopted to obtain $w(\omega,\theta)$.

The quantity $w(\omega,\theta)$ was first obtained, for the general case
of any two materials in contact, from the application of classical
acoustic or mechanical boundary conditions. As in the analogous
situation in optics, $w(\omega,\theta)$ then includes such phenomena as acoustic
Snell's laws, the critical angle, total internal reflection, interface (Stoneley) waves, and even frustrated total internal reflection
if some form of additional phonon absorption or "scattering" or
energy attenuation α is present in the opposing medium. All of
these phenomena have been studied using ultrasonic techniques [2].
If $w(\omega)$ is not a function of ω, $\dot{Q} \sim T^3$ and $R_B^{-1} \sim T^3$ just as for
phonon thermal conductivity in the presence of a frequency inde-
pendent mean free path. Less classical theoretical techniques are
needed and are being developed to obtain $w(\omega)$, but thus far these
have yielded only limited additional information [3,4,5].

The acoustic mismatch (AM) model described above, which incor-
porates the classical acoustic boundary conditions, works remarkably well, with no adjustable parameters, over a temperature range
of ~0.01-100 K and for a broad class of "ordinary" interfaces such
as gold-sapphire [6,7]. Somewhat surprising, however, is the fact
that roughly a decade passed after the introduction of the AM
model before it was generally recognized that the theory did provide a correct explanation of R_B. This is primarily because
various problems were encountered which beclouded the issue, and
which hopefully will be avoided in the future. In particular, in
order to compare theory with experiment, one must know $E(\omega)$ of
Eq. (1) in order to calculate R_B. That is, we must know the frequency dependence of the phonon populations incident on the interface from the two sides. Often the simple Debye approximation is
not appropriate. For example, an amorphous material such as a glue
or an epoxy acts like a frequency dependent, low-pass filter for
phonons. If $E(\omega)$ does not reflect this fact in some manner, the
calculated R_B may be in error by a factor of 10 or more for such
an interface [8]. As another example, dislocations often are
accompanied by localized phonons or vibrations. Resonant scattering of thermal phonons from these states creates a notch-filter
effect in thermal transport measurements [6]. A more general
problem, especially in nonmetallic materials, has been that a
damaged layer next to the interface may change $E(\omega)$ considerably
[6,9,10]. An alternative viewpoint is that the effective local
temperature of the phonons at the interface is not known. Failure
to appreciate the fact that $E(\omega)$ must represent the physical situation in question has caused some authors to state erroneously that
the AM model is not correct, or that Eq. (1) is wrong by a numerical
factor of 2.

Thus far I have emphasized that the AM model is valid over a
wide temperature range for a broad class of materials and a variety
of conditions provided one is careful in introducing approximations

into Eq. (1). There are two regimes, however, in which the AM
model in its present form fails miserably. This occurs when the
acoustic properties of the two media are either (i) very similar,
e.g. a grain boundary [11], or (ii) very dissimilar, e.g. a Cu-He
interface [1]. For very similar materials R_B is larger than
expected; for very different materials, it is smaller than expected.
Little work has been done on (i), which probably requires that the
AM model be extended to anisotropic materials for a proper application. I will therefore concentrate on (ii), to which most
current experimental and theoretical efforts are being directed.

     Let's review the experimental situation for an interface
between He and a solid cell material such as Cu. There is a decrease by an order of magnitude [12] in the quantity $R_B T^3$ as the
temperature is increased from ≈ 0.1 to 1 K. This implies that
$w(\omega)$ of Eq. (1) increases with increasing phonon frequency for
$10^{10} \lesssim \omega \lesssim 10^{11}$ Hz. Below ≈ 0.1 K, R_B is very sensitive to mechanical surface treatment or overlayers of condensed gases, but above
≈ 1 K, surface treatment, condensed gases, or even the cell material have much less influence on R_B [1,12,13,14]. Above ≈ 1 K, R_B
is essentially the same as liquid ^{4}He or ^{3}He, solid ^{4}He or
^{3}He, or even gaseous ^{3}He, while below 0.1 K the magnitude of R_B
to these several forms of He are quite different [12,15]. Thus
there appear to be two rather distinct temperature ranges in which
the physical properties of R_B are quite different, namely $T \lesssim 0.1$ K
and $T \gtrsim 1$ K.

     The AM model is most likely to be applicable in the range
below 0.1 K since the frequencies of thermal phonons are small,
the wavelengths large, and the classical boundary conditions therefore most relevant. The primary change within the model as the
acoustic mismatch between the two materials increases is that the
critical angle becomes small, roughly 5° for a liquid ^{3}He-Cu interface. Thus the frustrated total internal reflection mentioned
previously becomes important. That is, incidence of any phonon
from the He at $\theta \gtrsim 5°$ may result in energy transfer across the
interface if some phonon absorption or attenuation α exists in the
Cu, although a well-defined "critical" angle still remains [2].
Indeed for a carefully prepared surface, the measured R_B was very
close to the value obtained from the AM theory using no adjustable
parameters; the value of α was that measured for the absorption of
phonons by electrons [2]. This simply means that phonons incident
from the He at $\theta \gtrsim 5°$ interact with electrons in the Cu. (Alternatively the electrons experience an enhanced density of phonon
states near the interface. In other words, thermal energy can
flow in either direction across the boundary via this interaction.)
With less well-prepared surfaces, a single empirically determined
value of α permits one to calculate rather accurately R_B to liquid
or solid ^{4}He or ^{3}He at any pressure [16]. This implies that a
"damaged" surface serves to enhance the net Cu electron-He

phonon interaction. A complication does arise in low temperature experiments involving liquid [16] or solid [17] ^{3}He, apparently because of a non-phonon or magnetic energy exchange across the boundary [18], but this is not a topic for the present conference. The classical AM model is also somewhat awkward to use in the case of liquid ^{3}He at low temperatures since the coupling to "transverse" waves in the liquid must be accounted for [15].

But in brief, for a He sample at $T \lesssim 0.1$ K, the AM model is at least useful and may even be correct in the form described previously. This model also could account for the anomalously small magnitude of R_B at $T \gtrsim 1$ K by using a very large and probably unphysical value for α. However, it is generally believed that R_B at $T \gtrsim 1$ K is dominated by a mechanism not yet explicitly incorporated into the model described above, a mechanism which may operate in parallel with the phonons. This would be consistent with the experimental observations that have been listed.

The above mentioned measurements (including Refs. 7 and 9) were all performed using basically dc techniques. This information has been augmented at $T \gtrsim 1$ K by other techniques which will, I am certain, prove essential to our understanding of R_B to He. Pulses of thermal phonons (heat pulses) have verified that R_B is the same to solid, liquid, or gaseous ^{4}He [19,20]. Such pulses allow individual phonon modes to be recognized by observation of their time-of-flight. Using this technique it has been possible to demonstrate that transverse phonons in the cell material, incident nearly perpendicular to the interface, coupled energy very effectively to liquid He [19]. This is not expected from a classical viewpoint. By observing the angular distribution of phonons radiated into liquid ^{4}He, the existence of a "critical" angle at a value of θ appropriate to the interface has been detected which reinforces the suggestion that the acoustic mismatch mechanism is still present at $T \gtrsim 1$ K [21]. Furthermore these same measurements showed that there was in addition a significant "background" distribution of radiated phonons at all angles, and that this additional component increased in magnitude with increasing temperature such as to account for the decrease in the quantity $R_B T^3$ between 0.1 and 1 K. Also the internal reflection of monochromatic phonons [22] from a He-covered surface has shown that $w(\omega)$ increases with frequency in such a way as to account for the decrease in $R_B T^3$ between 0.1 and 1K. Transient measurements carried out on various thicknesses of He films show that the unknown mechanism "occurs" near the second monolayer of He at the interface [23]. Of similar importance is the evidence that R_B is also anomalously small to solid H_2 and D_2 at $T \gtrsim 1$ K, but that R_B to solid Ne may again be accounted for readily by the AM model [20, 24]. Additional details have been gleaned from both transient and dc measurements, but their significance in interpreting R_B is not obvious (to the author) at the present writing [1].

The qualitative picture that emerges for R_B to He is thus: (i) an AM mechanism operates at all temperatures just as for interfaces between "ordinary" materials, (ii) with increasing frequency above $\approx 10^{11}$Hz, phonons have an increasing probability of transferring energy across the interface by an additional mechanism, (iii) this additional mechanism spews phonons into the He at all angles, (iv) roughly the second monolayer of He plays a significant role in this mechanism, (v) the mechanism is relatively weakly influenced by the physical condition of the interface, and (vi) the mechanism appears to be important in R_B to liquid or gaseous ^{3}He or ^{4}He, or to solid ^{3}He, ^{4}He, H_2 or D_2, but not to solid Ne or other "ordinary" solids.

Many mechanisms [1] have been suggested to account for the anomalously small R_B to the light elements at $T \gtrsim 1$ K. Since AM or direct phonon coupling seems to be present, perhaps the AM model needs only to be modified. One such suggestion is that the dense layer of He at the interface due to the van der Waals' force serves to acoustically match the two media and thereby enhance the transmission of phonons. Combined with the effect of the phonon attenuation α in the cell material, this can account for the observed magnitude of R_B [25] at $T \gtrsim 1$ K, and indeed should be included in any discussion of the AM model for He at these temperatures. However, an α of the required magnitude in a nonmetallic cell material would reduce the thermal conductivity of that material sufficiently to mask R_B [10] underline the attenuation were confined to a very thin layer of a few atomic diameters at the cell surface. This conclusion is consistent with recent work on carefully prepared single crystal Cu samples [26], and it is compatible with the AM model [2,25]. However it is unlikely to explain the anomalously small R_B to He or H_2 as well as the normal R_B to Ne for the same surface.

A second suggestion is that a mechanical resonance [27,28] of some material at the interface enhances $w(\omega)$ at the frequency of resonance ω_o and thus reduces R_B at $T \gtrsim (4\hbar\omega_o/k)$. Indeed, a resonance associated with pump oil contamination has been detected [22] in $w(\omega)$ at $\omega \approx 10^{11}$ Hz, using monochromatic phonons, which would reduce R_B to this surface at $T \gtrsim 1$ K. However, since explicit efforts have been made in the past to avoid such layers [10,26,29], it is unlikely that resonances generally dominate R_B. One should also note that the halfwidth δT of such a resonance in a dc measurement is $\delta T/T \approx 1$. Therefore a resonance of this kind cannot be resolved within the usual experimental temperature range of 1-2 K.

A third mechanism which is again being revived is the absorption of a phonon at the interface and the consequent desorption or "evaporation" of a He atom out of the dense van der Waals' layer into the liquid. This mechanism can account for both the temperature dependence and the magnitude of R_B [12,30]. More impressive

is the fact that it predicted [12], prior to experimental verification, both the correct magnitude of the van der Waals' potential
and that the effect should occur in the second monolayer of He on
the surface. However, R_B is also anomalously small to <u>solid</u> He
(or solid H_2 or D_2), and thus the concept of "evaporation" would
have to take on the appearance of (virtual) vacancy formation.

Some authors have emphasized the possible importance of surface
motion or "waves" to the small magnitude of R_B. The role of this
motion from the viewpoint of the AM model has already been alluded
to. At a Cu-^{3}He interface, the depth of such waves in the Cu
caused by phonons incident from the He at $\theta \gtrsim 10°$ is only a couple
of atomic layers [2]. Thus the surface motion is very sensitive
to attenuation by the slightest of surface imperfections, or by
the enhanced anharmonicity experienced by the surface atoms. The
less classical theory involving transfer Hamiltonians should be
more appropriate to this regime, and this too suggests that attenuation of surface motion may be important [3]. Experimentally the
role of surface waves is implied by the results of a study of the
internal reflection as a function of angle of incidence, of phonons
incident on a surface in contact with He [31], and perhaps by the
observation that the coupling to He from a solid is greatest for
those phonons which in the absence of He scatter diffusively from
the surface or undergo "mode conversion" [20,23,31]. As yet,
however, there is no definitive evidence that surface motion is
associated with the small magnitude of R_B.

In summary, it has been suggested that R_B between two "ordinary"
solids is rather well understood in terms of the AM model, and that
this understanding may extend to R_B for He interfaces at $T \lesssim 0.1$ K.
On the other hand, no mechanism as yet convincingly explains the
anomalously small R_B to the light elements at $T \gtrsim 1$ K. This is a
tantalizing problem. We now know in some detail what the phonon
spectrum looks like as it approaches or leaves either side of the
interface, and the effects of changing or modifying the sample or
cell materials. But what takes place at the interface still eludes
us some 35 years after the phenomenon of the thermal boundary
resistance to liquid helium was first observed.

References

*This work was supported in part by the National Science Foundation
Grant DMR 72-03026.

1. A complete bibliography will not be presented. The following
    papers, and work cited therein, are intended as general
    references. See also the reviews by T. H. K. Frederking,
    Advances in Cryogenic Heat Transfer 64, 21 (1968); G. L.
    Pollack, Rev. Mod. Phys. 41, 48 (1969); N. S. Snyder, Cryo-
    genics 10, 89 (1970); J. D. N. Cheeke, J. Phys., Paris
    Colloque Suppl. C3, 129 (1970); L. J. Challis, J. Phys. C 7,

481 (1974); A. C. Anderson, Proc. Israel Phys. Soc. (to be published).

2. R. E. Peterson and A. C. Anderson, J. Low Temp. Phys. 11, 639 (1973).
3. T. J. Sluckin, G. A. Toombs, and F. W. Sheard, Solid State Comm. 14, 203 (1974).
4. K. R. Allen and J. D. Powell, Phys. Rev. B9, 1180 (1974).
5. W. M. Saslow, Phys. Rev. B11, 2544 (1975).
6. S. G. O'Hara and A. C. Anderson, J. Phys. Chem. Solids 35, 1677 (1974).
7. W. Kappus and O. Weis, J. Appl. Phys. 44, 1947 (1973).
8. C. L. Reynolds and A. C. Anderson, J. Low Temp. Phys. (to be published).
9. J. D. N. Cheeke and C. Martinon, Solid State Comm. 11, 1771 (1972).
10. J. D. N. Cheeke, Cryogenics 10, 463 (1970).
11. K. A. McCarthy, Low Temperature Physics LT9, edited by J. G. Daunt, D. O. Edwards, F. J. Milford, and M. Yaqub (Plenum Press, New York, 1965) p. 1155.
12. A. C. Anderson and W. L. Johnson, J. Low Temp. Phys. 7, 1 (1972).
13. M. F. Whelan and D. V. Osborne, J. Low Temp. Phys. 8, 449 (1972).
14. The dependence of R_B on the density and sound velocity of the cell material has been discussed in Ref. 1 (Challis), in Ref. 12, and in J. P. Harrison, J. Low Temp. Phys. 17, 43 (1974).
15. J. T. Folinsbee and A. C. Anderson, Phys. Rev. Lett. 31, 1580 (1973).
16. J. T. Folinsbee and A. C. Anderson, J. Low Temp. Phys. 17, 409 (1974).
17. L. E. Reinstein and G. O. Zimmerman, Phys. Rev. Lett. 34, 458 (1975).
18. D. L. Mills and M. T. Beal-Monod, Phys. Rev. A10, 343, 2473 (1974).
19. C.-J. Guo and H. J. Maris, Phys. Rev. A10, 960 (1974).
20. J. S. Buechner and H. J. Maris, Phys. Rev. Lett. 34, 316 (1975).
21. R. A. Sherlock, N. G. Mills, and A. F. G. Wyatt, J. Phys. C 8, 300 (1975).
22. E. S. Sabisky and C. H. Anderson (to be published).
23. H. Kinder and W. Dietsche, Phys. Rev. Lett. 33, 578 (1974).
24. J. T. Folinsbee, Ph.D. thesis, University of Illinois, 1974 (unpublished).
25. J. L. Opsal and G. L. Pollack, Phys. Rev. A9, 2227 (1974).
26. N. S. Snyder (to be published).
27. M. Vuorio, J. Phys. C5, 1216 (1972).
28. J. C. A. van der Sluijs, E. A. Jones, and A. E. Alnaimi, Cryogenics 14, 95 (1974).
29. R. C. Johnson and W. A. Little, Phys. Rev. 130, 596 (1963); R. C. Johnson, Bull. Am. Phys. Soc. 9, 713 (1964).
30. A. R. Long, J. Low Temp. Phys. 17, 7 (1974).
31. A. R. Long, R. A. Sherlock, and A. F. G. Wyatt, J. Low Temp. Phys. 15, 523 (1974).

REFLECTION COEFFICIENT OF PHONONS BETWEEN 15 AND 200 GHz AT A

SOLID-LIQUID HELIUM INTERFACE

C. H. Anderson, P. Call, and E. S. Sabisky

RCA Laboratories

Princeton, New Jersey

Results on the reflectivity of phonons between 15 and 315 GHz on very clean surfaces of cleaved SrF_2 immersed in liquid helium have recently been published.[1] In this report, we give preliminary results on changes in the reflectivity between 15 and 200 GHz at 1.4K induced by the addition of H_2, D_2, Ne and Ar to the interface.

In order to reduce surface impurities even farther than in ref. (1), the heater has been changed from a resistive element which is glued on with G.E. varnish to a 100Å thick evaporated chromium layer. The remaining surfaces are cleaved just before mounting the sample in a chamber set inside a superconducting solenoid. The chamber is immediately evacuated after the sample is introduced and generally kept as free as possible from contamination with materials other than pure gases. Before each run the crystal is heated to 200C under vacuum, usually overnight, and kept near this temperature during the transfer of liquid helium into the outer bath. After the outer walls of the sample chamber are cooled to a few degrees Kelvin (the sample has cooled to about 300K), the inner chamber is filled with helium and the heater then turned off. We believe this procedure provides a very clean surface, most of which is atomically flat.

The outer bath is stabilized at 1.4K and the heater is square wave modulated on and off at ~ 100 milliwatts and at a rate of a few cycles per second. The phonons within the crystal reach dynamic equilibrium within a few microseconds after the heater is turned on, where phonons emitted from the heater travel through the crystal scattering off the crystal surfaces, and impurities and imperfections inside the crystal until they are reabsorbed by

the heater or absorbed into the helium bath. We generally find
the data can be analysed by ignoring the possibility that the
energy of the phonons change anywhere except at the heater or in
the helium bath. The average occupation number of the phonons in
the crystal is related to the heater and bath temperatures through
the expression

$$N(T_S) = \varepsilon(\nu)N(T_H) + [1-\varepsilon(\nu)]N(T_B)$$

where $N(T_H)$ and $N(T_B)$ are the usual Einstein-Bose factors for the
heater and bath temperatures, respectively. We measure an average
temperature, T_S, for phonons of frequency ν through their resonant
interaction with trace amounts of the paramagnetic ion Tm^{2+}(2).
The factor $\varepsilon(\nu)$ which can then be derived from Eq. (1) provides a
measure of the coupling of the phonon modes at frequency ν to the
helium bath relative to that of the heater. It can have the value
1.0 if the surfaces in contact with the bath are perfectly
reflecting and a value ~ 0.1 in our particular experiment if the
surfaces are perfectly absorbing. For highly reflecting surfaces
$\varepsilon \approx R^n$, where R is an average reflectivity and n is the average
number of bounces the phonons make before being reabsorbed. In
particular, ε is monotonically related to the reflectivity.

The $\varepsilon(\nu)$ curve for clean surfaces in contact with helium with
this new heater displays the same rapid decrease in reflectivity
observed between 20 to 120 GHz in ref. (1), with no sign of a
resonant overshoot. We have, however, only recently become
convinced that this heater design does not emit a thermal or
blackbody spectrum, especially when it has a ~ 1 micron thick
layer on it of one of the pure gases we have been investigating.
Instead this thin metal film heater produces a spectrum which is
deficient in low frequency phonons. This has made our results to-
date qualitative in nature.

After the initial clean calibration run, which repeats
remarkably well day to day, the helium is pumped off and one of
the gases H_2, D_2, Ne or Ar is admitted down a heated tube. Film
thicknesses of a few atomic layers to 1 micron are deposited, then
the sample chamber is refilled with liquid helium. We believe the
experimental situation should be viewed in the following manner.
The outer surface of the frozen layer is probably very rough and
filled with dislocations and so the Kapitza conductance to the
helium at this interface should be very high.(3) The thermal
gradient across these thin layers for a heat flux of 100 milliwatts
cannot exceed a few millidegrees. Therefore, what $\varepsilon(\nu)$ primarily
measures is the reflectivity between SrF_2 and the disordered frozen
gas layers when the film thickness is ~ 1 micron. When the film
is one or two atomic layers thick, then it is the frozen gas-
helium interface which dominates.

In the classical acoustic mismatch theory, the reflectivity between SrF_2 and He, H_2, and D_2 should all be close to 1.0, but decreasing in that order, while for Ne and Ar it is roughly 0.7 and 0.65, respectively. The observations at low frequencies (20-40 GHz) at least keep the correct order. The Ne and Ar results were expected to be classical and so $\epsilon(\nu)$ should be a constant, but instead it was found to increase with increasing frequency. This is consistent with other evidence that the heater is non-thermal.

The H_2 and D_2 thick film results qualitatively confirm those of Buechner and Maris[4] in that at frequencies above 150 GHz the values of $\epsilon(\nu)$ are close to those of helium. However, both gases produce reflectivities which deviate from the classical values at lower frequencies than helium does. Also the shape of these curves can be radically altered by annealing at 5-8K. When the crystal is heated to above 15K without the helium present, most of the hydrogen evaporates leaving one or two atomic layers on the SrF_2 surface. The $\epsilon(\nu)$ curve, in this case, is similar to that observed with the clean surfaces.

We propose the following interpretation of these preliminary H_2 and D_2 results which attempts to maintain the essence of the suggestion by Anderson and Johnson[3] that the Kapitza conductance anomaly is associated with the deabsorption of He atoms in the second layer. This must be generalized to energy states at the interface associated with light molecules which interact in a highly non-linear fashion with the propagating modes. Kinder, in informal discussions with one of us (C. H. Anderson), discussed tunneling states over two years ago. The thick layers of hydrogen which are condensed onto the SrF_2 surfaces at a few degrees Kelvin must be highly disordered and thus contain a multitude of tunneling states covering a wide frequency range. Annealing the hydrogen changes the spectrum, but never produces the sharp threshold observed with helium. Helium is self-annealing so that the modes, or possibly single mode, can be well defined if the crystal surface is atomically flat.

References

(1) E. S. Sabisky and C. H. Anderson, Solid State Comm. (to be published).
(2) C. H. Anderson and E. S. Sabisky, Physical Acoustics 8 (Academic Press, New York, 1971).
(3) A. C. Anderson and W. E. Johnson, J. Low. T. Phys. 7, 1, (1972).
(4) J. S. Buechner and H. J. Maris, Phys. Rev. Letters 34, 316, (1975).

A MODEL OF INELASTIC HEAT TRANSFER MECHANISM FOR THE EXCESS

KAPITZA CONDUCTANCE

Hiroshi Namaizawa[†]

Fachrichtung Theoretische Physik, Universität des

Saarlandes, D66 Saarbrücken, West Germany

Observations of thermal boundary conductance (Kapitza conductance[1]) between a classical solid and quantum media (liquid and solid helium, solid H_2 and D_2[2]) show that, in addition to the elastic processes[3], inelastic processes are important[4,5]. We propose here a model of inelastic thermal heat transfer between a solid and liquid helium. We assume that a helium monolayer is formed between the solid filling the half-space $z<0$ and the bulk liquid He at $z>d$ (see Fig 1), due to an anisotropic field

$$W(z) = W^s(z) + W^\ell(z) \tag{1}$$

where W^s is the Van der Waals field from the solid and W^ℓ is the sum of Lennard-Jones potential from bulk He;

$$W^\ell(z) = \pi n^\ell r_0^3 v_0 \{(r_0/(d-z))^9/45 - (r_0/(d-z))^3/3\} \tag{2}$$

in which n^ℓ is the atom number density of He, r_0 and v_0 are Lennard-Jones parameters of He. We solve the equation

$$(-\hbar^2/2m \, d^2/dz^2 + W(z))\phi_m(z) = e_m \, \phi_m(z), \quad m=0,1,2... \tag{3}$$

If we further know the ground state property of the monolayer motion parallel to the layer we can describe collective excitations of the monolayer according to Jackson[6]. There are two kinds of them: One is longitudinal density wave running along the layer (ℓ-surfon) in which all the He atoms are in the ground state $\phi_0(z)$ with respect to z-motion, the other consists of superposition of propagating waves where one atom is excited in z-motion, corresponding to drum-like vibrations of the layer (t-surfon). The excitation energies of ℓ-surfons and t-surfons are given respectively by

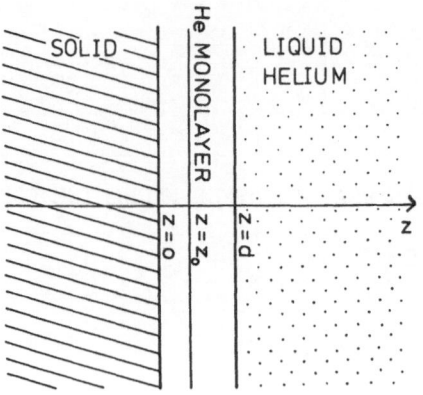

Fig 1

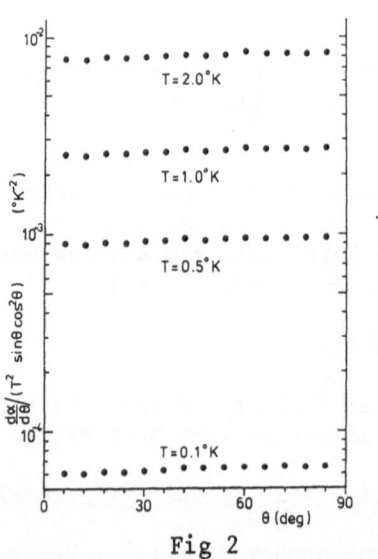

Fig 2

$$E^{\ell}(\underset{\sim}{k}) = \hbar^2 k^2/(2mS(\underset{\sim}{k})) \tag{4}$$

and

$$E_m^t(\underset{\sim}{k}) = \Delta e_m + \hbar^2 k^2/2m, \quad \Delta e_m = e_m - e_\theta \tag{5}$$

where $\underset{\sim}{k}$ is the surface wave-vector and $S(\underset{\sim}{k})$ is the static structure factor of the two-dimensional He^4[7].

Heat transfer from the solid to the liquid proceeds in the following way: A phonon impinging on the solid surface vibrates it and
causes displacement u_z^s in z-direction, and W^s changes by $-(\partial W^s/\partial z)u_z^s$.
This perturbation excites a collective mode on the monolayer, and

the excited mode, after running along the layer, gives its energy
via perturbation $(\partial W^{\ell}/\partial z)u_z^{\ell}$ to the liquid to create elementary
excitation. In the present study we choose argon as the solid and
we have

$$W^s(z) = \pi n^s r_1 v_1 \{(r_1/z)^9/45 - (r_1/z)^3/3\} \tag{6}$$

where n^s is the number density of Ar and, r_1 and v_1 are the Lennard-
Jones parameters of Ar-He[7]. In solving (3) we set d, the position
of liquid helium boundary, equal to the sum of z_0, the expectation
value of z calculated with the ground-state wave-function ϕ_0 of (1)
and d_0, the hard-core diameter of He atom ($d_0 \simeq 2.3$Å). At low temperatures there are essentially only ℓ-surfons in the initial and
the final states of the above scattering process and their dispersion relation is well approximated by a linear dispersion (sound velocity c). Similarly only the lowest t-surfon (m=1) plays an important role in the intermediate states and the excitations in liquid
helium are assumed to be phonons (sound velocity c^{ℓ}). Dividing the
flux of energy calculated by our model by the energy flux of incident
phonons we obtain the inelastic heat transmission coefficient at temperature T˚K

$$d\alpha/d\theta = \alpha_0 (k_B T/2\pi mc^{\ell} c^D)^2 F(\theta,T) \sin\theta \cos^2\theta \tag{7}$$

where θ is the phonon radiation angle normal to the interface and α_0
is a dimensionless number defined by

$$\alpha_0 = 40(2\pi)^{-6} \left(\frac{mn^{\ell}c^D}{Mn^s c^{\ell}}\right) |<\phi_0|\partial W^s/\partial z|\phi_1>/\Delta e_1|^2 |<\phi_1|\partial W^{\ell}/\partial z|\phi_0$$
$$x |<\phi_1|\partial W^{\ell}/\partial z|\phi_0>/\hbar c n^{\ell}|^2$$

$F(\theta,T)$ represents the reduced phase space integral of the corresponding scattering process, M and c^D are the atomic mass and the Debye
velocity of Ar, respectively. In Fig 2 we plotted $d\alpha/d\theta$ divided by
$(k_B T)^2 \sin\theta \cos\theta^2$ as a function of θ at T = 0.1,0.5,1.0, and 2.0˚K.
From this we find that θ-dependence of the transmission coefficient
is mainly determined by $\sin\theta \cos^2\theta$. In order to compare our results
with experiment[4] we show in Fig 3 the normalized transmission coefficient at T = 2˚K together with experimental results and $3\sin\theta \cos^2\theta$
(note that $\int_0^{\pi/2}\sin\theta\cos^2\theta d\theta=1/3$). Agreement of theory with experiment is well at angles $\theta>50°$. Since separation[4] of the background
signals from the central peak, corresponding to the critical cone of
the elastic theory, is not straightforward it is still difficult to
compare theory with experiment at small angles $\theta<40°$. The predicted
peak at $\theta \simeq \arcsin(1/\sqrt{3}) = 35.3°$ and the small angle behaviour of
$d\alpha/d\theta$ are to be checked by the future experiment. As is also shown
in this figure contribution from the longitudinal phonons (L) is
smaller than that from the transverse phonons (T). Finally we present in Fig 4 total inelastic transmission coefficient α as a function
of temperature along with a tentative fit.

† On leave of absence from University of Tokyo, Tokyo

1. L J Challis, J Phys C7, 481 (1974).
2. J S Buechner and H J Maris, Phys Rev Lett 34, 316 (1975).
3. I M Khalatnikov, 'Introduction to the Theory of Superfluidity'
   (Benjamin, New York, 1965).
4. N G Mills, A F G Wyatt and R A Sherlock, J Phys C8, 289 (1975)
   R A Sherlock, N G Mills and A F G Wyatt, ibid C8, 300 (1975).
5. H Kinder and W Dietsche, Phys Rev Lett 33, 578 (1974).
6. J D Jackson, Phys Rev 180, 184 (1969).
7. M D Miller and C-W Woo, Phys Rev A7, 1322 (1973) and a private
   communication.

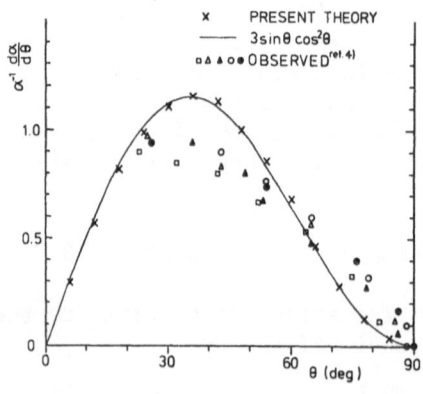

Fig 3

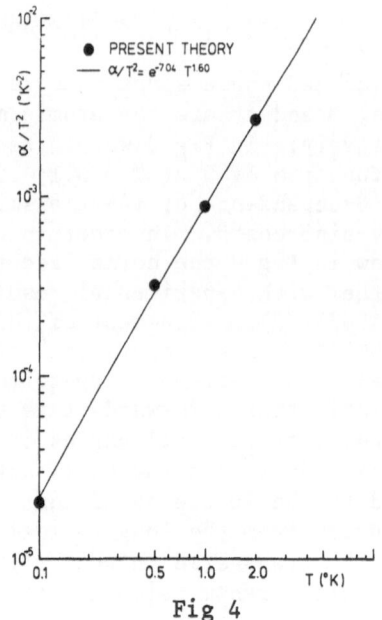

Fig 4

A NEW KAPITZA INVESTIGATION: DETECTING BACKSCATTERED PHONONS WITH TUNNEL JUNCTION AND BOLOMETER SIMULTANEOUSLY

W. Dietsche[+] and H. Kinder[+]

Institut für Festkörperforschung der KFA Jülich

At phonon frequencies above c.100 GHz the heat transfer across the interface between a solid and helium increases drastically over the values expected by the acoustical mismatch of the two materials.[1] In the acoustical model it is assumed that phonons are transmitted directly i.e. without changing their frequencies. In this paper, however, we will show that the phonons undergo frequency conversion in the first two atomic layers of helium at the solid.

For experimental proof it is necessary to analyze how the frequency distribution of the incident phonons is affected by the helium. Therefore, we used monochromatic phonons generated and detected by superconducting tunnel junctions.[2] For a qualitative frequency analysis we could use the detector junction in two states. In the "usual" junction state it was only sensitive to phonons with energies higher than the energy gap $2\Delta_D$ (c.120 GHz withAl:0). If the bias current was increased until one of the films became normal at a weak spot the detector behaved as a weak link and was now sensitive to the total energy regardless of frequency.

For the investigation described here we used phonon pulses of 290 GHz which were generated at one side of a Si crystal by a Sn junction. The phonons propagated along the 111 direction to the far side of the crystal where they were scattered. The backscattered phonons were detected on the generator side. The reflecting surface was either in vacuum or could be exposed to helium gas of various pressures.

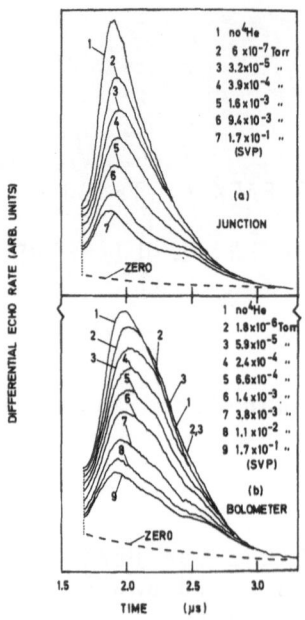

Fig.1 Transverse
echoes measured in
two detector states.

Fig.1 shows the transverse echo measured by the junction (a) and the
bolometer (b) together with the He
pressures. The traces 1 (without He)
were of different shape due to a
slightly higher time constant of the
bolometer. But this did not affect
the results. When He was introduced,
the echoes decreased strongly but
differently depending on the detector state. The bolometer showed an
additional time delay at low pressures where trace 2 and 3 even crossed trace 1. This delay was only weak
with the junction detector. In order
to compare the reflected intensities
we integrated over the total echo
areas. These areas were normalized to
the ones without helium to account
for the different sensitivities of
the two detectors. The results are
plotted in Fig.2 as a function of the
He pressure.

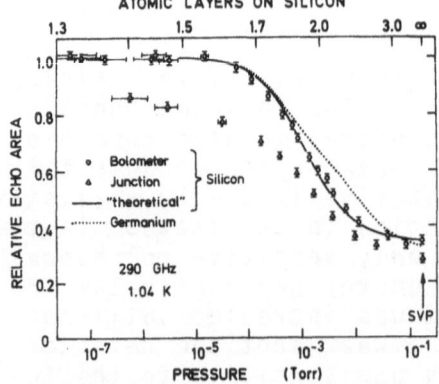

Fig.2 Relative echo in-
tensities as function of
pressure.

The bolometer shows that no
energy is lost up to a pressure of 10^{-4} Torr i.e. the
whole energy is reemitted into the solid. The junction,
however, exhibited a remarkable decrease already at a
pressure below 10^{-6} Torr or a
film of 1.4 atomic layers.[3]
It follows that the backscattered phonons were no longer
monochromatic but were down-
converted to frequencies below $2\Delta_D/h$.

At pressures above 10^{-4} Torr,
the bolometer intensity began also to decrease down to 35%
of its original value. In this pressure regime, the junction showed an additional decrease and its intensity was always less than that of the bolometer. The decrease of the
bolometer signal can be understood using simple gas kinetic arguments. Let us assume that a fraction Q_o of the in-

cident phonons is thermalized in the film. The energy is
then redistributed between the gas (Q_g) and the solid (Q_S).
The ratio depends on the ratio of the thermal resistances
$Q_S/Q_g = R_g/R_S$. R_g is given by the kinetic theory as $R_g = \rho/p$
with $\rho = 4x\sqrt{mT/k_B}/3 = 1.7 \times 10^{-4}$ sK/cm at 1K. Therefore,

Q_g (p)$= Q_0 \times p/(p + \rho /R_S)$. If we assume that Q_0 and R_S do not
depend on pressure or film thickness then $Q_0 = Q_g$ at high
pressures. With $Q_0 = 0.65$ and $R_S = 1.1 \times 10^{-4}$ sKcm^2/erg as a fit
parameter we plotted $1-Q_g$ in Fig.2. The agreement with the
experimental values is remarkable good and supports the
simple assumptions. In this picture one can also calculate
the time constant of the "cooling" of the film after it
was heated by the incident high frequency phonons. The
time constant τ is given by $\tau = C x (R_g^{-1} + R_S^{-1})^{-1}$ where C is
the heat capacity of the film. With $C \approx 0.03$ erg/cm^{2}K^4 we
get an upper limit of 3µs. The observed time constant is
smaller by about one order of magnitude. However, we used
for the heat capacity results measured by a stationary experiment while in our case the thermalization was probably
not yet complete at the low helium coverages. This leads
to a reduced effective heat capacity and a smaller time
constant.

Fig.2 shows also the intensities of a ST echo in Ge
measured by a junction detector.[5] Though a direct comparison with Si is not possible (bolometer data were not
taken) one can say that the decrease occurs at higher
pressures or, in the gas kinetic model, R_S is smaller.
This appears to be reasonable because the density of phonon states is higher in Ge than in Si. The smaller R_S
leads also to a faster "return time" to the solid. Therefore, a higher film thickness (stronger thermalization)
is needed for the frequency conversion to become noticeable. Besides, the observation of an only small frequency
broadening at low He coverages was very favourable at Ge.[5]

In conclusion we have shown that high frequency phonons (f≳100 GHz) are frequency converted in the first two
layers of He at the interface. This yields a smaller Kapitza resistance because the energy transmission is no
longer hindered by the different acoustical properties.

+ present address: Physik Department der Technischen
  Universität München, 8046 Garching, West-Germany
1. See for example: L.J.Challis,J.Phys.C7, 481 (1974).
2. H.Kinder,Phys.Rev.Lett.28, 1564 (1972).
3. E.S.Sabisky, C.H.Anderson,Phys.Rev.Lett.30,1122 (1973).
4. D.F.Brewer, J.Low Temp.Phys. 3, 205 (1970).
5. H.Kinder, W.Dietsche, Phys.Rev.Lett.33, 578 (1974).

THE ANGULAR DISTRIBUTION OF PHONONS RADIATED FROM THE CLEAVED

<100> FACES OF NaF, KCl AND MgO INTO LIQUID ^{4}He

G.J. Page, R.A. Sherlock, A.F.G. Wyatt, K.R.A. Ziebeck

Department of Physics, University of Nottingham

Nottingham, NG7 2RD, England

The angular distribution of phonons emitted from a <100> face of NaF into liquid ^{4}He at O.1K has been measured by Sherlock et al[1]. They observed a central peak, which arises from phonon transmission across the interface with the parallel component of the wave vector $(q_{||})$ conserved [2], superimposed upon an approximately cosine emission accounting for ∿90% of the energy transfer. The central peak was slightly wider than straightforward acoustic theory suggested and a modification including scattering in the solid was thought necessary[3]. Proving the generality of this result is important and we report similar measurements on MgO and KCl.

The experimental arrangement was similar to that used for NaF. The crystal was pulse heated to a few degrees K by a thin film heater glued to a cleaved <100> face with a thin layer of G.E. varnish. Finally in the experimental preparation the opposite face was cleaved. The crystal was mounted on a rotating frame inside a cell that was filled with ^{4}He and pressurised. All results were at 20 bar and .1K where the phonons travel in straight lines, hence those emanating from the surface were detected ∿10mm away with a graphite bolometer.

With all specimens the angular distribution was separable into background and peak, which after separation, was normalised to unity at $\theta = 0$, figs 1 and 2. The resolution of the source/detector system was 6° for MgO and 2° for KCl2, allowing structure to be seen. The intrinsic width of the peak should be $\sin^{-1}(C_{He}/C_S)$, C_{He} is the velocity of phonons in ^{4}He, C_S is the surface wave velocity, this was compared to the experimental results $(\Delta\theta^!)$ after subtracting the angular size of the crystal and balometer, fig. 3. The results lie on a line of slope 45°, but not passing through the origin. Sherlock et al suggested that this discrepancy came from broadening of the Rayleigh peak which arises if the phonons in the solid are

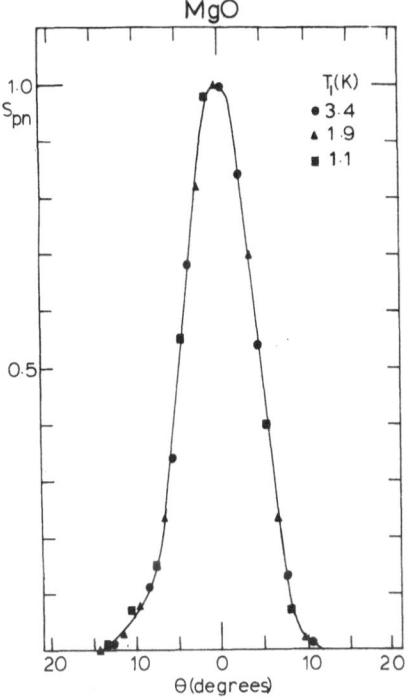

Fig 1: Normalised signal versus
angle for the peak.

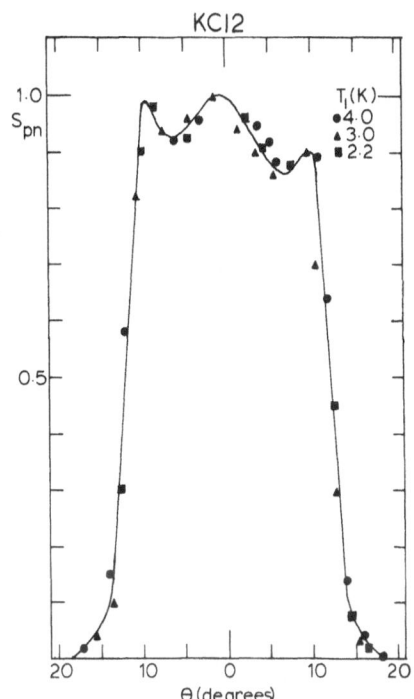

Fig 2: Normalised signal versus
angle for the peak.

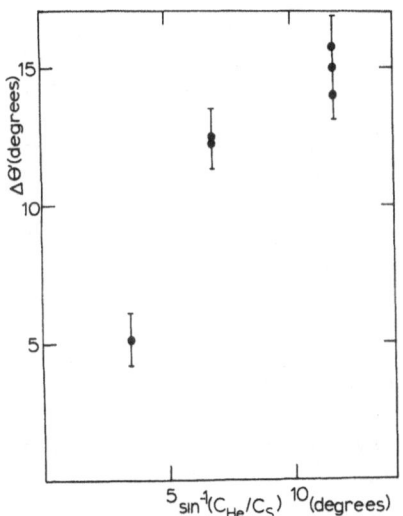

Fig 3: Measured versus theoretical peak width for MgO, NaF, and
KCl in ascending order.

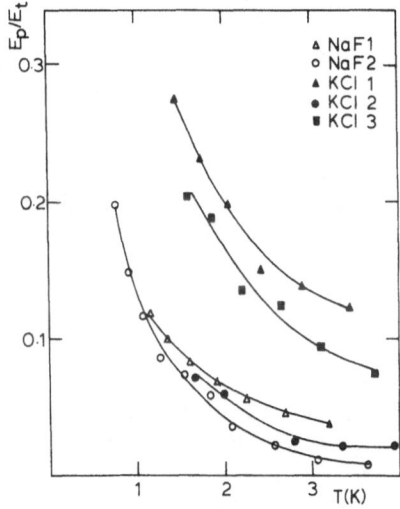

Fig 4: Ratio of the energy in the
peak to the total energy as a
function of source temperature.

strongly scattered via coupling between bulk and surface phonons[3,4].
This attenuation introduces an uncertainty in the surface wave vect-
or $\Delta K_s \approx L_s^{-1}$, L_s is the attenuation length in the solid, this broa-
dens the transmission peak by $\sim \Delta K_s / K_s$. However, to get $\sim 5°$ broadening at phonon frequencies of 100GHz requires L_s to be ~ 0.16 m, implying a density of 10^{10} dislocations per mm^2. This is extremely high
even for a surface stressed by cleaving, and strangely all specimens
require this large value. It is therefore by no means proved that
this mechanism is the cause of the broadening. Comparison between
experimental peak shape and a modified acoustic mismatch theory is
complicated with KCl, because it is not an isotropic solid and shows
strong focusing[5], for example the longitudinal [100] is enhanced by
23 times the isotropic case, adding more structure to the central
peak.

The ratio of the energy transmitted into the peak E_p and background E_b, is given by:-

$$E_p/E_b = \int_0^{\pi/2} S_p(\theta) \sin\theta \, d\theta \, (\int_0^{\pi/2} S_b(\theta) \sin\theta \, d\theta)^{-1}$$

because the angular dependences of S_p and S_b are temperature independent $E_p/E_b = R\, S_p(\theta = 0)/S_b(\theta = 0)$ where R is constant for each
sample. Thus the temperature dependence of E_p/E_b can be easily calculated. In fig 4 values of E_p/E_t ($E_t = E_p + E_b$ = total energy) for
NaF and KCl are shown, for MgO the ratio is a factor of 20 smaller
(this result is tentative as the crystal was diffuse to phonons),
but it exhibits the same qualitive behaviour, the energy in the peak
becomes a smaller fraction of the total energy with increasing temperature. This is consistent with the results from conventional Kapitza measurements, that as the temperature is increased the heat flow
becomes increasingly larger than predicted by acoustic theory.

The general features of the NaF result showing that the angular
distribution can be resolved into peak and background components app-
lies to MgO and KCl. The peak has been identified with the conservation of q_{11}, and the background is responsible for the excess Kapitza conductance. We would expect that these results are generally
applicable to all solid-^{4}He interfaces.

We gratefully acknowledge the Science Research Council for support under grants B/SR/9970 and B/RG/4231 and for a studentship
(G.J.P.) and post doctoral award (K.R.A.Z.).

References
1. Sherlock, Mills, Wyatt 1975, J Phys C: Solid State Phys Vol 8.
2. Little, 1959, Can J Phys 37, 334-49.
   Khalatnikov, 1965 Introduction to Superfluidity (N Y: Benjamin),
   pp 138-146.
3. Haug and Weiss, 1972, Phys Lett A40, 19-21.
4. Peterson and Anderson, 1973, J Low Temp Phys 11, 639-65.
5. Maris (1971) J Ac Soc Am 50, 812-8.

ENERGY TRANSFER BETWEEN A SOLID AND HELIUM BY EXCITED ATOMS

G. N. Crisp, R. A. Sherlock* and A. F. G. Wyatt

Department of Physics, University of Nottingham
University Park, Nottingham NG7 2RD, England
*Department of Physics, University of Waikato
Hamilton, New Zealand

In some recent experiments[1,2,3] it has been shown that the reflectivity of a crystal surface, for phonons incident on it from within the crystal, is sensitive to very thin layers of helium on the crystal surface. The reflectivity drops by ∿ 25% with only ∿ 3 atomic layers of helium and does not decrease further when the He thickness is increased to bulk values. This loss of energy on reflection must be accompanied by a transfer of energy through the interface into the helium. These experiments strongly suggest that the first few layers of He are all important in this energy transfer and they are of central importance to the Kapitza conductance problem. We suggested that single atom excitations were more likely than collective excitations in the thin ^{4}He film and we envisaged that acoustic energy excited He atoms out of bound states and the atoms transported the energy away as kinetic and potential energy[2]. The other hypothesis was that energy is converted into an interface excitation which is then reconverted into low frequency bulk phonons[3].

The direct test of whether atoms can be desorbed by phonons is to repeat the reflection experiment[2] with an additional bolometer mounted 6 mm away from the reflection surface of the crystal. The pressure of gas in the reflection side chamber was measured at 300 K, so thermomolecular pressure corrections have been made.

The crystal heater is pulsed with powers up to 0.35 W and 1 μs duration and the bolometer is monitored with an amplifier, transient recorder and signal averager. The reflection side chamber is initially evacuated before the cell is cooled, so that the reflection surface is free of He. As expected, the bolometer shows no response until He is admitted into the reflection side chamber.

From several runs the picture emerged that there is no detectable signal from the bolometer until 1 monolayer of He has formed.
The signal grows slowly until 2 layers are formed and then it grows
very rapidly to a peak at 2.6 layers.   The signal then decreases
and also begins to change its shape.   This behaviour is shown in
Fig. 1.   (Adsorption isotherms are notoriously prone to systematic
errors and so the film thicknesses which are derived from them
should be treated with some caution.)

The pulse shape for 1-2.6 layers is approximately the same.
There is quite a well defined delay of typically 44 μs after the
crystal heater pulse, the signal rises to a maximum and then decays
extremely slowly.   It has only fallen to 0.36 of its maximum by
1500 μs.   Above 2.6 layers the bolometer signal becomes very mcuh
shorter, which we believe is due to the energy propagating as a
collective mode.

We attribute the transmission of energy, from crystal to the
bolometer, to phonons travelling ballistically through the crystal
to the reflection face where they desorb atoms which travel to the
bolometer.   The only other He dependent mechanism is third sound
which at 50 m/s[5] would take 200 μs to reach the detector which is
much longer than the measured time of 44 μs.   To travel 6 mm in
this time the atoms must have kinetic energy equivalent to 10 K
which with a heater temperature of 7 K and a binding energy $\gtrsim$ 7 K
per atom is not unreasonable.   The increase in signal with coverage

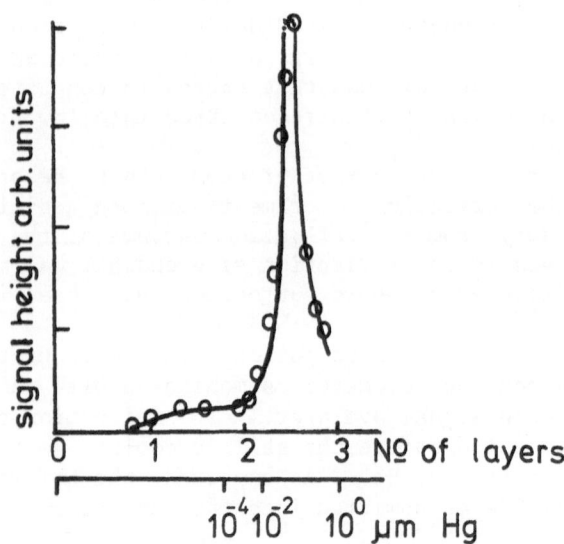

Figure 1: Peak height of energy flux associated with desorbed ^{4}He
atoms from a NaF crystal surface as a function of ^{4}He converage.

and with power implies that a phonon has a certain probability
(< 1) to desorb an atom in a particular layer.   As the number of
atoms in this layer increases and as the number of phonons increases then the number of atoms desorbed increases proportionately.
The probability of desorption is small in the second layer but an
order of magnitude larger in the third layer which is to be expected from the different binding energies in the two layers.   The
reason for the decrease in signal height above 2.6 layers is not
clear.   The possibilities that the detector sensitivity changes
with coverage, the change of mode of propagation alters the energy
flux or that less atoms are desorbed, cannot as yet be separated.

   At low pressures of He it might be expected that the atoms would
travel ballistically from the crystal face to the bolometer.   We
consider two models for desorption which together with ballistic
propagation predict a pulse shape.   In the first model we assume
that the desorbed atoms have an energy spectrum which is given by
shifting the phonon spectrum down by the atom's binding energy (E_s).
For a delta function input of phonons the bolometer signal as a
function of time is

$$S_1(t) \propto \left[\tfrac{1}{2}m(\tfrac{d}{t})^2 + E_b\right]\left[\tfrac{1}{2}m(\tfrac{d}{t})^2 + E_s\right]^2 \frac{d^3}{t^3} \cdot \left[\exp(\tfrac{1}{2}m\frac{d^2}{t^2} + E_s)/kT - 1\right]^{-1}$$

where E_b is the binding energy on the bolometer, T is the temperature of the heater and d is the separation of crystal face and bolometer.   In the second model we assume that the phonons heated the
He film up to a temperature T and then the atoms evaporated.   The
expected signal is then

$$S_2(t) \propto (d/t)^4 \left[\tfrac{1}{2}m(\tfrac{d}{t})^2 + E_b\right]\left[\exp(\tfrac{1}{2}m\frac{d^2}{t^2} + E_s)/kT - 1\right]^{-1} .$$

These models do not predict the very long pulses which are detected.
The long pulses could be due to diffuse rather than ballistic propagation or be an artefact of a bolometer with poor thermal grounding.

   In conclusion it is now clear that the loss in energy of phonons
reflecting off a crystal-He film interface is due to the desorption
of He atoms.   It is not unlikely that the same mechanism occurs at
interfaces of solids and bulk [4]He and the excited atoms rapidly
decay into phonons with a broad angular distribution[6].

   We gratefully acknowledge Dr. D.A. Jones for kindly giving us an
NaF crystal and C.J. Page for help with the early experiments and
the S.R.C. for support under grant B/SR/9970.

1. C.J. Guo and H.J. Maris,Phys.Rev.Lett.,29,855 (1972).
2. A.R. Long et al., J.Low Temp.Phys.,15,516 (1974).
3. H. Kinder and W. Dietsche, Phys.Rev.Lett.,10,578 (1974).
4. L. Meyer, Phys.Rev.,103,1593 (1956).
5. J.H. Scholtz et al.,Phys.Rev.Lett.,32,147 (1974).
6. R.A. Sherlock et al.,J.Phys.C,8,300 (1975).

PATH-INTEGRAL FORMULATION OF THE THEORY OF THERMAL BOUNDARY RESISTANCE

M C Phillips and F W Sheard

Department of Physics, University of Nottingham

University Park, Nottingham NG7 2RD, U K

The thermal contact resistance between two dissimilar insulating solids or a solid and liquid helium arises from the partial transmission of phonons at the interface. In the theories of Khalatnikov and Little it is assumed that there is a temperature drop across the interface but that the media on either side are in thermal equilibrium even though there must be a net heat flux in each medium. These theories also predict a finite temperature discontinuity at an interface between identical media which cannot be true in steady-state heat conduction[1,2]. We have used the Chambers path-integral method[3] of solving the Boltzmann equation to give a self-consistent solution of the transport problem in which this undesirable feature does not arise. In this method the phonon distribution function at a point is calculated by considering the various paths by which phonons may arrive at that point taking account of the probability of being scattered en route. The reflection and transmission of phonons at the boundary gives rise to a non-local thermal conductivity and the temperature distribution must be calculated self-consistently so as to conserve the heat current throughout the system in the steady state. The integral equation for the temperature distribution is difficult to solve even for simple models of the phonon transmission, but we have shown that in one dimension the temperature gradient is constant in both media apart from the discontinuity at the interface. However in general a nonlinear temperature variation near the interface is to be expected[4].

Rather than give the details of this calculation we shall describe a simplified one-dimensional treatment which contains all the essential physics but is not obscured by mathematical complexities. To illustrate the path-integral method we first consider thermal conduction in an infinite solid under a constant temperature gradient

$-dT/dx > 0$. On average phonons travel a mean free path ℓ correspon-
ding to a temperature rise or fall $\Delta T = |dT/dx|\ell$ and phonon populat-
ion difference $(\partial n^o/\partial T)\Delta T$, where $n^o(T)$ is the Bose-Einstein distribution. Phonons arriving at point x from the hotter regions $x' < x$
thus contribute a heat current (from left to right)

$$J_+ = \tfrac{1}{2}\Sigma \; (\partial n^o/\partial T)\hbar\omega v\Delta T$$

where the sum is over all modes, v is the phonon velocity and the
factor $\tfrac{1}{2}$ arises because in one dimension half of the phonons travel
in each direction. Introducing the phonon heat capacity C we have
$J_+ = \tfrac{1}{2}Cv\Delta T$. Similarly phonons arriving from the cooler regions
$x' > x$ give a contribution (from right to left) $J_- = -\tfrac{1}{2}Cv\Delta T$. The
net heat flux $J = J_+ - J_-$ is hence

$$J = Cv\Delta T = Cv\ell|dT/dx| = K|dT/dx|$$

which reproduces the usual formula $K = Cv\ell$ for the thermal conductivity in one dimension.

The analysis may now be extended to the problem of thermal boundary resistance. In figure 1(a) we show the various paths by which
phonons arrive at point A adjacent to the boundary but just inside
medium 2. For path A_1A the temperature drop includes both δT across
the interface and ΔT_1 in a mean free path within medium 1. Since
the path crosses the interface the heat flux is $\tfrac{1}{2}\alpha C_1 v_1(\delta T + \Delta T_1)$,
where the transmission coefficient α is illustrated in figure 1(b).
Those phonons from A_2 which undergo reflection also contribute to
the heat flux J_+ from left to right. Hence

$$J_+ = \tfrac{1}{2}\alpha C_1 v_1(\delta T + \Delta T_1) + \tfrac{1}{2}s C_2 v_2(-\Delta T_2)$$

Those travelling directly from A_2 to A give a reverse flux
$J_- = \tfrac{1}{2}C_2 v_2(-\Delta T_2)$ and so the net heat current $J_A = J_+ - J_-$ is

$$J_A = \tfrac{1}{2}\alpha C_1 v_1(\delta T + \Delta T_1) + \tfrac{1}{2}(1 - s)C_2 v_2\Delta T_2 \qquad (1)$$

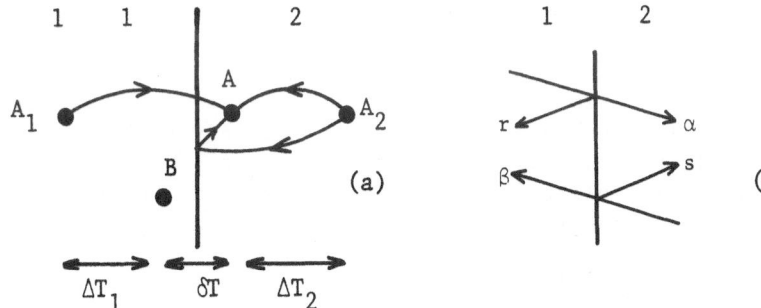

Fig 1(a) Paths for phonons arriv-   Fig 1(b) Definition of transmiss-
ing at A in medium 2.   ion and reflection coefficients.

By a similar argument the heat flux at B adjacent to the interface in medium 1 is

$$J_B = \tfrac{1}{2}\beta C_2 v_2 (\delta T + T_1) + \tfrac{1}{2}(1 - r)C_1 v_1 \Delta T_1$$

Since in the steady state these fluxes must be identical we have the relations $\alpha + r = 1$, $\beta + s = 1$, which express conservation of energy and the detailed-balance relation $\alpha C_1 v_1 = \beta C_2 v_2$. For a three-dimensional Debye solid at low temperatures $C_1/C_2 = (v_2/v_1)^3$ and thus $\beta/\alpha = (v_2/v_1)^2$. In the acoustic-mismatch model this relation between the average transmission coefficients is a consequence of the existence of the critical cone in the medium of lower sound velocity.

We now recognize that the same heat current must exist throughout the bulk media. Thus $J_A = J = C_1 v_1 \Delta T_1 = C_2 v_2 \Delta T_2$. Substituting into eq. (1) and using $1 - s = \beta$ gives

$$J = \tfrac{1}{2}\alpha C_1 v_1 \delta T + \tfrac{1}{2}\alpha J + \tfrac{1}{2}\beta J$$

The first term is just the heat current of the Khalatnikov-Little theory and the remaining terms are corrections arising from the non-equilibrium conditions in the bulk media. The thermal boundary conductance of the interface between media 1 and 2 defined by $h_{12} = J/\delta T = h_{21}$ is thus

$$h_{12} = \frac{\tfrac{1}{2}C_1 v_1 \alpha}{1-\tfrac{1}{2}\alpha-\tfrac{1}{2}\beta} = \frac{\tfrac{1}{2}C_2 v_2 \beta}{1-\tfrac{1}{2}\alpha-\tfrac{1}{2}\beta} \tag{2}$$

We note that for identical media $\alpha = \beta = 1$, $h_{11} = \infty$ and the boundary resistance $R_{11} = h_{11}^{-1}$ vanishes as it should.

An interesting application of this theory is to calculate the thermal resistance of a thin metal foil immersed in liquid helium for the case when the phonon mean free path in the foil is much larger than the thickness. There is then no temperature gradient or thermal resistance inside the foil itself but we have to take into account the multiple reflections of the phonons in passing through the foil. We may then regard the foil as a composite interface separating two baths of liquid helium (medium 2) but with an effective transmission coefficient $\alpha_{eff} = \beta/(2 - \alpha)$, which is obtained by summing the intensities of the multiply reflected beams of phonons[5]. Thus from eq. (2) applied to the case $C_1 = C_2$, $v_1 = v_2$, $\alpha = \beta = \alpha_{eff}$, the effective thermal resistance of the foil is $R_{eff} = (1 - \alpha_{eff})/\tfrac{1}{2}C_2 v_2 \alpha_{eff}$ which, with a little algebra, gives

$$R_{eff} = \frac{2(1-\tfrac{1}{2}\alpha-\tfrac{1}{2}\beta)}{\tfrac{1}{2}C_2 v_2 \beta} = 2R_{12}$$

Here $R_{12} = h_{12}^{-1}$ is the Kapitza resistance at the metal-He interface

given by eq. (2). Despite the multiple reflections inside the foil
the total thermal resistance is therefore just the sum of the boundary resistances at the two interfaces as if each behaved independently.

1. Phillips M C, 1974, Ph.D. thesis, University of Nottingham.
2. Simons S, 1974, J Phys C 7, 4048.
3. Chambers R G, 1952, Proc Phys Soc A65, 458.
4. Budd H and Vannimenus J, 1971, Phys Rev Lett 26, 1637.
5. Challis L J and Sherlock R A, 1970, J Phys C 3, 1193.

THE ROLE OF ELECTRONS IN THE KAPITZA RESISTANCE OF LEAD

L. J. CHALLIS and M. C. PHILLIPS

Department of Physics, University of Nottingham

University Park, Nottingham NG7 2RD, England

In a pure metal, thermal conduction is very largely by conduction electrons so that the Kapitza conductance is determined by the rate at which energy can be transferred to these electrons from the excitations in the helium. Two routes are possible in principle. In one, energy passes first to the phonon system in the metal presumably by processes similar to those for a dielectric. It is then transferred to the electrons. In the other a more direct interaction between the electrons and the helium is envisaged. The first process introduces an additional electron-phonon contact resistance which is not of course present in a dielectric while the second provides an additional conductance. The earlier experimental investigations of these effects were made by comparing the conductances in the normal and superconducting states of a metal. Much of this work is now held to be unreliable because it is thought that surface strain induced perhaps by differential contraction reduces the thermal conductivity in the surface region. The effect should be less in thin foils and measurements on these[1] show that the conductance is somewhat larger in the superconducting state for lead, tin and indium. This suggests that the electron-phonon contact resistance is important: an earlier interpretation of these data as a size effect[1] now seems less likely in view of recent theoretical work[2].

Information on these electronic processes can also be obtained from measurements in high fields. In the present investigation measurements from 1 to 2 K have been made using the AC technique[1] in fields up to 50 kOe. The principal work was on two single crystal discs cut from the same sheet of lead of resistance ratio 3800 ($\rho_{293}/\rho_{4.2}$). This corresponds to an electron mean free path of 30 µm. The plane of the discs is the (111) crystallographic

plane and they were chemically polished to 230 μm (Pb1) and 110 μm
(Pb2).   Measurements were also made on Pb1 after it had been immersed in dilute sulphuric acid for 10 minutes to provide a PbSO$_4$ coating 0.2 μm thick on each surface.

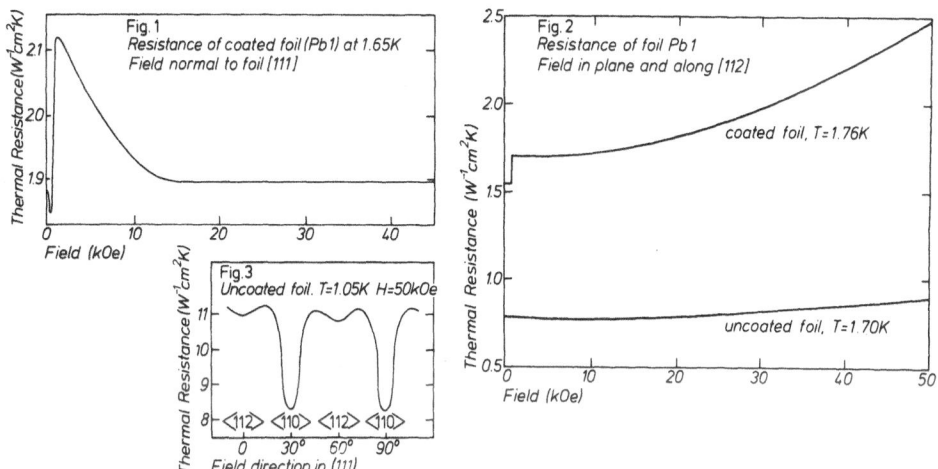

We first describe the results on the coated sample.   The
2000 Å coating is much larger than the dominant phonon wavelengths
in helium and lead of $\lesssim$ 40 and 200 Å respectively and we assume
there is no direct exchange to the electrons from the helium.   The
foil resistance in the normal state at low fields is larger than
that in the superconducting state by $\sim$ 1.5T$^{-3.7}$ W^{-1} cm^2 K.   This
may be compared with the resistance W $\sim$ 6/C$_{vc}$ $\sim$ 0.5T^{-3} W^{-1} cm^2 K of
two layers,each a phonon mean free path thick (3T^{-1} μm) at each surface in which only lattice conductivity can occur.   It seems possible therefore that the additional resistance in the normal state
is due to electron-phonon contact resistance.   The resistance in
the normal state is decreased substantially by fields normal to the
foil, e.g. 12% by 50 kOe at 1.65 K, Figure (1).   This cannot be
due to a reduction in bulk thermal resistance since this is only
0.5% of the resistance at the critical field and we conclude it is
a reduction in the electron-phonon contact resistance.   Measurements in transverse fields show pronounced anisotropy.   The changes
are broadly consistent in magnitude, temperature and field dependence with those expected from the bulk thermal magnetoresistance of
the lead[3].   The main difference is the small decrease in resistance
that occurs in many directions at low fields.   It would seem possible that this is again due to a decrease in the electron-phonon
contact resistance with field.

The increase in resistance of the uncoated foil with transverse
field is much less at the higher temperatures than that of the

coated foil for most field directions (e.g. Figure (2)).   If the
increase for the coated foil is indeed a bulk effect, this indicates that the Kapitza resistance of the uncoated foil decreases
markedly with field.   However, we cannot wholly exclude size
effects (the electron mfp is $\sim$ 1/8 of the foil thickness) which
would be affected by changes in the surface and more work is needed
to clarify this.   For the moment we assume there is a decrease in
Kapitza resistance with field.   The amount of the decrease is not
very well defined because of uncertainties of $\lesssim$ 10% in the absolute
values of the foil resistances, etc., but it seems likely that the
Kapitza resistance is at least halved by a field of 50 kOe at 1.7 K.
The decrease is smaller at lower temperatures and by 1.0 K, the
field dependence of the foil resistance is not significantly changed
by coating.   The behaviour shown in Figure (2) is representative
of other field directions in the (111) plane except those near the
[110] direction.   In these directions, the resistance of the coated
foil does not show the marked increase at high fields apparent in
Figure (2) but is very nearly field independent.   This is consistent with the saturation of magnetoresistance that occurs for directions near to [110].   The resistance of the uncoated foil and so
presumably the Kapitza resistance also varies only weakly with field
in this direction, i.e. the field dependence of the Kapitza resistance is also anisotropic.   The anisotropy of the total resistance
is shown in Figure (3).   Finally we note that measurements at higher temperatures on a second uncoated foil (Pb2) also show much
smaller changes with field, for example in the <112> direction, than
those expected from bulk magnetoresistance and again suggest that
the Kapitza resistance decreases with field at the higher temperatures.

     In summary then our data on a coated sample suggest that there
is an electron-phonon contact resistance in series with the phonon
part of the Kapitza conductance and that it is decreased by magnetic fields.   On an uncoated sample (or rather a sample whose surface is only coated by the effects of chemical polishing and expos-
ure to air), the conductance appears to increase with field at least
at the higher temperatures.   Because of possible size effects,
more work is needed to confirm these conclusions.   Finally, we
should note that field dependent foil resistances have been observed
previously in polycrystalline lead[1] and also in a detailed investigation on single crystal gallium[4].

1. L.J. Challis and R.A. Sherlock, J.Phys.C,3,1193,1970, and Proc.
     11th Int.Conf. on Low Temp.Phys.,St. Andrews 1968, p.571.
2. M.C. Phillips and F.W. Sheard, these proceedings.
3. L.J. Challis, J.D.N. Cheeke and P. Wyder, Proc. 9th Int.Conf. on
     Low Temp.Phys., Columbus 1965, p.839.
4. F. Wagner, F.J. Kollarits and M. Yaqub, Phys.Rev.Letters, 32,
     1117,1974.
     We are grateful to Dr. F.W. Sheard for helpful discussions.

INFLUENCE OF DISLOCATIONS ON THE KAPITZA CONDUCTANCE OF CLEAN

COPPER SURFACES

K.C. Rawlings and J.C.A. van der Sluijs

School of Physical and Molecular Sciences
University College of North Wales
Bangor, Gwynedd, Great Britain, LL57 2UW

An experiment is described in which the variation of the Kapitza conductance of copper with the electropolishing depth of the surface is investigated in the temperature range of 1 to 2K. Preliminary results show a dependence and the role of dislocations in the heat transfer mechanism is discussed.

1. SURFACE PREPARATION AND RESULTS

The surface preparation method developed by Al Naimi and van der Sluijs has been described before (1,2) and consists of keeping annealed, electropolished samples under high vacuum for several days, and doing the experiments without reexposing them to the atmosphere. This treatment produces Kapitza conductance results which reproduce to within 5% for different samples. In the present work the samples consisted of 99.999% pure polycrystalline copper machined to disks 6mm in diameter and 3mm thick. All samples were annealed under vacuum for 1 day and then electropolished in a 30% solution of orthophosphoric acid in water. The electropolishing liquid was rinsed off with distilled water and the samples put to dry in a vacuum dessiccator. After drying the samples were mounted in the sample cell described previously (3) and the cell evacuated to a pressure of 10^{-7}mm Hg, measured at the ion pump. The cell was at the end of a capillary of 1 m long and a bore of 1 mm for the first two samples and of 6 mm for the others.
    Some preliminary results are shown in fig. 1 as a function of the temperature. It appears that there is a significant change due to variation of the electropolishing depth. All results have been obtained after sufficiently long vacuum treatment to make

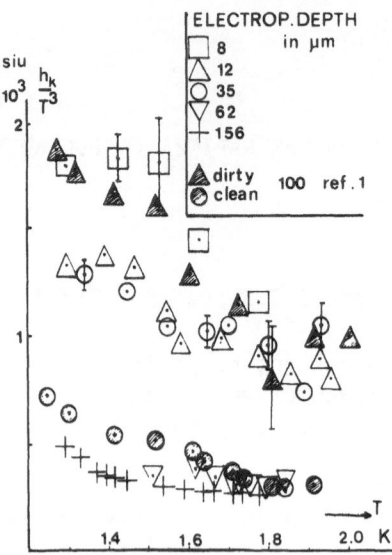

Fig.1. h_k/T^3 as a function of the temperature

subsequent measuring runs reproduce. The results for deep electropolishes agree rather well with earlier work at a depth of 100 μm,
and the results for shallow electropolishes look similar to curves
obtained on samples before vacuum treatment after a 100 um electro-
polish. These curves have been included (1).

2. COMPARISON OF THE RESULTS WITH TRANSFER MODELS.

    In fig.2 the h_k/T^3 has been plotted as a function of the elec-
tropolishing depth for three temperatures. The dependence is one of
a decreasing transfer coefficient for increasing electropolishing
depth and if it is desired to do so a roughly exponential dependence
could be read from the curves. This would be rather nice because the
dislocation density near to the surface falls exponentially with
increasing depth on an unpolished sample and electropolishing leaves
the dislocation density in each layer unchanged from the original
one. Dislocations caused by strains introduced by mounting the sample
with an indium seal would not affect this conclusion as these would
not be concentrated near to the surface. Therefore these cause only
scatter of results between different samples, which is not found.
    It appears that there is one heat transfer process, also opera-
ting at small dislocation densities (4) and one additional process
proportional to the dislocation density. The temperature dependence

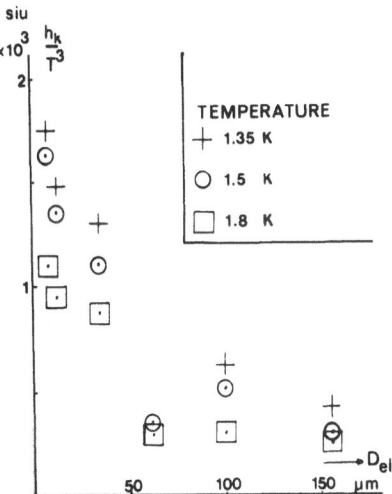

Fig.2. h_k/T^3 as a function of the electropolishing depth.

of the conductance of the former can be worked out to be $T^{1.6}$ in agreement with earlier results (1) and that of the latter $T^{2.6}$.

It would be tempting to equate the second process with the heat transfer by scattering of evanescent waves against dislocations near to the interface(5,6,7). However the similarity with the curves obtained on dirty samples suggests a second possibility. There may be preferred sites for adsorbed impurities on dislocations and grain boundaries. Therefore these are the ones most difficult to remove. This would explain all features of the curves. By consequence the dislocation scattering process may not become dominant at these temperatures unless the cleaning method has been improved further.

This research is supported by an SRC research grant, and an SRC CASE studentship (KCR).

REFERENCES

1. A.E.Al Naimi and J.C.A.van der Sluijs, Cryogenics 13, 722 (1973).
2. A.E.Al Naimi and J.C.A.van der Sluijs, Cryogenics 14, 599 (1974).
3. E.A.Jones and J.C.A.van der Sluijs, Cryogenics 13, 535 (1973).
4. J.C.A.van der Sluijs and A.E.Al Naimi, 2nd Phonon Scatt. Conf.
5. H.Haug and K.Weiss, Phys.Lett. 40A, 1 (1972).
6. A.C. Anderson and W.L. Johnson, J.Low Temp.Phys. 7, 1 (1972).
7. J.L. Opsal and G.L. Pollack, Phys.Rev. 9A, 2227 (1974).

INFLUENCE OF SURFACE CONTAMINATION ON THE KAPITZA CONDUCTANCE OF

COPPER, SILVER, MOLYBDENUM AND BERYLLIUM

J.C.A. van der Sluijs and A.E. Al Naimi*

School of Physical and Molecular Sciences, University
College of North Wales, Bangor, Gwynedd, Great Britain
*Present address: University of Technology, Baghdad,Iraq

From experimental data published previously(1,2) the behaviour
of the Kapitza conductance of various types of surfaces is analysed
in terms of contamination, dislocations and strains.

1. **General approach and summary of the experimental results.**
An analysis is given of experimental work by our group which is
aimed at obtaining reproducible Kapitza conductance results at temperatures between 1 and 2K, finding the cause of bad reproducibility
and studying conductance of surfaces showing reproducibility.

Methods and results have been described before (1,2) and only a
short summary will be given. Because of the apparent influence of
contamination the design of the vacuum system is crucial. The
experimental results are summarised in figure 1.

The diagrams show that the h_k/T^3 of dirty samples, where h_k is
the Kapitza conductance, have a structure which can be enhanced by
depositing material and reduced by vacuum treatment. There is no
significant dependence upon electropolishing depth, which should vary
the dislocation density near the interface and the results obtained
on samples subject to extensive vacuum treatment become reproducible.
Results are shown for only one metal at the time, but insofar checked
all four metals behave similarly.

2. **Comparison with theories and conclusions.** Vacuum treatment
of annealed and electropolished samples produces surfaces whose
Kapitza conductance between 1 and 2K behaves differently from that
found before, because untreated surfaces have a bad reproducibility
a difference of two orders of magnitude with the acoustic mismatch
theory and a non uniform temperature dependence of h_k around T^3,
whereas the h_k for treated surfaces is fairly reproducible, differs
one order of magnitude from the acoustic mismatch, and has a fairly

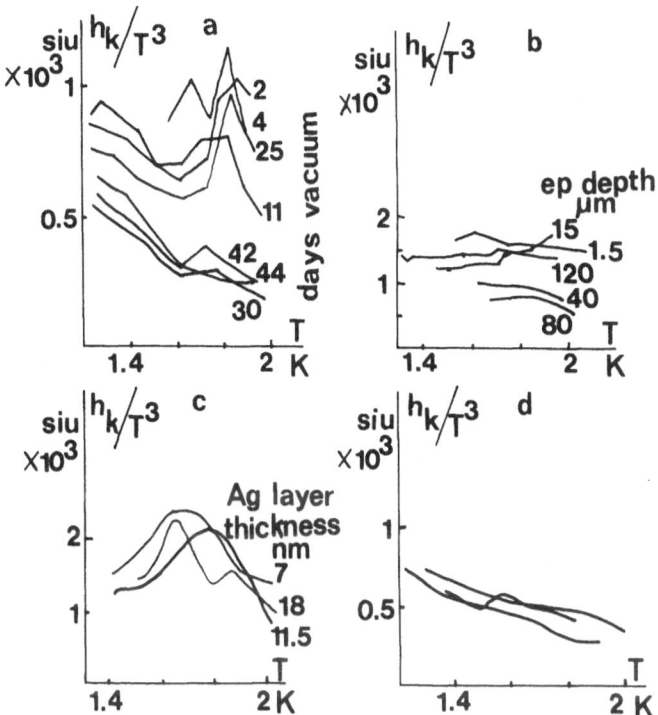

fig.1. Experimental h_k/T^3. a. For silver and varying pumping time. b. For dirty copper and varying electropolishing depth. c. For copper sample with silver coating. d. Lowest for 3 copper samples.

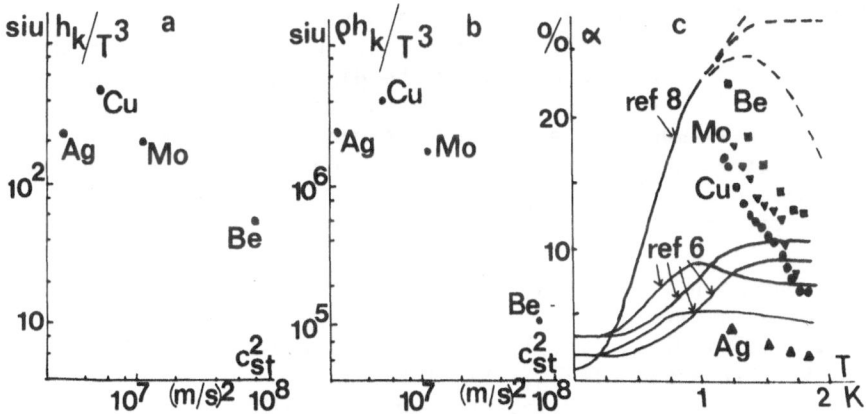

fig.2. Comparison with theory.   a and b: Material constants. c: Temperature dependence.

uniform temperature dependence of about $T^{1.5}$.

The following models will be compared with these two cases.
1. Acoustic mismatch(3) requiring reproducibility, h_k one or two orders of magnitude below the observed one, a T^3 temperature dependence and a $1/\rho c_t^3$ dependence on the material constants of the solid, with ρ the density and c_t the transverse velocity of sound.
2. Static dislocations or strains near the surface require a T^3 temperature dependence(4,5), with a dense helium layer added non uniform(6). The dependence on the properties of the solid is $1/c_t^x$ with X < 2.
3. Adsorbed impurities give a non uniform temperature dependence and a dependence upon properties of the solid of $1/\rho c_t^3$ or less (7).

The order of magnitude of the experimental data rules out the acoustic mismatch and the temperature dependence the acoustic mismatch and the "bare" dislocation models. For untreated surfaces the adsorbed impurity model is the only one to describe the data qualitatively. For the treated surface the choice between the dislocation + dense helium layer and the adsorbed impurity model must be made by further analysis.   The analysis on the basis of the material constants is given in fig.2a and b. Excluding a ρ dependence the result is $h_k \sim 1/c_t^{1.3}$ and with ρ dependence $\sim 1/ c_t^{2.6}$. From the first result follows a prediction for copper of h_k/T^3 of $32W/m^2K^4$, not far from the acoustic mismatch and much below the clean surface results.   The $1/ c_t^{2.6}$ is quite compatible with the adsorbed impurity model.

In fig.2c data have been recalculated as a transfer coefficient α (8) together with curves extrapolated from those of Anderson and Johnson(5) by Challis(8) and theoretical curves from copper by Opsal and Pollock(6). It appears that all our curves suggest a falling branch above 1K compatible with the low temperature data. The theoretical curves based on the dislocation plus dense layer model behave qualitatively similarly, but the detailed shape is quite different.

Therefore it may be concluded that the untreated surfaces are described quite well by the adsorbed impurity model, and that such surfaces are instable leading to bad reproducibility. The treated surfaces may be described either by the dislocation plus dense layer model or the adsorbed impurity model, but in either case there are some flaws which make the argument unconvincing. Further experimental work is required and some of this will be described in another paper at this meeting.

This research is supported by an SRC research grant.

References
1) E.A. Jones and J.C.A. van der Sluijs, Cryogenics 13, 535 (1973).
2) A.E. Al Naimi and J.C.A. van der Sluijs, Cryogenics 13, 722 (1973); ib. 14, 599 (1974); ib. 15, in press (1975).
3) I.M. Khalatnikov, ZETF 22, 687 (1952).
4) H. Haug and K. Weiss, Phys.Lett. 40A, 1 (1972).
5) A.C. Anderson and W.L. Johnson, J.Low Temp.Phys. 7, 1 (1972)
6) J.L. Opsal and G.L. Pollack, Phys.Rev. 9A, 2227 (1974).
7) J.C.A. van der Sluijs, E.A. Jones and A.E. Al Naimi, Cryogenics 14, 95 (1974).
8) L.J. Challis, J.Phys.C. 7, 481 (1974).

A DIRECT APPROACH TO FINITE PHONON TRANSPORT

(REALISTIC TRANSFER AT BOUNDARIES)

W. Schneider and M. Wagner

University of Stuttgart, Institute of

Theoretical Physics, 7000 Stuttgart 80

Recently phase-transition like phenomena in phonon transport have gained some interest [1,2]. However, in these systems the phonon flux is finite and therefore the conventional transport assumptions have to be discussed carefully. One of the unsolved problems is the energy transfer at the boundaries. Conventionally an externally fixed energy transfer is postulated. In our approach a more realistic view is taken. We let external particles collide with the end atoms of a linear chain. In this way an energy flux is established. The chain may also contain an impurity molecule at some site. The flux created in this manner not only depends on the state of the external particles, but also on the reaction of the system itself. The external particles arrive in a stochastic sequence which is governed by a modified Maxwell-Boltzmann-distribution.

For our mathematical procedure a special application of the classical response-formalism turns out to be very appropriate [3]. The relevant stochastic process exhibits great difficulties. Therefore, we have to calculate each sample function exactly, and in succession to average over several sample functions.

For a test of our method the following well-known results have been verified: (1) The equipartition in equilibrium. (2) The connexion between fluctuation and dissipation. (3) The temperature dependence of the lattice conductivity. This may be compared with measurements which have been made at crystals of small dimension (cca 10 Å), lacking both symmetry and periodicity [4] (fig. 1,2). In contrast to earlier explanations our calculation is of first principle nature.

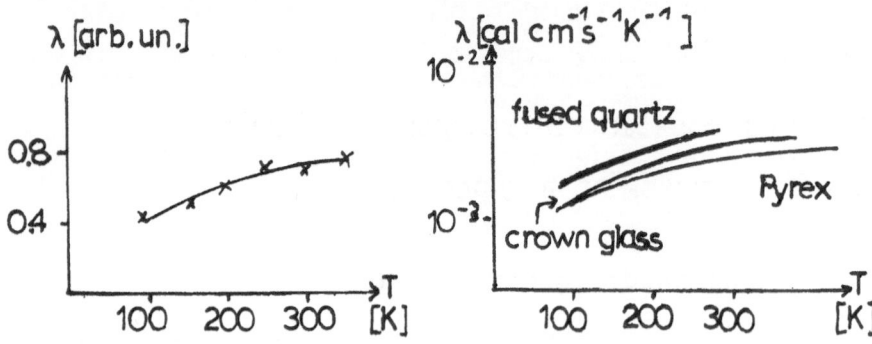

Fig. 1: theoretical curve   Fig. 2: exp. curve [4]
Temperature dependence of the lattice conductivity.

       From these findings we are confident that also the
data on more microscopic properties are very reliable.
We have performed a sequence of studies. (a) time scale
problems of the flux and energy, (b) fluctuations and its
dependence on temperature, temperature gradient and
internal parameters, (c) relaxation, (d) spacial energy
structure, (e) local transport excitations. For lack of
space we only present here two results in detail:
(1) The behaviour of the flux, viewed within several
time scales (fig. 3). The large time scale behaviour,
i.e. a mean value behaviour, appears after about 10^4
external collisions. However, within smaller scales the
fluctuations may produce interesting cooperative phenomena in time and/or in space. For example the flux may
temporarily go into the "wrong" direction. (2) In fig. 4
we show the energy structures with an ideal chain and a
disturbed one containing a molecular impurity (local
mode). At the end of the chain thermal equilibrium at two
different temperatures is assumed. We recognize that in
contrast to the ideal case in the disturbed chain there
is no monotonic change of the "temperature". This is a
strong violation of one of the transport postulates. The
energy indentation appears in the neighbourhood of the
defect and does not even disappear if the temperature is
much larger than the characteristic local mode temperature $\theta = \hbar \omega_\ell /k$.
       We should like to note two additional advantages of
our algorithm. There is no difficulty (1) to include
phase correlated external forces, (2) to extend the
formalism to real crystals, provided the eigenvectors of

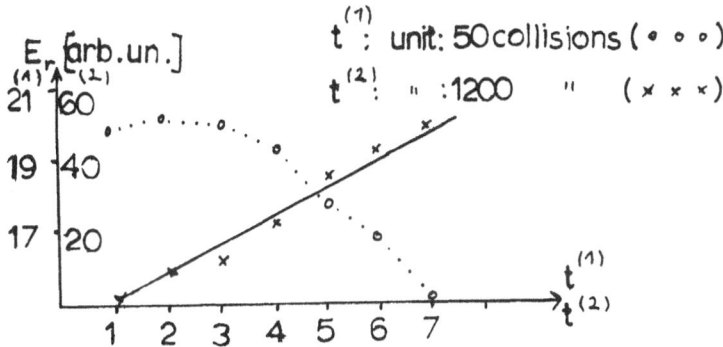

Fig. 3: Large (solid line) and small (dotted line) scale
behaviour of the flux. E_r is the energy, transferred from
the boundary of higher temperature to that of lower one.

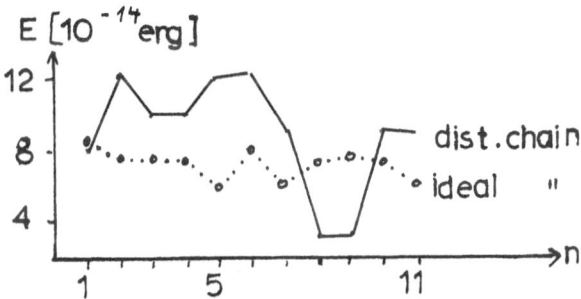

Fig. 4: Energy structure between two points of local
equilibrium with different temperature (T(n=1)>T(n=11)).
n: chain index.

them are known (see e.g.[6]). For the detail of our theoretical method we refer to a forthcoming paper [5]. It
appears that our direct approach to the phonon transport
problem is well suited to give valuable insight for the
discussion of the validity of conventional transport
postulates.

References
[1] M. Wagner, in "Cooperative Phenomena", ed. H. Haken
    and M. Wagner, Springer, Berlin (1973)
[2] F. Kaiser, M. Wagner, Sol. State Comm. 15, 945 (1974)
[3] M. Wagner , Z. Phys. 206, 131 (1967)
[4] C. Kittel, Phys. Rev. 75, 972 (1949)
[5] W. Schneider, M. Wagner, to be published
[6] W. Bluthardt, W. Schneider, M. Wagner,
    phys. stat. sol. (b) 56, 453 (1973)

REFLECTION OF PHONONS IN LIQUID ^{4}He FROM SOLID SURFACES

A F G Wyatt, N A Lockerbie, K R A Ziebeck and R A
Sherlock

Department of Physics, University of Nottingham
University Park, Nottingham NG7 2RD, U K

The interface between a solid and liquid helium has interesting properties and affects phonons in a way which is not yet understood. The system is being probed by transmitting and reflecting pulses of phonons. There are four possible arrangements of the generator and detectors and these correspond to a) reflecting from the solid side[1], b) reflecting from the liquid side, c) transmitting from the solid to the liquid[2], and d) from the liquid to the solid[3]. The results of experiments using a) and c) have been reported. At Nottingham we are using b) and d) and in this paper we should like to concentrate on b) and report on our first experiments.

If we reflect a pencil of phonons from a flat surface with no surface states then there is translational symmetry which ensures that the angle of reflection is equal to the angle of incidence. If the angle of incidence is outside the critical cone (i.e. $\sin^{-1} c_{\text{He}}/c_s$) then the reflectivity is 1 and even if it is inside the critical cone the reflectivity is ~ 0.997 because of the large acoustic mismatch. So we expect in this case a specularly reflected pencil with the same intensity (to within 0.3%) as the incident pencil.

In contrast if the surface is rough but again without surface states then we expect phase shifts in the reflected waves of $2hk_\perp$ where h is a measure of the depth of the roughness and $k_\perp$ is the perpendicular component of the incident wave vector. If the phase shifts are large then reflection will appear diffuse rather than specular. So we should expect that a surface will show more specular reflection for low frequency phonons than for high and at large angles of incidence rather than normal incidence. However the phonon frequencies are unchanged.

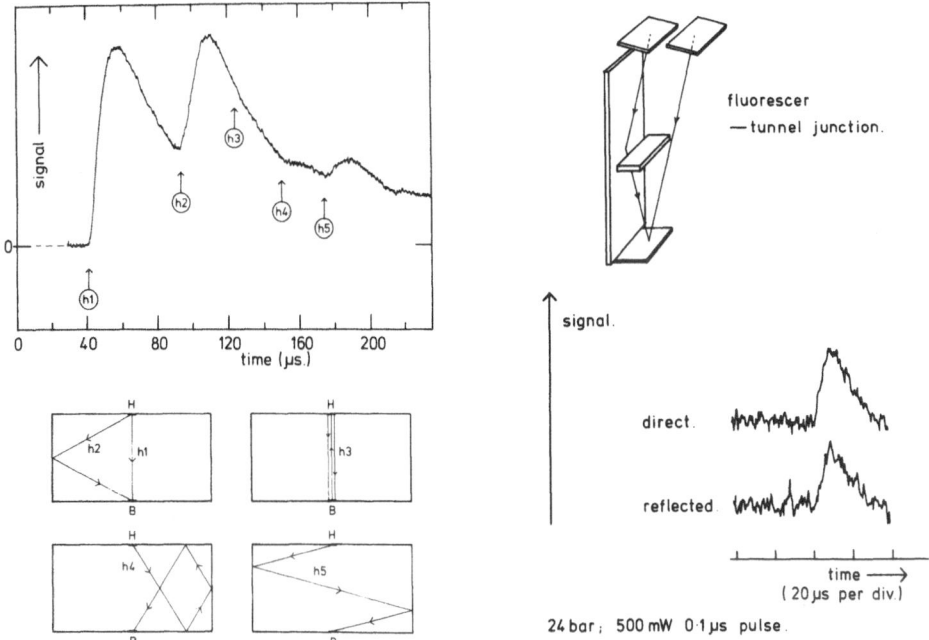

Figure 1: Heater-bolometer signal at 0 bar pressure for 1μs 0.5mW pulse. Below: deduced pulse trajectories within the cylindrical cavity of liquid helium.

Figure 2: Comparison between direct and reflected (from a glass microscope slide) fluorescer-tunnel junction signals. The propagation length is 5 cms.

For a flat surface with surface states the reflected phonons will depend on the inelastic scattering from these states. The surface states may be extended interface waves or localised states which break the translational symmetry and change energy and momentum. The incoming phonon can interact with these states raising them from their ground state if they are thermally unoccupied and both increasing and decreasing their energy if they are thermally excited. We shall then expect that the reflected phonons will have a different frequency and angle to the incident ones and by observing these reflected phonons we should be able to investigate these surface states.

Our preliminary experiments are at nearly normal and oblique incidence with high and low phonon frequencies. For low frequencies we use liquid ^{4}He at 0.1K and at SVP. At this pressure all the phonons with $\hbar\omega/k < 9.5$K decay to low frequencies via the 3 phonon decay process[4]. A thin film heater on a glass slide is held opposite a broad band, carbon film bolometer by a section of glass tube, 1 cm long and 2 cm diameter. The heater and detector were along the symmetry axis and were immersed in liquid helium which filled the system. Phonons emitted from heater during the 1μs pulse rapidly deg-

rade in frequency and travel with the ultrasonic velocity. The signal which was detected showed not only phonons which propagated directly but other pulses which arrived later at well defined times.
These we attribute to reflections off the glass surfaces and the
identified paths are shown in figure 1 together with their computed
flight times shown by arrows on the received signal. From this result we conclude that low frequency phonons mainly reflect specularly
at all angles of incidence because diffuse reflection would give
quite different pulse shapes as all paths would then contribute to
the detected signal.

At high pressures we can study phonons of high frequency as
they propagate without decay over many centimetres[4]. With an Al/Al
oxide/Al superconducting tunnel junction we can study 2Δ phonons with
$\hbar\omega/k$ = 4.2K. For oblique incidence we use the arrangement shown in
figure 2. Two identical heaters are arranged so that the direct path
from one is identical to the specularly reflected path from the other
to the detector. As can be seen the two signals are very similar
in shape and the one reflected off the microscope slide is $\sim$ 80% of
the direct beam. The pulses have been broadened by their 5 cm path
in the helium probably by small angle 4 phonon processes. At normal
incidence we only have the result for reflection off a metal film
evaporated on to a glass slide and it's apparent that the specular
reflection of these high frequency phonons is small. For a wavelength of 40Å the surface would only have to rough on the scale of 10Å
to account for this.

The deviations from perfect specular reflection must be examined in more detail to see how surface states affect the reflectivity
as they surely must from the evidence of the transmission experiments[2,3].

1. C-J Guo and H J Maris, Phys Rev Lett 29, 855 (1972)
   A R Long, R A Sherlock and A F G Wyatt, J Low Temp Physics 15,
   523 (1974)
   H Kinder and W Dietsche, Phys Rev Lett 10, 578 (1974)
2. R A Sherlock, N G Mills and A F G Wyatt, J Phys C 8, 300 (1975)
3. A F G Wyatt and G J Page to be published
4. R C Dynes and V Narayanamurti, Phys Rev Lett 33, 1195 (1974)
   A F G Wyatt, N A Lockerbie and R A Sherlock, Phys Rev Lett 33,
   1425 (1974)

THE THERMAL BOUNDARY RESISTANCE AT EPOXY-RESIN/METAL INTERFACES AT LIQUID HELIUM TEMPERATURES

F. F. T. de Araujo and H. M. Rosenberg

Clarendon Laboratory, University of Oxford

Oxford, OX1 3PU, U.K.

Abstract

The thermal boundary resistance, R_B, in the range 1.4 to 20 K has been measured between an epoxy resin (Epikote 828) and Cu, Ag, Au, Al, Sn and Pb. In all cases the magnitude of R_B was about an order of magnitude higher than that predicted by the theory of Little (1959) and its temperature dependence instead of being T^{-3}, ranged from $T^{-1.5}$ to T^{-2}. Various possible reasons for the extra resistance are discussed. ------------

In order to interpret some experiments on the heat conductivity of metal-powder/epoxy composite materials[1] we have measured the thermal boundary resistance at the interface of epoxy-resin/metal surfaces. The temperature range covered was 1.4 to 20 K and the metals investigated were Cu, Ag, Au, Al, Sn and Pb. The resin was Epikote 828, with Epikure NMA hardener and BDMA accelerator used in the proportions 100:90:0.5 by weight respectively. The specimens were made in the form of a sandwich of two identical metal disks separated by a thin layer of ($\sim$30 μm) of epoxy. Au-Fe:Chromel thermocouples were inserted in the disks in order to measure the temperature gradient across the sandwich when a known quantity of heat was generated electrically in a heater fixed to one disk. The other disk was anchored to a cold sink in the apparatus.

Although the thermocouples were embedded in the metal disks, the thermal resistance of the metals could be neglected since it was much less than that of the resin film and the boundary resistance. However, in order to calculate the boundary resistance a value for the thermal conductivity of the resin had to be assumed.

Experimental Results

Sets of experimental values for the thermal boundary resistance between resin and metal are shown in the figure. The main
features are that the resistance is quite small (compared with
that of the resin itself) above 4 K but that it increases below
that temperature and in all cases its temperature dependence lies
between $T^{-1.5}$ to T^{-2}. A similar result for copper has been
reported by Schmidt[2].

Discussion

The theory of Little[3] shows that the boundary resistance
between two media is due to an acoustic mismatch effect at the
boundary which gives rise to an extra phonon scattering. The
theory predicts that the boundary scattering should vary as T^{-3}.
Calculations based on the theory for the epoxy/metal boundaries
which we have investigated show that not only is the T^{-3} temperature dependence in disagreement with experiment, but that the
actual magnitude of the effect is appreciably larger than the
theory would predict. At 4 K the experimental value is about 10
times larger than the theoretical one. Only when the temperature
approaches 1 K do the theoretical and experimental values tend to
agree with one another. Peterson and Anderson[4] have shown that
quite good agreement can be achieved in some cases as $T \rightarrow 0$ K.

Whilst the discrepancy between theory and experiment could be
accounted for by experimental error e.g. in measurements of the
temperature gradient across the resin or in the actual value used
for its thermal conductivity, this would appear to be unlikely
since the disagreement occurs for several different metals and in
addition, as already stated, it is confirmed by an independent
determination[2].

There seem to be three possible causes for the discrepancy:
(a) The conductivity of the resin film may be different from that
of the bulk resin due to extra strain caused by differential
thermal expansion between the resin and the metal. Experiments
are in progress to check whether the resin conductivity is strain
dependent.
(b) Since the agreement between theory and experiment for
dielectric/epoxy boundaries is much better than that which we have
observed between metal/epoxy boundaries, the discrepancy in the
case of metals might be due to some electron/phonon scattering
mechanism - although this should only be important if one dimension
of the metal is small compared with the phonon mean free path in
the metal[5], and indeed, our results for superconducting Sn (below
3.7 K) and Pb show no significantly different trends from those for
the other metals.
(c) There might be inhomogeneities in the contact between the

resin and the metal interface. If these were of the order of, or
greater than, the phonon wavelength then the effective area of
contact would be frequency dependent. This would also lead to a
higher value of the boundary resistance as the temperature is
increased.

It would appear, both from our experiments and those of other
workers, that the mechanism controlling the thermal boundary
resistance for metals is by no means fully understood and that
further investigations would be desirable.

References

(1) F.F.T. de Araujo & H.M. Rosenberg, Proc. Intern. Conf. on
     Composite Materials (1975) to be published.
(2) C. Schmidt, Cryogenics 15, 17 (1975).
(3) W.A. Little, Can. J. Phys. 37, 334 (1959).
(4) R.E. Peterson and A.C. Anderson, J. Low Temp. Phys. 11, 639
     (1973).
(5) S.G. O'Hara and A.C. Anderson, J. Phys. Chem. Solids 35, 1677
     (1974).

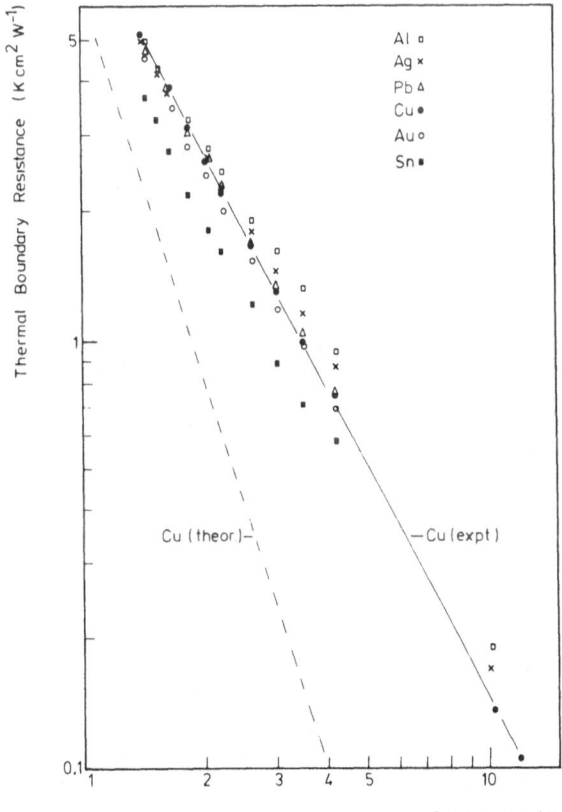

DETERMINATION OF KAPITZA CONDUCTANCE FROM THE VELOCITY OF DESTRUCTION OF SUPERCONDUCTIVITY BY CURRENT*

W. C. Overton, Jr. and H. Weinstock[†]

Los Alamos Scientific Laboratory

Los Alamos, New Mexico   87544, USA

Kapitza conductance, so strongly dependent upon surface preparation, is virtually impossible to determine for thin conducting wires by direct measurement. Here, we describe a technique for determining this conductance indirectly for superconducting wires, and report on results obtained for Ta wires immersed in liquid helium and embedded in Stycast 2850FT epoxy.

The physical measurement required is the velocity of propagation of a normal-superconducting (N-S) interface as it moves down a wire carrying a current I. This current, imposed prior to the onset of the N-S interface propagation, must be greater than a threshold value (I_i) for the initiation of such propagation and less than the superconducting critical current (I_c) appropriate to the ambient bath temperature (T_B) and magnetic field (which is zero for the present investigation). While a normal region may be produced by various methods, we have done so by the application of a heat pulse to a spot near one end of the wire. The local temperature at this spot exceeds the superconducting critical temperature corresponding to the current I. The subsequent Joule heating $\rho_N J^2$ (where J = current density and ρ_N = residual resistivity) causes expansion of the N-region and subsequent contraction of the two S-regions on either side of the original hot spot. The velocity of N-S interface propagation remains constant, for constant I, until the normal state prevails over the entire length of wire.

The velocity of interface propagation is measured by a transit time method. The voltage to ground (which is fixed at the end of the wire farthest from the initial hot spot) is monitored by pickoff probes at two points along the wire. Photographs of a dual trace oscilloscope screen are taken of voltage vs. time, with the arrival of the interface at each probe denoted by a sharp change in slope of the corresponding trace.

To relate the interface velocity to Kapitza conductance, it is necessary to match solutions of the heat balance equations at the interface boundary and account for heat absorbed at the interface. Specifically, this first requirement involves finding solutions to two second order heat balance equations in a moving coordinate frame $z \equiv x - vt$, for which $z = 0$ locates the interface position:

$$\frac{d}{dz} K_N [T_N(z)] \frac{dT_N}{dz} + v C_N [T_N(z)] \frac{dT_N}{dz} + \rho_N J^2 + \frac{2H_N}{r} (T_N^4 - T_B^4) = 0, \qquad (1)$$

$$\frac{d}{dz} K_S [T_S(z)] \frac{dT_S}{dz} + v C_S [T_S(z)] \frac{dT_S}{dz} + \frac{2H_S}{r} (T_S^4 - T_B^4) = 0, \qquad (2)$$

where v = interface velocity, H_N, H_S = Khalatnikov "radiation" constants, K_N, K_S = longitudinal thermal conductivities, and C_N, C_S = heat capacities per unit volume. Equation 2 may be solved exactly for $T_S(z)$, but certain approximations must be introduced to find a suitable solution to Eq. 1. Eq. 1 already omits the Thomson heat term; this is justified because it is $O(10^{-4} \rho_N J^2)$.

Accounting for the heat absorbed at the interface, we use the first order equation

$$[-K_N(T)(dT_N/dz)]_0 + [K_S(T)(dT_S/dz)]_0 = vQ_{eff} , \qquad (3)$$

where Q_{eff} is the heat absorbed at the interface. In type I superconductors Q_{eff} is the ordinary thermodynamic latent heat, and in type II materials it involves the entropy corresponding to specific heat anomaly.

A complete analysis of the simultaneous solution of the above three equations will be presented elsewhere.[1] The results of this analysis will, however, be stated for the limiting case for which $v = 0$. In this limit, the right-hand side of Eq. 3 is zero, so that the exact form of Q_{eff} is of no importance. Although we have taken into account the fact that in general $H_N \neq H_S$, the present space limitations require that we consider only data near $T_c - 4.495$ K for our Ta samples — for which $H_N \approx H_S \equiv H$.[2]

For the zero velocity limit, the solution is reduced to two equations which must be evaluated simultaneously to find a value for H:

$$r\rho_N J_i^2 / [2H(T_t^4 - T_B^4)] = (< T_N^2 > + T_B^2)/(T_t^2 + T_B^2)$$

$$+ [K_S(< T_N^2 > + T_B^2)/(2K_N T_t^2)]^{1/2} \qquad (4);$$

$$T_{-\infty}^2 = T_B^2 + r\rho_N J_i^2 / [2H(< T_N^2 > + T_B^2)] \qquad (5);$$

where $< T_N^2 > \equiv (T_\infty^2 + T_{-\infty}T_t + T_t^2)/3$, and J_i is the current density associated with the threshold current I_i.

Although there is a minimum interface velocity (~ 10 cm/sec) which can be measured due to limitations of the medium-bandwidth electronics used, extrapolating to $v = 0$ the experimentally determined I vs. v data at a fixed T_B, provides a reasonably accurate ($\sim 2\%$ precision) value for I_i. Alternately, I_i has been determined by direct static dc measurements. In these, we merely monitor the voltage developed between the pick-off probes after application of a heat pulse. I_i is found to be that current for which the measured voltage goes somewhat discontinuously from zero to a sizable finite value. For Ta embedded in epoxy, the I_i's found in this manner agree quite well with those obtained using the I vs. v extrapolation technique. For Ta immersed in liquid helium, there is considerable uncertainty in locating the center of the resistive transition. The problem arises because there are values close to I_i for which partial propagation occurs, i.e., the N-S interface travels part way between the probes and becomes stationary in that region. Nevertheless, the (less precise) static dc method for determining I_i is far simpler and less time consuming than the dynamic method we have described above.

Once I_i has been evaluated, H may be found from the solution of Eqs. 4 and 5 for given values of K_N, K_S and T_t. T_t can be found utilizing the Sibsbee relation $I = 5rH(T_t)$ where I is in amps, r in cm and $H(T_t)$ is in Oe. $H(T_t)$, in turn depends on whether the wire is type I or type II material. For type I, $H(T_t) = H_0(1 - T_t^2/T_c^2)$, and for type II, the precise relationship must be determined from independent experimental data. For Nb, and also type II Ta, the specific heat anomalies occur at T_t values corresponding to H_c (thermodynamic). Thus we choose $T_t = T_c(1 - J_i/J_0)^{1/2}$. Then, using independently-measured K_N, K_S in (4), (5), we solve for $H = H_N = H_S$. We obtain the results: $H(0.025$ cm diam. Ta in LHe at 4 K$) = 4.9 \times 10^{-5}$; $H(0.0178$ cm diam. Ta in epoxy at 4 K$) = 1.2 \times 10^{-4}$; $H(0.0127$ cm diam. Ta in epoxy at 4 K$) = 1.1 \times 10^{-4}$. Above units are W cm^{-2} K^{-4}.

These results can be expressed in W cm^{-2} K^{-1} upon multiplying by $< T^3 >$. The corresponding values are h = 0.015, 0.036, 0.033, and 0.085, respectively. The equivalent value h(Ta — LHe at 4 K) = 0.085 is smaller than that quoted for any metal by Cheeke.[2] This indicates that the nonequilibrium dynamical process, with large ΔT and "short time constant, comes closer to agreement with Khalatnikov theory.

*Work performed under the auspices of the U.S.E.R.D.A.

†Visiting Staff Member; permanent address: Physics Dept., IIT, Chicago, IL 60616, USA

1. W. C. Overton, Jr., to be published.
2. L. C. Challis, Proc. of the VIIth Intl. Conf. on Low Temp. Phys, G. M. Graham and A. C. Hollis Hallett eds. (U. of Toronto Press, 1961) p. 476; and J.D.N.Cheeke, Cryogenics 10, 463 (1970).

COMPUTER-SIMULATED SCATTERING OF ENVELOPE SOLITON FROM IMPURITY

AND INTERFACE IN A ONE-DIMENSIONAL NONLINEAR LATTICE

Tetsuro Sakuma, Tsuneyoshi Nakayama and Fumio Yoshida

Department of Engineering Science, Hokkaido University

Sapporo 060, Japan

In 1970, Tappert and Varma[1] showed that the behavior of heat pulse propagation in solids is governed by the nonlinear Schrödinger equation under certain conditions on phonon dispersions and lattice anharmonicity in a continuum limit. The solitary wave (soliton) solution resulting from the equation is called envelope soliton (E soliton). The real crystals, however, contain some impurities or interfaces, so that the soliton propagating in this anharmonic crystals will suffer considerable perturbation from them. In this paper, we investigate numerically the interaction of E soliton with a mass impurity and an interface in a one-dimensional nonlinear lattice.

In order to obtain the nonlinear Schrödinger equation including a mass impurity, we start with a one-dimensional nonlinear lattice which consists of N-1 identical particles and one mass inpurity. Assume that the potential contains the terms up to quartic in the displacement of each particles, then the dynamical equation of this lattice may be written by

$$m_i \frac{d^2 y_i}{dt^2} = \varkappa [(y_{i+1} - 2y_i + y_{i-1}) + \alpha\{(y_{i+1} - y_i)^2$$
$$- (y_i - y_{i-1})^2\} + \beta\{(y_{i+1} - y_i)^3 - (y_i - y_{i-1})^3\}], \quad (1)$$
$$i = 1, 2, \cdots\cdots, N$$

where the subscript i refers to the i-th particle, y the displacement of the particle and $\varkappa$ the linear spring constant. α and β are the parameters directly related to the relative weakness of anharmonic forces. Suppose that there exists a mass impurity located at j-th site.

$$m_i = m_o + \Delta m \delta_{ij} \,. \tag{2}$$

m_o is the mass of N-1 identical particles.

Substituting (2) into (1) and taking a continuum limit, we obtain

$$[1 + \Delta \rho h \delta(x - x_o)]u_{tt} = (1 + \epsilon P u_x + \epsilon^2 Q u_x^2)u_{xx}$$
$$+ (h^2/12)u_{xxxx} \,, \tag{3}$$

where h is the equilibrium lattice separation, $\Delta \rho = \Delta m/m_o$, $P = 2\alpha h^2$, $Q = 3\beta h^4$ and the values of $\chi h^2/m_o$ is normalized to unity. In deriving (3), we have introduced the dimensionless amplitude u by

$$\epsilon u = y/h \,, \tag{4}$$

where ϵ is a small parameter, since we are considering a small but finite amplitude.

Now we use the reductive perturbation method developed by Taniuti and Yajima[2] to obtain the nonlinear Schrödinger equation for interacting E solitons.

$$u = \sum_{m=o}^{\infty} \sum_{l=-\infty}^{\infty} \epsilon^m \phi(1,m)_l \, \mathrm{i}1(kx-\omega t) \,. \tag{5}$$

Assuming the smallness of impurity mass difference by

$$\Delta \rho = \epsilon \gamma \,, \tag{6}$$

and defining the slowly varying variable ξ and τ by the following relation

$$\xi = \frac{2\epsilon}{h}(x - \lambda t) \,,$$
$$\tau = \epsilon^2 kt \,, \tag{7}$$

where λ plays the role of group velocity, we obtain the modified nonlinear Schrödinger equation for the amplitude $\phi^{(1,0)} \equiv \phi$ after some calculations.

$$i\phi_\tau = \frac{1}{2}\phi_{\xi\xi} + [\,|\phi|^2 - \gamma \delta(\xi - \xi_o)\,]\phi \tag{8}$$

Since (8) has the well-known E soliton solution for the case $\gamma = 0$, we see that (8) describes the scattering of E soliton with the mass impurity located at $\xi = \xi_o$. Time development of this interacting E soliton is numerically investigated and one example of the results is shown in Fig.1. Needless to say, the behavior of an E soliton interacting with the impurity depends upon the mass difference parameter γ. Further results and discussions will be given elsewhere.

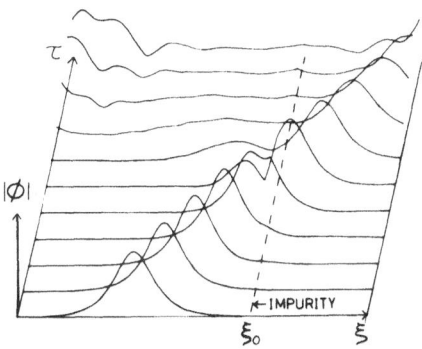

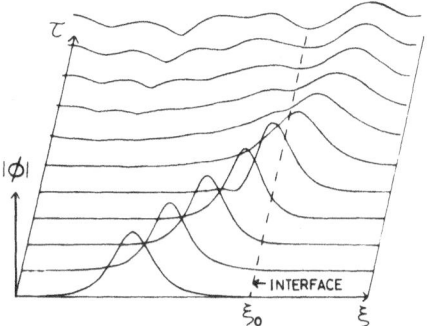

Fig.1. Example of E soliton   Fig.2. Example of E soliton
       interaction with a mass   interaction with an inter-
       impurity : $\gamma = 8$.   face : $\bar{\gamma} = 150$.

     Next consider the interaction of an E soliton with the inter-
face in a one-dimensional nonlinear lattice. The derivation of
modified nonlinear Schrödinger equation in the presence of a mass
interface is quite similar to the case of mass impurity. That is,
the modified nonlinear Schrödinger equation with mass interface is
written as :

$$i\phi_\tau = \frac{1}{2} \phi_{\xi\xi} + [|\phi|^2 - (\bar{\gamma}/2)\theta(\xi - \xi_0)]\phi \quad , \tag{9}$$

where $\theta(\xi - \xi_0)$ is Heaviside's step function and $\bar{\gamma}$ defined by

$$\Delta\rho = \varepsilon^2\bar{\gamma} \quad . \tag{10}$$

We see that (9) describes the interaction of E solitons with the
mass interface. It is noted that the mass difference of the interface should be smaller than that of an impurity by the order of ε
to reduce the modified nonlinear Schrödinger equation consistently.
Equation (9) can also be solved numerically and one example of the
results is shown in Fig.2. We see that when the soliton enters
into the heavier mass region, the soliton width becomes wider and
its velocity is slightly reduced. Only a small fraction of it is
reflected at the interface but these characters depend largely on
the sign and the magnitude of $\bar{\gamma}$. Detailed results and discussions
will also be given elsewhere.

References

[1] F. Tappert and C. M. Varma : Phys. Rev. Letters, <u>25</u> (1970)
     1108.
[2] T. Taniuti and N. Yajima : J. Math. Phys., <u>10</u> (1969) 1369.

SURFACE PHONON SCATTERING BY DENSITY FLUCTUATION ON SOLID SURFACES

Tsuneyoshi Nakayama and Tetsuro Sakuma

Department of Engineering Science, Hokkaido University

Sapporo 060, Japan

Since the real surface of solids is rather rough, surface phonons are scattered considerably by the surface irregularities. The present authors calculated the frequency and correlation length dependence of the damping rate of surface phonons due to the density fluctuations on solid surfaces in a previous work[1], hereafter referred to I. It is assumed there that the density fluctuation is localized on solid surfaces. In this report we extend I to allow the density fluctuations with finite depth from solid surfaces. The frequency and correlation length dependence of the rate is calculated on the basis of Gaussian distributed density-fluctuation.

We can expand the displacement $\vec{u}(\vec{x}, t)$ at a point $\vec{x} = (\vec{r}, z)$ and a time t in terms of eigenmodes, [2]

$$\vec{u}(\vec{r}, t) = \sum_J (8\pi^2 \rho \omega_J)^{-\frac{1}{2}} [a_J(t) \vec{u}^J(z) e^{i\vec{k}\cdot\vec{r}} + h.c.] ,\qquad (1)$$

where ρ is the density of the crystal, J the set of quantum numbers, and a_J the annihilation operator of mode J. J is composed of three quantum numbers, that is, $J = \{\vec{k}, c, m\}$: k the wave vector in the x-y plane, c the propagation velocity in the x-y plane, and m which stands for the five eigenmodes of elastic waves in a half-space.

Now we separate the density function $\rho(\vec{x})$ into a non-random part or the average density ρ, and a random part with zero mean:

$$\rho(x) = \rho + \Delta\rho(\vec{x}) \qquad : \qquad \langle \Delta\rho(\vec{x}) \rangle = 0 .\qquad (2)$$

Following Bucaro and Flax[3] we take the density fluctuation to be exponentially decreasing in the vicinity of solid surface like ge^{-gz} $\Delta\rho(\vec{r})$. In the limit of $g \to \infty$, this is the same density function with that previously considered in I, i.e., $\Delta\rho(\vec{r})\delta(z)$. We take the correlation function to be statistically uniform and Gaussian as follows:

$$\left\langle \Delta\rho(\vec{r})\Delta\rho(\vec{r}')\right\rangle = (\hat{\Delta\rho}^2 a^2/\pi^2 1_c^2)\exp(-|\vec{r} - \vec{r}'|^2/1_c^2) \quad , \tag{3}$$

where the anguler bracket denotes ensemble average and a is the characteristic length parameter. From (3), in the zero range limit $1_c \rightarrow 0$, one has two dimensional δ-function for the correlation function which represents the so-called white noise case. The infinite range limit, $1_c \rightarrow \infty$, it is trivial, that is, the correlation function vanishes. Then for small $\hat{\Delta\rho}$ or large 1_c the correlation function (3) will be small, implying that the Hamiltonian containing $\Delta\rho(\vec{r})$ may be treated as a perturbation. Thus we obtain the following perturbed Hamiltonian,

$$H_1 = \sum_{J,J_1} V_{J, J_1} [a_J \exp(-i\omega_J t) - h.c.][a_{J_1} \exp(-i\omega_{J_1} t) - h.c.]. \tag{4}$$

Here the vertex function V_{J,J_1} is of the form,

$$V_{J,J_1} = [(\omega_J \omega_{J_1})^{1/2}/2\rho]\int d\vec{r} \int dz\{\exp i(\vec{k} + \vec{k}_1)\cdot r\}\Delta\rho(\vec{r})e^{-gz}g$$
$$\times [\vec{u}^J(z)\cdot\vec{u}^{J_1}(z)] \quad . \tag{5}$$

We can obtain from (4) an expression for the damping rate of a J-mode surfon by means of a standard Green's function:

$$\Gamma_J(\omega) = \sum_{J_1} \frac{\pi}{\rho^2} \delta(\omega_J^2 - \omega_{J_1}^2)\omega_J \omega_{J_1}^2 |\int dz \vec{u}^J(z)\cdot\vec{u}^{J_1}(z)e^{-gz}g|^2$$
$$\times\left\langle|\Delta\rho(\vec{k} + \vec{k}_1)|^2\right\rangle \quad . \tag{6}$$

Since the density fluctuation is assumed to be localized near the surface, it would be enough to take into account only two modes, the Rayleigh mode and the mode with total reflection, among five modes which form a complete orthogonal set of the eigenmodes. The other three modes, having oscillating wave functions of a distance z from the surface, will contribute less to surfon scattering. Thus we may regard J to be Rayleigh mode and J_1 to run through Rayleigh and total reflection modes. Now introducing the explicit forms of Rayleigh and total relfection modes into (6), we can obtain the following expression for the damping rate of the Rayleigh mode wave:

$$\Gamma(\omega) = \Gamma_1(\omega) + \Gamma_2(\omega) \quad . \tag{7}$$

Here the first term coming from the decay process into other Rayleigh modes is given as follows:

$$\Gamma_1(\omega) \sim \omega^5 \exp(-x)[f_1^2\{I_0(x) + I_2(x)\} - 4f_1 f_2\gamma^2 I_1(x)$$
$$+ 2\gamma^4 f_2^2 I_0(x)] \quad , \tag{8}$$

where

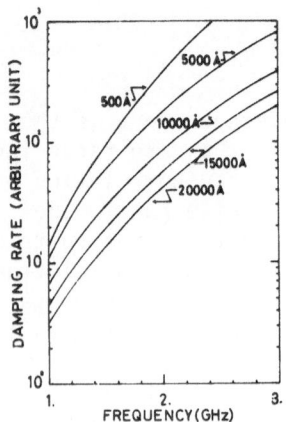

DAMPING RATE (ARBITRARY UNIT)

5000Å
500Å
10000Å
15000Å
20000Å

FREQUENCY (GHz)

DAMPING RATE (ARBITRARY UNIT)

1.8 GHz
1.6 GHz
1.4 GHz
1.2 GHz
1.0 GHz

CORRELATION LENGTH (×250Å)

Fig.1 (a) Frequency dependence   (b) Correlation length dependence
of damping rate for $1/g = 1000A$.   of the damping rate for $1/g = 1000$ A.

$$f_1 = 1/2(2\gamma\omega/c_R + g) + P_1^2/(2\eta\omega/c_R + g) - 2P_1/[(\gamma + \eta)\omega/c_R + g], \quad (9)$$

$$f_2 = f_1(P_1 \to P_2) \quad , \tag{10}$$

and

$$x = 1_c^2\omega^2/2c_R^2 , \quad P_1 = 2\gamma\eta/(1 + \eta^2) , \quad P_2 = 2/(1 + \eta^2) , \tag{11}$$

and I_n is the modified Bessel function of n-th order. The expression of the second term coming from the decay process into total reflection modes is lengthy but general feature is essentially the same as (8), so that it will not given here.

It should be noted that (9) and (10) contain the frequency variable ω in the denominators. Thus the frequency dependence of the rate depends on the relative magnitude of g and ω. For g large compared with the frequency of the incident waves ω is less important in the denominators, while for small g, the contrary is the case. In Fig.1(a) and (b) the predicted numerical results of the correlation length and frequency dependence of the damping rate are shown. The finite depth effect makes the damping rate quite different from the case of the density fluctuation localized on the solid surfaces.

References

[1] T. Nakayama and T. Sakuma : J. appl. Phys. 15,(1975) 2445.
[2] H. Ezawa : Ann. Phys. 67,(1971) 438. In this article, Ezawa
     named the quantum of elastic waves in a half space "surfons".
     Hereafter we use this term for surface phonon.
[3] J. A. Bucaro and L. Flax : J. appl. Phys. 45,(1974) 765.

THEORY OF RAYLEIGH WAVE SCATTERING BY MASS DEFECTS AND SURFACE ROUGHNESS *

A. A. Maradudin, D. L. Mills and R. F. Wallis

Department of Physics, University of California, Irvine

Irvine, California 92664

In the presence of a mass defect situated at the point $\vec{x}_0 = (0,0,x_{03})$ in a semi-infinite, isotropic, elastic medium occupying the half-space $x_3 > 0$, the equations of motion of the medium can be written in the form

$$\rho \sum_{\mu} L_{\alpha\mu} (\vec{x},t) u_{\mu} = \Delta m \delta (\vec{x} - \vec{x}_0)(\partial^2/\partial t^2)u_{\alpha} \tag{1}$$

where $u_{\alpha}(\vec{x},t)$ is the α Cartesian component of the displacement field at the point $\vec{x}$ at the time t, ρ is the mass density of the medium, Δm is the increase in the mass of the medium due to the introduction of the defect, and $L_{\alpha\mu}(\vec{x},t)$ is a second-order differential operator involving the elastic moduli which have a step-function behavior at the surface.

We now introduce a Green's function $G_{\alpha\beta}(\vec{x},\vec{x}';t-t')$ as the solution of the equation

$$\sum_{\mu} L_{\alpha\mu} (\vec{x},t) G_{\mu\beta} (\vec{x},\vec{x}';t-t') = \delta_{\alpha\beta} \delta(\vec{x}-\vec{x}') \delta(t-t') \quad . \tag{2}$$

We can write $u_{\alpha}(\vec{x},t) = u_{\alpha}^{(0)}(\vec{x},t) + u_{\alpha}^{(s)}(\vec{x},t)$ where $\vec{u}^{(0)}(\vec{x},t)$ is a solution of the corresponding homogeneous equation,

$$\sum_{\mu} L_{\alpha\mu} (\vec{x},t) u_{\mu}^{(0)} (\vec{x},t) = 0 \quad , \tag{3}$$

and in the present context represents a Rayleigh wave propagating along the surface $x_3 = 0$ of the semi-infinite elastic medium. With the Fourier decompositions

$$u_{\alpha}(\vec{x},t) = u_{\alpha}(\vec{x},\omega) e^{-i\omega t}, u_{\alpha}^{(0)} (\vec{x},t) = u_{\alpha}^{(0)} (\vec{x},\omega) e^{-i\omega t} \tag{4}$$

$$G_{\alpha\beta}(\vec{x},\vec{x}';t-t') = (1/2\pi)\int d\Omega \, G_{\alpha\beta}(\vec{x},\vec{x}';\Omega) e^{-i\Omega(t-t')} \tag{5}$$

$$G_{\alpha\beta}(\vec{x},\vec{x}';\Omega) = (1/2\pi)^2 \int d^2 k_{\parallel} e^{i\vec{k}_{\parallel}\cdot(\vec{x}_{\parallel} -\vec{x}_{\parallel}')} g_{\alpha\beta}(\vec{k}_{\parallel}\Omega|x_3 x'_3) \tag{6}$$

where $\vec{x}_{\parallel} = (x_1, x_2, 0)$ and $\vec{k}_{\parallel} = (k_1, k_2, 0)$, we can write the scattered field $u_\alpha^{(s)}(\vec{x}, \omega)$ in the form

$$u_\alpha^{(s)} = - \frac{\Delta m \omega^2}{4\pi^2 \rho} \int d^2 k_{\parallel} e^{i\vec{k}_{\parallel} \cdot \vec{x}_{\parallel}} \sum_\beta \frac{g_{\alpha\beta}(\vec{k}_{\parallel} \omega | x_3 x_{o3})}{D_\beta(\vec{x}_o; \omega)} u_\beta^{(o)}(\vec{x}_o, \omega) \tag{7}$$

where $D_\alpha(\vec{x}_o; \omega) = 1 + (\Delta m \omega^2 / \rho) G_{\alpha\beta}(\vec{x}_o, \vec{x}_o; \omega)$.

The nonzero elements of the tensor $g_{\alpha\beta}(\vec{k}_{\parallel} \omega | x_3 x_{o3})$ have been calculated recently [1] for a semi-infinite, isotropic, elastic-medium occupying the upper half-space $x_3 > 0$, with a stress-free surface. For the incident wave we assume a Rayleigh wave propagating in the positive x_1-direction with wave vector k_o.

From a knowledge of the amplitudes of the incident and scattered elastic waves the scattering efficiency of Rayleigh waves due to scattering by mass defects or surface roughness can be calculated in the following way. The time average of the energy in an elastic wave crossing unit area in unit time is given by the real part of the complex elastic Poynting vector $\vec{\zeta}$, whose Cartesian components are given by

$$\zeta_\alpha = - \frac{1}{2} \sum_{\beta\mu\nu} C_{\alpha\beta\mu\nu} \dot{u}_\beta^* \frac{\partial u_\mu}{\partial x_\nu} . \tag{8}$$

The energy per unit time stored in the incident wave is

$$\frac{dE^{(o)}}{dt} = L_2 \int_0^\infty dx_3 \bar{\zeta}_1 = L_2 \int_0^\infty dx_3 |A|^2 \rho \omega^2 F_o(\omega, x_3) \tag{9}$$

where L_2 is the dimension of the crystal in the x_2-direction, A is an arbitrary amplitude, $F_o(\omega, x_3)$ decreases exponentially with increasing x_3 with an exponent proportional to the wave vector of the incident Rayleigh wave, and $\bar{\zeta}_1$ is the real part of ζ_1.

From Eq. (7) and using the explicit form [1] for the Green's function $g_{\alpha\beta}(\vec{k}_{\parallel} \omega | x_3 x_{o3})$, one can evaluate the asymptotic form of $u_\alpha^{(s)}(\vec{x}, \omega)$ for large $|\vec{x}|$ using the method of steepest descents and the theory of residues. The amplitude of the scattered wave can be decomposed into contributions corresponding to scattering of the incident Rayleigh wave into bulk waves and into other Rayleigh waves. The latter contribution arises from the pole in the Green's function when $k_{\parallel}$ takes on the Rayleigh wave value. The general form of the Poynting vector for the scattered wave obtained by using Eqs. (7) and (8) can be written as

$$\zeta_\alpha^{(s)} = -[(\Delta m)^2 \omega^5 / 32\pi^2 \rho] F_\alpha(k_{\parallel}, \omega; \vec{x}, \vec{x}_o) \tag{10}$$

where the $F_\alpha(k_{\parallel}, \omega; \vec{x}, \vec{x}_o)$ are rather complicated functions.

For the case of scattering into bulk waves, the energy per

unit time scattered from the mass defect into solid angle range $d\Omega$, $d^2E^{(s)}/d\Omega dt$, can easily be obtained from $\varsigma_\alpha^{(s)}$. The ratio of $d^2E^{(s)}/d\Omega dt$ to $dE^{(0)}/dt$ is a measure of the scattering efficiency which we shall call p. One can verify that the function $F_\alpha(k_\parallel, \omega; \vec{x}, \vec{x}_0)$ in Eq. (10) is proportional to the frequency ω and independent of the mass change Δm. Also, $dE^{(0)}/dt$ as given by Eq. (9) is proportional to ω, so we see from Eq. (10) that the scattering efficiency p is proportional to $(\Delta m)^2 \omega^5$, the result previously obtained by Steg and Klemens (2) and by Sakuma. (3)   For the case of scattering into other Rayleigh waves, we also find that the scattering efficiency is proportional to $(\Delta m)^2 \omega^5$.   A further point of interest concerns the quantity $D_\beta(\vec{x}_0, \omega)$ which appears in the denominator of Eq. (7). This quantity possesses a zero for a particular value of ω, and this leads to a resonance in the scattering efficiency. (3)

We have studied the scattering produced by surface rough-ness by the same method. One begins by presuming the shape of the free surface is defined by the statement

$$x_3 = f(x_1, x_2) \qquad (11)$$

where $f(x_1, x_2)$ measures the height of a point on the surface above the $x_1 x_2$ plane. One then writes down the equations of motion for a medium with position dependent elastic constants $C_{\alpha\beta\mu\nu}(\vec{x})$ given by

$$C_{\alpha\beta\mu\nu}(\vec{x}) = C_{\alpha\beta\mu\nu} \theta(x_3 - f(x_1, x_2)) \qquad . \qquad (12)$$

The scattering problem is solved in Born approximation by the Green's function method described above. We have obtained expressions for the contribution to the attenuation rate from roughness-induced radiation of energy into bulk transverse and bulk longitudinal waves, as well as roughness-induced radiation into Rayleigh waves. If the wavelength of the incident Rayleigh wave is large compared to the transverse correlation length of the roughness, a limit one expects appropriate for Rayleigh waves at microwave frequency which propagate on a well-prepared substrate, then as in the case of mass defect scattering, we find an ω^5 variation of all contributions to the attenuation length. The physical origin of the ω^5 variation is the same as for the mass defect scattering.

References:

1.   A. A. Maradudin and D. L. Mills, to be published.
2.   R. G. Steg and P. G. Klemens, Phys Rev. Lett. 24, 381 (1970).
3.   T. Sakuma, Phys. Rev. B8, 1433 (1973).

*   Supported in part by the U. S. Air Force Office of Scientific Research under Grant No. AFOSR 76-2887 and by the U. S. Office of Naval Research under Contract No. N00014-69-A-0200-9003.

PHONON DIFFUSION IN A TEMPERATURE GRADIENT

R A Lucas

Ecole Sup. de Physique et Chimie

10 rue Vauquelin, 75231 Paris Cedex 05, France

The diffusion of phonons of energy $\hbar\omega$ and momentum $\hbar k$ is studied by taking account of the temperature dependence for wave velocities V_S and V_p in isotropic solids.

Generally wave velocity V in a uniform gradient of temperature decreases as T increases. For large mean free path phonons, the trajectory is indicated on fig 1.

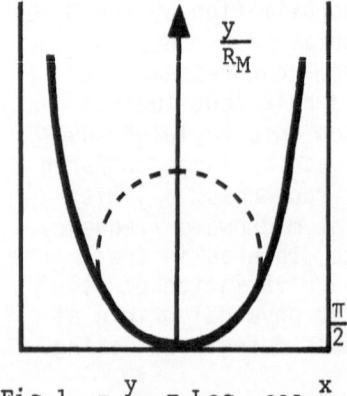

Fig 1 $\quad -\dfrac{y}{R_M} = \mathrm{Log}_e \cos \dfrac{x}{R_M}$

With $1/R_M = (1/V)(\partial V/\partial T)(\partial T/\partial y)$ the equation of a tangent trajectory in $x = 0$ is:

$$- y/R_M = \mathrm{Log}_e \cos(x/R_M)$$

and shows that phonons may travel towards hot regions. In calculating the flux through plane $y = 0$, we shall consider the mean free path ℓ and use equipartition with trajectories along Ox, Oy and Oz.

A: Phonon diffusion in the y direction (no mirage):

Calculation of flux of phonons coming from the dv element between $y = +\ell$ and $y = -\ell$ leads to

$$F'_y = -\frac{1}{3} C \, V \, \bar{\ell} \left(1 + \frac{1}{2} \frac{T}{V} \frac{\partial V}{\partial T}\right) \frac{\partial T}{\partial y}$$

(C: heat capacity per unit volume).

B: Flux of phonons travelling along Oy and Oz, suffering mirage effect.

If $\partial V/\partial T < 0$, only phonons coming from the region $y < 0$ give a result if the path from dv to the plane $y = 0$ is $< \overline{\ell}$. Taking account of the fact that $\overline{\ell} \ll R_M$ it is found:

$$F''_y = -\frac{1}{3} C V \overline{\ell} \frac{2T}{V} \frac{\partial V}{\partial T} \frac{\partial T}{\partial y}$$

The total flux becomes

$$F_y = -\frac{1}{3} C V \overline{\ell} \left(1 + \frac{5}{2} \frac{T}{V} \frac{\partial V}{\partial T}\right) \frac{\partial T}{\partial y}$$

instead of the classical value which is obtained for $\partial V/\partial T = 0$.

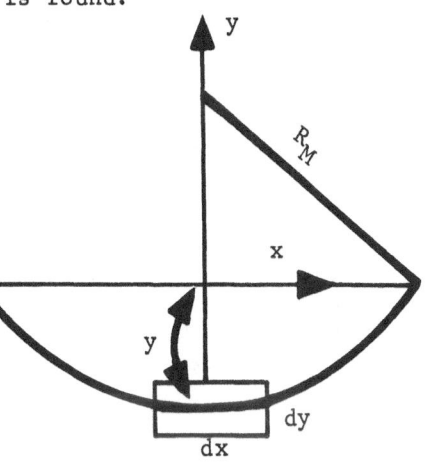

Fig 2

Taking into account two kinds of phonons of velocities V_S and V_p, we suppose the transverse phonon energy to be twice that of longitudinal phonons $V_D = 1/3 (2V_s + V_p)$ we have

$$F_y = -\frac{1}{3} C V_D \overline{\ell} \left(1 - \frac{5}{2} mT\right) \frac{\partial T}{\partial y}$$

with $m = -\frac{1}{V_D} \frac{\partial V_D}{\partial T} = -\frac{d \log V_D}{dT}$

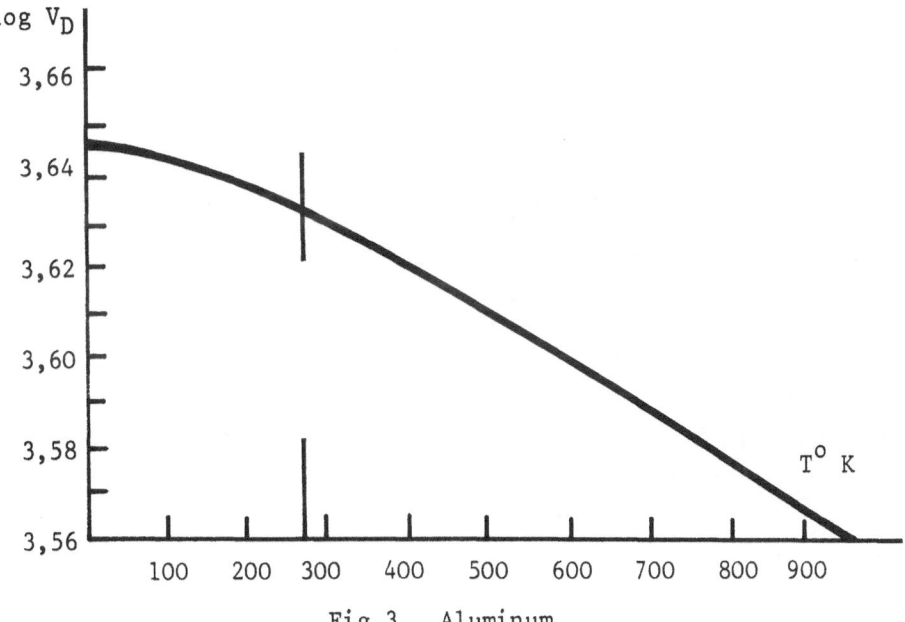

Fig 3   Aluminum

Temperature Dependence of V_D

For some solids of cubic system (Al, Cu, Ag), values of V_S, V_p (then V_D) are known between $0°K$ and $850°K$. The curve Log $V_D = f(t)$ shows a linear region between 250 and $850°K$. So, for Al, Cu, Ag we have the values m = 2.57 x 10^{-4}, 1.44 x 10^{-4}, 1.65 x 10^{-4}.

Retrodiffusion Flux

Terms depending on m in F_y correspond to a phonon flux going from cold to hot regions. With preceeding values of m it represents at $700°K$, 30 to 50% of the flux without mirage.

If m remains constant until the melting point T_f, the term $(1 - 5/2\ m\ T_f)$ has to remain positive so that, statistically, the phonon flux flows towards decreasing temperatures. This condition holds for preceeding solids.

New Equation for Heat Transfer

The classical expression $F_y = -\frac{1}{3} C V \bar{\ell} \frac{\partial T}{\partial y}$ leads to Fourier equation

$$\frac{\partial T}{\partial t} = -\frac{1}{3} V \bar{\ell} \left(\frac{\partial^2 T}{\partial y^2}\right)$$

by using energy conservation.

With the new expression of F_y one obtains an equation with terms $\partial^2 T/\partial y^2$ and $\partial^2 T^2/\partial y^2$ which is not linear. These results make possible the comprehension of experimental facts due to J Jacq and reported by L Kaiser.

REFERENCES

R Lucas: (1) J de Phys Coll C 6; 33, 1972, 76
           (2) C R Acad Sc B 252, 1961, 852
           (3) C R Acad Sc B 270, 1973, 575
           (4) 5th Intern Heat Transfer Conf, 1974,
                Tokyo I, 157 and VII, 21

Single Crystal Elastic Constants and Calculated Aggregate Properties. The M I T Press, Cambridge, Massachusetts and London, England (2nd edition).

PHONON DIFFUSION AND TEMPERATURE FIELD IN METALS

Louis KAISER

Laboratoire de Physique Statistique

Centre Universitaire. 66000 Perpignan (France)

According to J.Jacq's experiments /I/, Fourier's theory does not predict the suitable temperature distribution, near the interface air-metal (fig.I).
R. Lucas has explained this surprising fact by suggesting that the thermal gradient gives rise to a curvature of the phonon trajectories by variation of the phase velocity (mirage effect). This produces a retrodiffused flux /2/.

The mirage may be formulated by a phonon with continuously variable wave vector k. The gradient of temperature acts on k (mirage), but phonon retrodiffusion does alter the temperature profile : with such a reciprocal interaction, the temperature field must itself behave as a set of quantized oscillators that might be called thermons.
The action of a phonon in state k on a thermon at the point r is a quantized oscillatory process, let us say a phonothermon $\Phi(r,k)$ proportional to $\exp(jkr)$ as a vibration (with $j^2 = -I$), γ being the coupling constant. So, we have $\Phi(r,k) = \gamma \exp(jkr)$.
Statistically, the phonothermon contributes to the macroscopic temperature by its probability density $P(k)$ in k space. This quasi-particle does not satisfy usual distribution laws, for many reasons : (i) irreversible process, (ii) non-uniform temperature, (iii) phonothermon being itself a component of the temperature concept. A general principle for random processes (the lowest an energy level is, the greatest its population) suggests $P(k) = \exp(- |k|/k_o)$ where

k_0 is a constant > 0. For all configurations, we have

$$\int_0^{k_D = \text{Debye Limit}} \gamma \,\Theta(rk)P(k)dk = \int_0^{k_D} \gamma \, e^{-\frac{|k|}{k_0}} e^{jkr} dk$$

Retrodiffused phonons are progressively set up in the high temperature space till the steady state for which the configuration multiplicity reaches its greatest value m, so that the action of retrodiffusion on the thermal field at the point r is

(I) $\quad \Theta(r) = \int_0^\infty \gamma \, e^{-\left(\frac{|k|}{k_0}\right)^m} e^{j(kr)^m} d(k^m) = \boxed{\dfrac{\gamma k_0^m}{1+(k_0 r)^{2m}}}$

because, for Iron (with $m = 4$, as we shall see) :

$$\int_{k_D}^\infty \gamma \, e^{-\left(\frac{|k|}{k_0}\right)^m} e^{j(kr)^m} d(k^m) < 140 \, \exp(-10^{24}) \sim 0$$

From (I), the initial value (γk^m) of $\Theta(r)$ is reduced to $(\gamma k^m)/2$, when $r = r^* = I/k_0 = \lambda_0/(2\pi)$. Hence the characteristic wave length

(2) $\qquad \boxed{\lambda_0 = 2\pi \, r^*} \qquad\qquad$ (r^*, for point I)

Let r^{**} be the depth for which the Fourier's profile $\varphi(r)$ becomes true, the temperature distribution $T(r)$ is

$$T(r) = \Theta(r) + \Psi(r)$$
$$\Psi(r) = \begin{cases} \varphi(r^{**}) = \text{Const.}, & \text{if } 0 \leqslant r \leqslant r^{**} \\ \varphi(r) & \text{if } r > r^{**} \end{cases}$$

For the Iron, the Jacq experimental profile /I/ gives $m = 4$, and (I) may be written

(3) $\quad \boxed{\Theta(r) = \dfrac{140}{1+(r/80)^8}} \quad k_0 = \frac{1}{r^*} = \frac{1}{80} \, (\mu^{-1}), \, (\gamma k_0^m = 140\,°c)$

which is compared (fig.I) with experimental data/I/.

Experimentally, Wolf and Kostenko /3/ have revealed a parallelism between heat transfer and propagation of ultrasonics of frequency $\boxed{\nu = 7 \times 10^6 \text{ Hz.}}$ According to (2), $\lambda_0 = 502 \, \mu$ at the point $r^* = 80 \, \mu$ and for the mean velocity $c = 3614$ ms^{-I} of sound in Iron, conform with the Debye definition :

$$\frac{3}{C^3} = \frac{2}{C^3_{Trans}} + \frac{1}{C^3_{Longit}} \;;\quad \begin{cases} C_{Longit} = 5960 \text{ ms}^{-1} \\ C_{Trans} = 3240 \text{ ms}^{-1} \end{cases}$$

we find the frequency

$$\nu_o = c/\lambda_o = \boxed{7.2 \times 10^6 \quad \text{Hz}}$$

in perfect agreement with experiment.

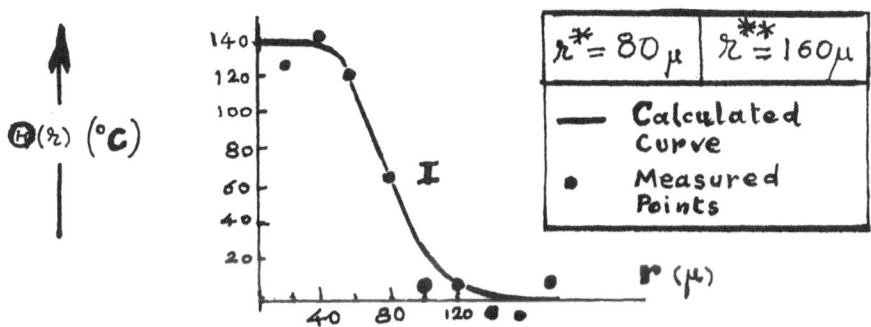

−fig.I− Temperature Profile in Iron.

CONCLUSION. − Expression (I) shows that temperature distribution and phonon distribution are essentially Fourier transform of each other. The experimental temperature profile /I/ involves the existence of phonons with variable velocity suggested by Lucas /2/ and, at the point r * is associated an ultrasonic phonon of a frequency which is conform to experiment.

/I/ J.Jacq, Thèse Docteur-Ingénieur, n° 327I, I970, C.N.R.S., 3 nov. I969.
/2/ R. Lucas, C.R.Ac.Sc.,Paris, 252, I96I, p.852 and 276,série B,I973, p.575.− 276,série B, I973,p.659. Heat Transfer, I974, Tokyo, I, p.I57 and VII,p.2I.
/3/ L.Wolf Jr. and C.Kostenko, Ultrasonic Measurements of Thermal Conductance of Joints in Vacuum, IIT, Research Institute, Chicago, I97I, p. 769−772.

ROLE OF UMKLAPP IN ACOUSTIC BOUNDARY RESISTANCE

N S Shiren

Thomas J Watson Research Center, Yorktown Heights

New York 10598, U S A

We consider the transmission and reflection of ultrasonic waves at the interface between a real periodic solid and a continuum liquid. If the sound velocity in the liquid is considerably smaller (as e.g. in He) than that of the solid then at sufficiently high frequencies the wavelength in the liquid can be smaller than a lattice spacing of the solid. In that case the periodicities in the boundary plane become important and there is an additional transmission with the modified Snell's law,

$$\vec{k}_T = \vec{k}_i + \vec{G}$$

where $\vec{k}_T$ and $\vec{k}_i$ are the projections on the boundary plane of the transmitted and incident wave vectors, respectively, and $\vec{G}$ is a reciprocal lattice vector in the boundary plane.
The transmitted and reflected amplitudes are calculated from the usual boundary conditions on stress and particle velocity with the modified Snell's law. The results show that for a single such umklapp process, from either a longitudinal or transverse incident wave, the transmission into the liquid varies from 2% to 30% depending on the angle of incidence and this transmitted power is primarily taken from the reflected mode conversion signal. It is these two facts which make the calculation of interest in connection with the problem of the thermal boundary resistance, since they are in accordance with the results of recent heat pulse investigations.
In its present form, however, the calculation cannot provide a final explanation of the experimental results for the following reasons:

1. The required frequencies are too high.
2. The assumption of a continuum liquid and continuum boundary conditions imply that the coupling is the same for normal

64

and umklapp processes[1], whereas the latter should include
a scattering form factor analogous to the atomic form fact-
or in X-ray Bragg scattering.

3. The calculation is not self consistent in that each process
    has been treated separately.

This is probably a good approximation if only one umklapp process
is allowable since the normal transmission is so small. However,
if more than one umklapp is allowed the approximation is expected
to be poor.

1. This method of handling the coupling was suggested by Dr B D
    Silverman.

DISCUSSION ON KAPITZA CONDUCTANCE AND SURFACES

The Thermal Boundary Resistance   A C Anderson   Page 1

H Namaizawa: Is there any experimental attempt to measure heat transfer from a quantum solid, such as solid H_2 or solid He to another quantum medium, like liquid He?

A C Anderson: The heat transfer between liquid and solid He was, I believe, reported at an LT conference. No definitive results have yet been reported for solid H_2 to liquid He.

L J Challis: In view of the present emphasis on processes that may be occurring in the first one or two layers of the helium, do you think anyone has done measurements on ^{3}He pure enough that ^{4}He contamination of the layers is avoided?

A C Anderson: Yes, for example in the measurements of Folinsbee.

C H Anderson: You did not emphasise the growing sense of non linearity in the subject.

A C Anderson: By non linearity you presumably mean non frequency conserving processes. There is evidence that they are taking place and one such source of such a process is anharmonicity at the interface.

H M Rosenberg: As you have said electrons in a metal do couple to surface waves and hence to the phonons in the helium, could you mention any more about the electron problem in Kapitza conductance?

A C Anderson: Consider the particular problem of dislocations at the boundary which shorten the phonon mean free path. The surface wave excited by the helium gives its energy to the solid if the solid attenuates phonons. Then the problem is to get the energy away from this surface layer and this can be done more effectively by electrons than by phonons with short mean free paths. Therefore the enhancement of thermal transport due to surface damage can only be seen with metals.

Reflection Coefficient of Phonons Between 15 and 200 GHz at a Solid-
Liquid Helium Interface   C H Anderson, P Call and E S Sabisky
                                                        Page 8

W E Bron: I thought you said that you expected a black-body phonon
spectrum emitted from the chromium heater film. What experimental
evidence do you actually have in this regard?

C H Anderson: Our main purpose for changing to the chromium layer
was to improve the cleanliness of the experiment and we did not seriously consider the spectral properties of the heater. In fact this
heater turned out to be non-thermal, namely deficient in low frequency phonons.

H Maris: (a) What is the lowest frequency at which the acoustic-
mismatch theory fails for hydrogen?
(b) Buechner and I have made some preliminary measurements of the
reflection coefficient at a silicon-solid hydrogen interface as a
function of the temperature of the heat-pulse generation. We found
that the results appeared to tend to the acoustic-mismatch value as
the temperature was lowered to $1^{\circ}K$. This temperature is significantly higher than the corresponding temperature for helium.

C H Anderson: (a) It depends on the state of the hydrogen layer,
but it could be less than 20 GHz.
(b) It is possible that our results are strongly effected by the
backing of liquid helium we have on our layer and is not present
on yours.

H Kinder: Your curves show that the reflection coefficient is falling off around 40-50 GHz and this energy is too low for the desorption model which would need at least 7K. What is your opinion of the
explanation?

C H Anderson: I don't have a model, but if you notice the H_2 and D_2
results the break is at even lower frequencies and if we anneal these
surfaces with H_2 and D_2 then we can change the spectrum dramatically
at the low frequency end. We start off by freezing those layers on
at low temperatures so they are very disordered and have a large
number of tunnelling modes with a wide frequency range and then annealing (up to $T \sim 6K$) changes these modes. This is consistent with
the model of quantum tunnelling states associated with light mass
particles.

                        ****

A Model of Inelastic Heat-Transfer Mechanism for the Excess Kapitza
Conductance   H Namaizawa   Page 11

H Maris: We have carried out some similar calculations. I should
like to mention one dangerous trap in theories of this type. One is

dividing the physical system into 3 parts: 1. the classical solid, 2. the 'interface' atoms or quasi-particles, 3. the bulk liquid. If one treats the bulk liquid as a source of <u>static</u> potential, one obtains an unreasonably large energy absorption simply because the interface atoms are being 'squeezed' between the solid and a rigid wall. It appears essential to use a model which does not have this unphysical property.

H Namaizawa: You might be quite right at a low temperature at which the thermal phonon frequency is not so high that your three parts can make simultaneous movements adiabatically. But at high temperatures this adiabatical situation becomes no longer true and the 'static' potential approach turns out to be not so unphysical, since the vibration frequency is large. I am sorry that I cannot tell you at which temperature adiabatic movement breaks down, but it might be very high actually. As a matter of fact we see in fig 4 that the total transmission coefficient α is still rather small since the couplings are not large enough. In this respect I cannot understand why you obtained, in your calculations, such large couplings.

L J Challis: I have not really understood how your transverse surfons play such an important role if their lowest frequency is $\sim$ 50K.

H Namaizawa: There are essentially no transverse surfons at low temperatures in the initial and the final state of the scattering process due to the large excitation energy you pointed out. But it appears only in the intermediate states of the scattering processes. It is true that its large energy makes the energy denominator of the intermediate states rather big, but there is no need to excite it thermally for it to give a finite contribution to heat transfer.

A F G Wyatt: Your background phonon distribution peaks at 35° so that if we add to it a critical cone of basal width 20° then we get a minimum at $\sim 20^\circ$. However our measurements show no signs of such a minimum.

H Namaizawa: There is no need to have a dip at angle between $25^\circ \lesssim \theta \lesssim 35^\circ$. If my theory is true, the angular distribution plotted in fig 3 <u>defines</u> the background part. If we separate this 'theoretical' background from your measurement, then we <u>obtain</u> the central peak.

****

A New Kapitza Investigation: Detecting Backscattered Phonons With Tunnel Junction and Bolometer Simultaneously   W Dietsche and H Kinder   Page 15

A F G Wyatt: Another interpretation of your results is that without helium on the reflecting surface you get specular reflections. This is likely because you have a polished surface and the phonon wavelength is $\sim 100\text{Å}$. Then when you introduce helium the reflections

become diffuse due to inelastic scattering from the helium. The phonons returning to the detectors then have lower energy and are time
delayed because of their longer path length in the crystal.

W Dietsche: We calculated the pulse shape expected for diffusive
reflection and found good agreement with the observed shape even
at a bare surface. Therefore we assume that the reflection is completely diffusive at all helium coverages.

J K Wigmore: Did you bias the junction with a magnetic field to get
the transition temperature down to 1.5K?

W Dietsche: No, we biased it with a large current, so that a weak
spot close to the junction becomes normal. This was at the same temperature as before, i.e. nothing else was changed.

K Dransfeld: Could you say something about the polarisation of the
phonons you used?

W Dietsche: The effect was much stronger for the transverse polarisations than for the longitudinal ones and so the accuracy of measuring is too low to do the same measurements on them. Only about
5-10% of the energy is in the longitudinal modes.

W Eisenmenger: This is an experimental point; did you check the
linearity of your bolometer?

W Dietsche: For frequency?

W Eisenmenger: No, for amplitude because quite often bolometers do
not show as good linearity as tunnel junctions due to their transition.

W Dietsche: I did not check this in detail but I had the impression
that the linearity was better for our bolometers than for our junction, because we had a wide range of currents which would bias it
with the same sensitivity, which is impossible for a junction.

L J Challis: A numerical estimate I made of the thermal resistance
of a low pressure gas between two plates suggested it was rather
large compared with typical values of Kapitza resistance at the pressures you were using; two orders of magnitude at 10^{-3} mm.

W Dietsche: I think this is only true at very low pressures.

****

The Angular Distribution of Phonons Radiated from the Cleaved {100}
Faces of NaF, KCl and MgO into Liquid ^{4}He   G J Page, R A Sherlock,
A F G Wyatt and K R A Ziebeck   Page 18

C H Anderson: Have you considered the possibility that the forward elastic scattered phonons are intrinsically modified by the presence of the large presumably inelastic, process which is the cause of the large Kapitza conductance, My guess is that these elastic phonons have an uncertainty in their energy and momentum, which will account for your observation that the forward peak is broader than the calculated critical cone.

A F G Wyatt: We think that a phonon incident on the crystal-helium interface has a small but finite probability of transmission with q_{11} almost exactly conserved and a large probability of being inelastically scattered into the helium. We had considered that these two channels were independent of each other but it may be that the system that causes the inelastic scattering does also modify the interface for q_{11} conserving process.

H Kinder: Have you considered that the dense layers of helium on the surface could be the cause of the broadening of the central peak?

A F G Wyatt: An intermediate layer which has translational symmetry does not change q_{11} conservation.

A C Anderson: The dislocation density would only have to be 10^7 per mm^2 if the dislocations were not static.

G J Page: The peak shape is independent of temperature so this implies a scattering rate $\propto \omega^{-1}$ which would be correct for static dislocations.

****

Energy Transfer Between A Solid and Helium by Excited Atoms
G N Crisp, R A Sherlock and A F G Wyatt   Page 21

P G Klemens: Were these experiments done at a series of temperatures because it seems to me that the desorption should be very temperature dependent.

G N Crisp: No, these were all carried out at $\sim$ 1K.

A C Anderson: Have you tried these experiments with ^{3}He?

G N Crisp: No we have not, but in going from ^{4}He to ^{3}He should change the detected pulse shape.

L J Challis: I believe you find that the signal goes on for a long time, order of 1 ms.

G N Crisp: Yes, this is correct. It was a lot longer than we expected from a Maxwellian distribution for example.

L J Challis: Would you not get this if there was a process involving 3 levels in which you first pump the system to an upper level and then it decays to an intermediate level which is metastable, then this would go on giving energy out over a long time.

H Kinder: You would get this behaviour if the detector took a long time to get back to equilibrium.

G N Crisp: Yes, this is a possibility that we have considered but not yet checked experimentally.

H Kinder: I believe you are suggesting that there is a direct process or quantum process when one phonon of the solid desorbs one atom in the film and kicks it out into the gas and that process should exhibit a strong temperature dependence as Professor Klemens has pointed out. We have seen, much more directly, that there is no such threshold as in the photoelectric effect by changing the frequency from $\sim$ 5K up to 40K and we saw no dependence on the energy of the incident phonons. Whereas you might imagine that you would need a threshold energy of more than 7K. Perhaps we should leave this for the discussion tomorrow.

A F G Wyatt (reply made in discussion group): I do not think it is clear yet, from any experiment, whether the atoms leave by a single or a multiphonon process. However there may not be a sharp threshold for single phonon process as the helium atoms will be in a spectrum of potential wells on the solid surface, especially on a polished surface which will have many different crystal orientations exposed. Also the temperature of the atoms is not negligible compared with the desorption energy and so any threshold will be thermally smeared. To look for quantum desorption I think one should look in transmission rather than reflection as there are other possible mechanisms, such as a change in the degree of diffuse reflection, which would obscure seeing a threshold in reflection. We see very hot atoms coming from the film; a significant number which would have an energy of 17K in the film. I find it hard to imagine that these atoms have long enough lifetimes in the film for them to be in thermal equilibrium.

****

Path-Integral Formulation of the Theory of Thermal Boundary
Resistance   M C Phillips and F W Sheard   Page 24

A C Anderson: Isn't this sort of calculation rather academic until someone can do a measurement on this type of system?

F W Sheard: I think these concepts do arise when measurements are done on thin metal foils in a second sound cell and where the electrons can be taken out of the conductance by making the film superconducting. This theory can be generalised to that situation so this analysis does come into practical measurements.

****

The Role of Electrons in the Kapitza Resistance of Lead
L J Challis and M C Phillips   Page 28

H M Rosenberg: Has anyone measured the longitudinal magnetoresistance of lead?

L J Challis: No but theoretical analysis by Pippard indicates that
in all metals, the resistance should increase slightly and then saturate. This would then produce an <u>increase</u> in the foil resistance
in longitudinal fields of say 1 to 2%. Let me stress also that even
if the magnetoresistance were negative this could only reduce the
resistance of the foil by $\sim$ 0.5% - the bulk contribution at low
fields.

A C Anderson: I seem to remember that in the Ohio State work there
was a change in the phase of the second sound wave in the detection
cavity, as the field was changed.

L J Challis: No in fact they didn't see a phase change and argued
that this was evidence that the changes they observed in foil resistance were not due to bulk effects since these would cause large
phase changes. We also saw no significant phase changes but this does
not conflict with our conclusion that bulk effects are important; in
our work these would cause phase changes too small for us to see.

W C Overton Jr: Can you make field measurements at higher temperatures?

L J Challis: One of the problems of looking at field effects is to
keep the bulk effects reasonably small compared with the Kapitza
resistance. This is the advantage of the second sound foil technique
but it is of course limited to HeII. If a conventional technique
were used, the thermometer would have to be $\sim$ 100 µm from the interface to obtain a comparable situation.

H Maris: You mention that it is not possible to use the second
sound resonator technique to measure the Kapitza resistance above
the λ-point. It might be possible to make measurements using ordinary sound resonators. Because of the large thermal expansion of
helium ordinary sound has a large temperature fluctuation associated
with it. The only problem one might worry about is the <u>mechanical</u>
energy transfer between the resonators due to the non-rigid behaviour of the sample foil.

L J Challis: It's an interesting idea. We like to use temperature
oscillations $\sim$1mk in the driving cavity to get a reasonable signal in
the detector cavity but I suppose that might be possible to achieve.

Influence of Dislocations on the Kapitza Conductance of Clean
Copper Surfaces   K C Rawlings and J C A van der Sluijs Page 31
and
Influence of Surface Contamination on the Kapitza Conductance of
Copper, Silver, Molybdenum and Beryllium   J C A van der Sluijs
and A E Alnaimi   Page 34

H M Rosenberg: Why do you assume that the dislocation density varies exponentially with the depth?

K C Rawlings: The damage is introduced at the surface and falls off
as the depth increases. An exponential decrease seemed the most
likely.

A C Anderson: Does the temperature dependence increase or decrease
as you remove the surface damaged layer?

K C Rawlings: The conductance of the final surfaces, i.e. after we
have removed most of the surface damage, varies as $T^{1.6}$. If we suppose that the surface damage provides a parallel process of conductance, the temperature dependence of this process (h_K with damage) −
h_K (after maximum electropolishing) $\propto T^{2.6}$.

                    ****

A Direct Approach to Finite Phonon Transport (Realistic Transfer at
Boundaries)   W Schneider and M Wagner·   Page 37

W E Bron: Have you considered anharmonicities in your formalism?

W Schneider: Till now we have considered no anharmonicities within
the system. But as for the interactions with the external particles,
anharmonicities have been taken into account because of the assumptions of collision at the boundaries. Beyond that, our algorithm is
especially established to also include anharmonicities within the
system, provided the anharmonic problem is of low dimension.

                    ****

Reflection of Phonons in Liquid ^{4}He From Solid Surfaces   A F G
Wyatt, N A Lockerbie, K R A Ziebeck and R A Sherlock   Page 40

J P Harrison: Is it obvious why the phonons do not thermalise when
the helium is at the high pressure?

A F G Wyatt: At high pressures the dispersion curve bends downwards
continuously from the origin and so 3 phonon decay processes are not
allowed because energy and momentum cannot be conserved. This is
not so at low pressures where the curve initially bends upwards.

H Kinder: I notice that your received pulses are long compared with
the drive pulse, can you explain them?

A F G Wyatt: The propagation path is very long, 5 cm, and we think
that there may be a lot of small angle scattering processes still
going on at 24 bar, probably due to 4 phonon processes, which would
spread out the pulse.

H Kinder: With such long pulses it is very difficult to say whether
your reflection is diffusive or specular as you lose time resolution.

A F G Wyatt: I think that would be a valid point if we had not had
measured the direct signal at the same time. The two signals were
similarly broadened out in time and so it appears as though it is
an effect of the bulk helium rather than that of the reflection process. The fact that we received $\sim$ 80% of the energy after reflection indicates that most of the energy is specularly reflected.

****

The Thermal Boundary Resistance at Epoxy-Resin/Metal Interfaces
F F T de Araujo and H M Rosenberg   Page 43

J P Harrison: In calculating the acoustic mismatch theory for the
epoxy-copper interface, how were the epoxy lattice parameters chosen?

H M Rosenberg: From published experimental values of density and
sound velocity.

L J Challis: How big is the contribution of the epoxy resin to the
total resistance you measure?

H M Rosenberg: It depends on the temperature, but in the helium
range they are of the same order of magnitude.

****

Determination of Kapitza Conductance from the Velocity of Destruction of Superconductivity by Current   W C Overton Jr and
H Weinstock   Page 46

L J Challis: How important is the thermal resistance of the ^{4}He I
next to the wire in comparison with the Kapitza resistance?

W C Overton Jr: The temperature increases rapidly in this phenomenon
from He bath temperature T_B to an N-state value between 0.5K and
about 1K above T_c (for Ta this is 4.5K). This increase occurs rapidly (several nanoseconds to a few microseconds) over a distance of
only a few hundred coherence lengths ($\xi \sim 10^{-5}$ cm). Nucleate boil-
ing of He does not have time to form. Consequently it seems most
reasonable to assume that the transverse heat flow will encounter a
thermal boundary resistance. However, the usual approximation h $\sim$
$HT^3 \sim Q/\Delta T$ is not valid for large ΔT. Therefore, the full form Q =
$(2H/r) (T^4 - T_B^4)$ must be used when it is assumed that Khalatnikov
theory holds. No other physical mechanism is presently known.

Theory of Rayleigh Wave Scattering by Mass Defects and Surface
Roughness   A A Maradudin, D L Mills and R F Wallis   Page 55

Y Korczynskyj: Could you compare your analysis with that of the
previous paper (T Nakayama and T Sakuma)?

R F Wallis: The previous paper was concerned with scattering by
density fluctuations and we have considered a special case of this,
that is a surface we have called rough. This implies a randomness
in the density in the vicinity of the surface. The ω^5 dependence
was also found by Drs Nakayama and Sakuma.

****

Phonon Diffusion in a Temperature Gradient
R A Lucas (read by Professor L Kaiser)   Page 58

J A Rayne: Could you explain the experimental set-up for the data
you have quoted?

L Kaiser: I have reported at the International Heat Transfer Conference in Tokyo (1974)(see Proceedings). When iron samples were
immersed in an molten aluminium bath the temperature decreases ins-
tantly instead of rising which it would do according to classical
theory. We can interpret this by the retrodiffused flux which is
set up towards high temperatures that is in the molten aluminium
bath; so the temperature <u>decreases.</u> Shortly after, the classical
flux gives rise to an increasing temperature recording.

P G Klemens: Is this effect a reduction of effective conductivity
in the presence of high temperature gradient?

L Kaiser: It is, certainly. According to Lucas' expression of
thermal conductivity

$$\lambda = -\frac{1}{3}\,\overline{CV\Lambda}\,(1 + \frac{5}{2}\,\theta\,\frac{\partial LV}{\partial\theta})\,\frac{\partial\theta}{\partial y}$$

the term

$$\frac{5}{2}\,\theta\,\frac{\partial LV}{\partial\theta}$$

decreases as θ increases, and it is multiplied by the temperature
gradient.

F Michard: Are these experiments made in other materials than
metals?

L Kaiser: Harrington (U S A) has made experiments in metals and
other materials, and he found an anomalous heat conduction in the
interface region.

Phonon Diffusion and Temperature Field in Metals   L Kaiser
                                                      Page 61

J P Maneval: Is there experimental evidence that retrodiffusion of
heat takes place at the velocity of sound?

L Kaiser: We have few experiments, but in the experiment I have
shown in Tokyo, when iron is immersed in an aluminium bath, the
lattice suffers a dilatation and we have an elastic wave propagating at the sound velocity.

****

Role of Umklapp in Acoustic Boundary Resistance   N S Shiren
                                                     Page 64

J Jackle: How do you determine the relative magnitude of the amplitudes for the momentum-conserving and the umklapp transition processes?

N S Shiren: On the present calculation which is not self consistent,
each process has been calculated separately; i.e. as if it were the
only possible process, and utilising the appropriate Snell's law and
the continuum boundary conditions. This is similar to calculations
of acoustic attenuation for example in which each 3-phonon process
is calculated in the absence of all the other similar processes.

J P Harrison: It is known from standard Kapitza resistance measurements that the transmission above that given by acoustic mismatch
theory increases with θ_D. Is your model in accord with this?

N S Shiren: I have not considered this question previously and am
not prepared to answer it.

P G Klemens: If the surface is skew with respect to a crystal plane,
the surface perturbation will have a longer periodicity than the atomic one, and the 'inverse lattice vector' will be reduced. It
would thus be possible to enhance the transmission of phonons to
lower frequency.

N S Shiren: I agree with your statement. In fact, in some sense,
the umklapp is a special case of the rough boundary, in that the
latter may be Fourier analysed with a distribution of $\vec{k}$ vectors
which are available for umklapp processes at lower frequencies.

PHONON PROPAGATION IN LIQUID HELIUM

Humphrey J. Maris

Physics Department, Brown University

Providence, Rhode Island 02912, U.S.A.

1. INTRODUCTION

In this paper we will review recent research on phonon propagation and scattering in superfluid helium. The study of phonons
in helium has been a very active research area in the last few
years for two reasons. Firstly, phonon experiments have been a useful way to study the low-energy part of the dispersion relation.
It is difficult to study this part of the dispersion curve in other
ways. Secondly, some of the experiments that have been performed
provide striking confirmation of the most basic concepts of phonon
propagation and phonon interactions. While such experiments are
possible in principle in other materials, in practice the effects
are much more clearly observed in helium. This is because experiments in other materials are nearly always complicated by crystalline anisotropy, defects, impurities, etc. These are effects which
either do not exist or which can be safely ignored in helium.

In this review we will concentrate our attention on the second
class of experiments, i.e., for the most part we will discuss phonon propagation and scattering with the assumption that the phonon
dispersion relation is known. We therefore begin by stating what
is currently known about the dispersion relation and then go on to
describe the experiments.

2. PHONON DISPERSION RELATION

The general features of the dispersion curve for excitations
in superfluid helium-4, as measured by neutron scattering[1], are
shown in Fig. 1. At low temperatures (T<0.5°K) most thermal excitations are of the phonon type, i.e., come from the approximately
linear part of the dispersion curve near to the origin. To understand phonon-phonon scattering in helium, it turns out to be important to know the details of the deviations of the dispersion

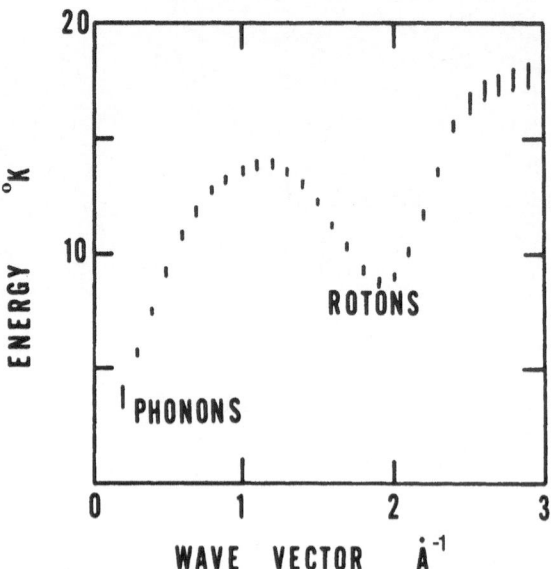

Fig. 1. Dispersion relations for excitations in helium-4 as measured by neutron scattering.

curve from linearity. For small wave momentum p we may approximate
the phonon energy by

$$\varepsilon(p) = c_o p[1 + \gamma p^2 + \ldots] \tag{1}$$

It was originally assumed[2] that the dispersion curve was convex
upwards, i.e., that γ was negative. However, in the last five
years much evidence has accumulated[3-11] which now shows unambiguously that γ is in fact positive. This is surprising because it
means that the phase velocity of an excitation is c_o at p=0, increases for finite p, reaches a maximum and then decreases as p approaches the roton region. This is referred to as anomalous dispersion. Although the experiments show that γ is definitely positive, there is still substantial uncertainty in the precise form of
$\varepsilon(p)$ for small p. One simple parameterization is[7]

$$\varepsilon(p) = c_o p \{ 1 + \gamma p^2 \frac{1 - p^2/p_A^2}{1 + p/p_B^2} \} \tag{2}$$

with $\gamma \hbar^2 = 1.1 \; \overset{o}{A}{}^2$, $p_A/\hbar = 0.54 \; \overset{o}{A}{}^{-1}$, $p_B/\hbar = 0.33 \; \overset{o}{A}{}^{-1}$. This is probably a reasonable approximation to $\varepsilon(p)$ up to wave numbers of
around 1.0 $\overset{o}{A}{}^{-1}$, but should be treated with some caution since at
present γ, p_A, and p_B are all uncertain by possibly as much as
± 20%. In Fig. 2 we show the phonon phase velocity c and the group

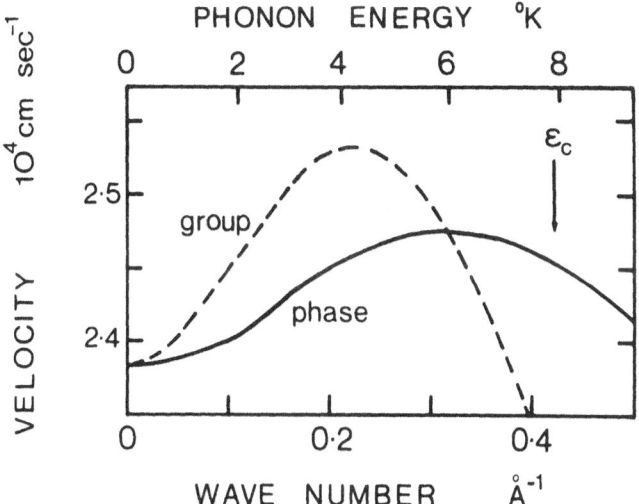

Fig. 2.   Phonon group and phase velocity as a function of wave number.

velocity v as a function of wave number.

A unique property of helium is that the dispersion curve can be modified substantially by the application of pressure. The velocity of sound c_o increases with increasing pressure by 50% before the liquid freezes. In addition, γ decreases and changes sign. The first evidence for this was from specific heat measurements[4]. These data indicated that γ went through zero at around 5 atmospheres. More accurate information is now available from the phonon scattering experiments that we will discuss below.

3.   SCATTERING EXPERIMENTS

In a continuum one expects that when phonons 1 and 2 collide to produce phonon 3 the following conservation laws must be satisfied:

$$\epsilon_1 + \epsilon_2 = \epsilon_3 \tag{3}$$

$$\vec{P}_1 + \vec{P}_2 = \vec{P}_3 \tag{4}$$

For the decay of phonon 1 into phonons 2 and 3 we must have

$$\varepsilon_1 = \varepsilon_2 + \varepsilon_3 \tag{5}$$

$$\vec{p}_1 = \vec{p}_2 + \vec{p}_3 \tag{6}$$

The peculiar dispersion relation of helium permits a very direct verification of these conservation laws. Consider first the scattering of a very low energy phonon (e.g., an ultrasonic phonon) by thermal phonons. In this case we can make the approximations $p_1 \ll p_2$ and $\varepsilon_1 = c_o p_1$. Equations (3) and (4) can then be used to derive the condition

$$c_o = \vec{v}(p_2) \cdot \hat{p}_1 \tag{7}$$

where $\hat{p}_1$ is a unit vector in the direction of $\vec{p}_1$. Thus the component of the group velocity of phonon 2 in the direction of $\vec{p}_1$ must be equal to the phase velocity of phonon 1. (This is the condition for the so-called "surf-riding resonance"). For helium at temperatures less than 0.5°K most of the thermal phonons have energies below 2°K. From Fig. 2 it can be seen that all of these thermal phonons have group velocities larger than c_o and so there are many phonons which can satisfy condition (7). Thus the attenuation of the ultrasonic phonons should be large. This is in agreement with the experimental results of Roach et al[12]. Under pressure the peak in the phonon group velocity shifts from its zero-pressure position at $0.2\text{\AA}^{-1}$ to a lower wave number. Thus, at a given temperature the number of phonons that have group velocity greater than c_o decreases as the pressure goes up, and hence the scattering of the ultrasonic phonons also decreases[13]. Above, a critical pressure P_c (about 20 atmospheres) the attenuation is essentially zero. This is because γ has changed sign and all thermal phonons now have velocity less than c_o.

Another striking consequence of the conservation laws shows up when one considers the scattering rate for thermal phonons as a function of energy. As shown above, low energy phonons can be scattered when the dispersion is anomalous. As the energy ε increases the scattering rate increases, and in addition the probability of spontaneous decay (see Eqs. 5 and 6) increases rapidly (approximately as ε^5). However, at high energies the phonon phase velocity starts decreasing rapidly and eventually at some energy ε_c the conservation laws (5) and (6) prevent the decay process from occurring. For the dispersion curve Eq. (2) one finds

$$\varepsilon_c = c_o p_B \left(\frac{5}{2}\right)^{1/2} \left[(1 + 16\ p_A^2 / 25\ p_B^2)^{1/2} - 1\right]^{1/2}$$

$$= 7.9°K \tag{8}$$

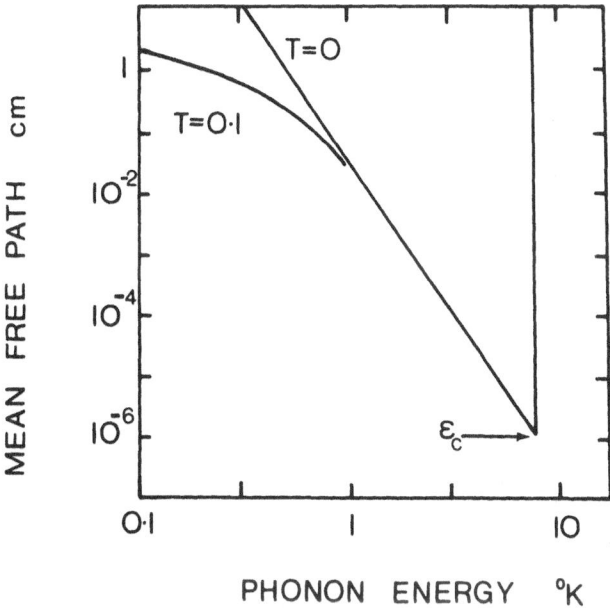

Fig. 3. Mean free path of phonons as a function of energy.

One then obtains a phonon mean free path which varies with energy as shown in Fig. 3. As the energy increases the mean-free-path drops to a minimum value at ε_c which is only 100A. It then suddenly increases to a macroscopic value! Results consistent with this remarkable prediction have been obtained by Dynes and Narayanamurti [9], and by Wyatt, Lockerbie and Sherlock [10]. These experiments constitute a striking confirmation of the conservation laws on which the mean-free-path calculation is based, and also provide a verification of the assumed form for the dispersion relation.

4. PHONON HYDRODYNAMICS

The effects described so far may be called "single-scattering processes" in the sense that one collision is sufficient to remove a phonon from a beam and thus limit the phonon mean free path. In hydrodynamics, on the other hand, one considers phenomena in which many phonons take part and a very large number of collisions occur. The hydrodynamic behaviour is determined by the same conservation laws that govern the collisions between elementary excitations. In particular there is a general connection between the number of conserved quantities and the number of propagating "sound-like" modes that can occur in a given system. This comes about in the following way. Let us suppose that at some time t we disturb a gas so that the distribution function becomes

$$n(\vec{p},\vec{r},t) = \bar{n}(\vec{p}) + \delta n(\vec{p},\vec{r},t) \tag{9}$$

Here $\bar{n}(\vec{p})$ is the distribution function for the gas when it is at
some reference temperature T_o, and δn represents a small disturbance. After a few collision times τ the distribution will relax
to the form

$$n(\vec{p},\vec{r},t) \simeq \bar{n}(\vec{p}) + \frac{\partial\bar{n}}{\partial T}\,\Delta T(\vec{r},t) + \frac{\partial\bar{n}}{\partial N}\,\Delta N(\vec{r},t) + \frac{\partial\bar{n}}{\partial\vec{v}}\cdot\vec{v}(\vec{r},t)$$

This is for a gas of particles. At this stage of its time evolution the gas is described by a distribution function which at each
point in space is completely specified by a local temperature var-
iation $\Delta T(\vec{r})$, a local fluctuation in number density $\Delta N(\vec{r})$, and a
local drift velocity $\vec{v}(\vec{r})$. All other features of the initial distribution function have been "washed-out" by the collisions. The
quantities ΔT, ΔN, and $\vec{v}$ are called hydrodynamical variables (HV).
They do not tend to zero in times of the order of τ because collisions between gas atoms always conserve energy E, number of particles N, and momentum $\vec{p}$. To derive equations for the rate of change
of the HV one has to consider the specific nature of the system.
Generally[14], these hydrodynamic equations express the rate of
change of one HV in terms of spatial derivatives of the set of HV.
Suppose now that one tries to find plane wave solutions of these
equations, i.e., solutions in which each HV depends upon space and
time according to

$$e^{i(kz - \Omega t)}$$

where K is real, but Ω may be complex. The number of different
solutions that can be found is equal to the number of equations,
which in turn is equal to the number of HV. Now some of the solutions correspond to propagating modes ($\text{Re}\,\Omega\neq 0$) and must therefore
appear in pairs. Using the fact that the number of HV is equal to
the number of conserved quantities it follows that:

> # of pairs of propagating modes $\leq$ 1/2 # of conserved quanti-
> ties. (10)

In a gas there are 5 conserved quantities, namely E, N, and p_x,
p_y, p_z. There is one pair of propagating modes (ordinary sound),
two damped transverse viscous waves (Re $\Omega=0$), and a damped temperature wave (again Re $\Omega=0$). For the phonon system in an ordinary
solid the only conserved quantity is the energy, and so from (10)
above it is clear that there can be no propagating modes. In superfluid helium or in a solid where only N-process collisions occur
there are 4 conserved quantities, i.e., E, p_x, p_y, p_z. In this
case there are two propagating modes (second sound) and two damped

shear waves.

The above discussion has been given to emphasize the connection between conserved quantities and hydrodynamic modes. This connection has an interesting application to helium at low temperatures[15]. This is because below about 0.5°K there exist quantities which are "nearly conserved" in addition to the exactly-conserved quantities E and $\vec{p}$. These extra conserved quantities come about in the following way. Collisions between phonons in helium at low temperatures are all small-angle collisions. To see this, note that if the dispersion relation were exactly linear ($\varepsilon = c_o p$), Eqs. (3) and (4) would become

$$c_o p_1 + c_o p_2 = c_o p_3 \tag{11}$$

$$\vec{p}_1 + \vec{p}_2 = \vec{p}_3 \tag{12}$$

Then $\vec{p}_1$, $\vec{p}_2$, and $\vec{p}_3$ must be exactly parallel. If the dispersion relation deviates from linearity by a small amount in the way given by Eq. (1), the angles between the vectors $\vec{p}_1$, $\vec{p}_2$, and $\vec{p}_3$ are of the order of

$$\alpha \sim (\gamma p_3^2)^{1/2} \tag{13}$$

Using $\gamma/\hbar^2 = 1.1 \overset{\circ}{A}{}^2$, we find that for thermal phonons at 0.5K the typical collision angle α is of the order of $10°$.

Consider now the way a phonon gas in which only small-angle collisons can occur comes to equilibrium after it has been disturbed. Let us define a "temperature" $T(\vec{p},\vec{r},t)$ for the phonons of momentum $\vec{p}$ at point $\vec{r}$ at time t by the relation

$$n(\vec{p},\vec{r},t) \equiv \frac{1}{e^{\varepsilon_p/kT(\vec{p},\vec{r},t)}-1} \tag{14}$$

where $n(\vec{p},\vec{r},t)$ is the distribution function. If the disturbance occurred at t=0, then for $t \gtrsim \tau_{||}$ ($\tau_{||}$ =time for one small-angle collision to occur) the temperatures of all phonons at the same point in space and with momenta in the same <u>direction</u> will have become equal. Thus for $t \gtrsim \tau_{||}$ we can write

$$T(\vec{p},\vec{r}.t) \approx T(\theta, \phi, \vec{r}, t)$$

$$= \sum_{\ell m} Y_{\ell m}(\theta,\phi) A_{\ell m}(\vec{r},t) \tag{15}$$

where θ,ϕ are the angles specifying $\vec{p}$, $Y_{\ell m}(\theta,\phi)$ are spherical harmonics, and $A_{\ell m}(\vec{r},t)$ are coefficients depending on space and time.

Further relaxation occurs on a much longer time scale. The effect of collisions is to make the amplitude $A_{\ell m}$ of each spherical harmonic decay exponentially with a characteristic time τ_ℓ (independent of m). To work out how τ_ℓ depends upon ℓ it is convenient to draw a polar plot of $T(\theta,\phi)$ assuming that only one spherical harmonic $Y_{\ell 0}(\theta,\phi)$ is present in addition to the equilibrium distribution, i.e., assuming

$$T(\theta,\phi) = \text{constant} + Y_{\ell o}(\theta,\phi) \tag{16}$$

This is done in Fig. (4). For $\ell=0$ the distribution is isotropic and therefore collisions produce no relaxation. Hence $\tau_o = \infty$. For $\ell=1$ the distribution has a net momentum in the z-direction. Since collisions between particles cannot change the momentum it follows that this distribution cannot decay and therefore that $\tau_1 = \infty$ also. For $\ell=2$ the temperature in the +ve and -ve z-directions are greater than the temperatures near the equator. Since each collision between phonons changes the direction of a phonon by only the small angle α, many collisions are needed to bring the poles and the equator into equilibrium. One can show that

$$\tau_2 \sim \frac{\tau_{\parallel}}{\alpha^4} \tag{17}$$

For higher values of ℓ the hot and cold regions are closer together in direction and so τ_ℓ decreases rapidly as ℓ increases. One finds

$$\tau_\ell \sim \frac{\tau_{\parallel}}{\ell^4 \alpha^4} \tag{18}$$

Finally for large ℓ (>20 at 0.2°K) successive maxima and minima of $Y_{\ell 0}(\theta\phi)$ are so close together that one collision can take a phonon from a hot direction to a cold direction. Thus

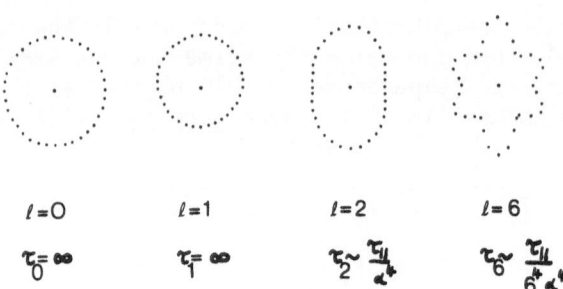

$\ell=0$ $\ell=1$ $\ell=2$ $\ell=6$

$\tau_o = \infty$ $\tau_1 = \infty$ $\tau_2 \sim \frac{\tau_{\parallel}}{\alpha^4}$ $\tau_6 \sim \frac{\tau_{\parallel}}{6^4 \alpha^4}$

Fig. 4. Polar plots of direction dependent temperature $T(\theta,\phi)$ when only one spherical harmonic is present in addition to the equilibrium distribution.

$$\tau_\ell \sim \tau_{11} \tag{19}$$

Quantitative calculations give the following results at 0.2°K: $\tau_2 = 1.8 \times 10^{-3}$, $\tau_3 = 3.7 \times 10^{-4}$, $\tau_6 = 3.1 \times 10^{-5}$, $\tau_{12} = 4.1 \times 10^{-6}$ secs.

Suppose now that we consider some oscillation of the phonon gas that occurs at frequency Ω. If

$$\Omega\tau_2 \gg 1$$

the amplitude of the $Y_{20}(\theta,\phi)$ component in the distribution function will relax very little during one cycle of the oscillation. Thus this amplitude may be considered as a conserved quality in the same right as the energy or the momentum. Hence at higher frequencies the number of nearly conserved quantities increases as $\Omega\tau$ becomes greater than 1 for more values of ℓ. One might expect, therefore, that more propagating modes should appear as Ω goes up. Detailed calculations show that this does indeed happen[15]. These extra modes are called second second sound, third second sound, etc. They have lower velocity than the ordinary (first) second sound. Although the attenuation is larger than for ordinary second sound, the wave is nevertheless definitely a propagating wave since $\Omega_R \gg \Omega_I$. The physical nature of these waves is quite complicated, since the distortion of the phonon distribution function has components with Y_{20}, Y_{30}, etc., symmetry in addition to Y_{00} and Y_{10}. For a discussion and some explicit results for the distribution function see reference 15.

These extra waves have not yet been observed experimentally. The principle difficulty is that any conventional second sound generator (such as a heated film) will generate ordinary (first) second sound and the higher second sounds with comparable efficiency. In a pulse experiment the higher second sound components should arrive after the main pulse because of their lower velocity. However, detailed calculations[15] show that because the velocity difference is not large and because the pulses spread due to dispersion the higher components appear in the tail of the main pulse. Thus distinct separate pulses are not observable. Experiments in which the temperature of the second sound generator is varied sinusoidally appear to be more attractive. A measurement of the magnitude and phase of the temperature fluctuation induced in the liquid at several distances from the source could then be analyzed to yield the velocity and attenuation of the various second sound modes. Measurements of this type would be very worthwhile. In addition to the intrinsic interest of detecting new modes of energy propagation in helium, an observation of these waves would constitute an extremely interesting confirmation of the theory of phonon-phonon interactions.

ACKNOWLEDGMENT

This work was supported in part by the National Science
Foundation.

REFERENCES

1. A. D. B. Woods and R. A. Cowley, Phys. Rev. Lett. 24, 646
   (1970); Can. J. Phys. 49, 177 (1971).
2. This was assumed by Landau and Khalatnikov in their calculations
   of kinetic processes in helium. This work is reviewed in I. M.
   Khalatnikov, Introduction to the Theory of Superfluidity, (Benjamin, New York, 1965).
3. H. J. Maris and W. E. Massey, Phys. Rev. Lett. 25, 220 (1970).
4. N. E. Phillips, C. G. Waterfield and J. K. Hoffer, Phys. Rev.
   Lett. 25, 1260 (1970).
5. C. H. Anderson and E. S. Sabisky, Phys. Rev. Lett. 28, 80 (1972).
6. H. J. Maris, Phys. Rev. Lett. 28, 277 (1972); Phys. Rev. 8, 2629
   (1973).
7. H. J. Maris, Phys. Rev. 8, 1980 (1973).
8. N. G. Mills, R. A. Sherlock and A. F. G. Wyatt, Phys. Rev. Lett.
   32, 978 (1974).
9. V. Narayanamurti, K. Andres and R. C. Dynes, Phys. Rev. Lett.
   31, 687 (1973); R. C. Dynes and V. Narayanamurti, Phys. Rev.
   Lett. 33, 1195 (1974) and Phys. Rev., to be published.
10. A. F. G. Wyatt, N. A. Lockerbie and R. A. Sherlock, Phys. Rev.
    Lett. 33, 1425 (1974).
11. W. R. Junker and C. Elbaum, Phys. Rev. Lett. 34, 186 (1975).
12. P. R. Roach, J. B. Ketterson and M. Kuchnir, Phys. Rev. 5,
    2205 (1972).
13. J. Jackle and K. W. Kehr, Phys. Rev. Lett. 27, 654 (1971).
14. This is not always true. For example, in a plasma the presence
    of long range forces makes the rate of change of the current at
    $\vec{r}$ depend upon the number density in a more complicated way.
15. H. J. Maris, Phys. Rev. Lett. 30, 312 (1973); Phys. Rev. 9,
    1412 (1974).

A FIRST STEP TOWARDS ZERO SOUND IN IONIC CRYSTALS

F. MICHARD[+], F. SIMONDET[+], L. BOYER[x], R. VACHER[x]

DRP-Université Paris VI[+], LPEC-Université Montpellier[x]

[+]Tour 22, 4 Place Jussieu-Paris 75230 Cédex 05

Experimental studies of the ultrasonic absorption in crystals and its dependence on the frequency are important from the standpoint of further development of the theory of phonon-phonon interactions .

In the present paper we report the results of an investigation of the absorption of ultrasound in some ionic crystals: alkali halides (NaCl, KCl, KBr) and isomorphous nitrates $(Sr(NO3)2, Ba(NO3)2, Pb(NO3)2)$ over the frequency range from 100 MHz to 1 GHz for longitudinal elastic waves propagating in the directions $[100]$ and $[11\bar{0}]$.

The measurements were performed by using the opto-acoustic interaction between the light beam of a He-Ne laser and acoustic pulses generated by a lithium-niobate transducer . In the frequency range investigated the interaction can be described as BRAGG diffraction ; for low acoustic power the intensity of the diffracted light is proportional to the acoustic power (1) .

The measurements of the diffracted light intensity as a function of the distance allow us to obtain the intrisic attenuation in the crystal, getting rid of spurious effects due to the bonding of the sample and to the non parralelism of its faces .

In the alkali halides the attenuation of the elastic waves was found to be proportional to the square of the frequency, as predicted by AKHIESER'S theory (2) . The measurements that we have made by BRILLOUIN scattering show that this dependence on frequency is still valid at about 30 GHz .

Incidentally, our results are in nominal agreement with the results previously obtained at lower frequency : 30. 200 MHz (3) . Regarding the results obtained for strontium and lead nitrates the most important fact is that there is a frequency dependence which

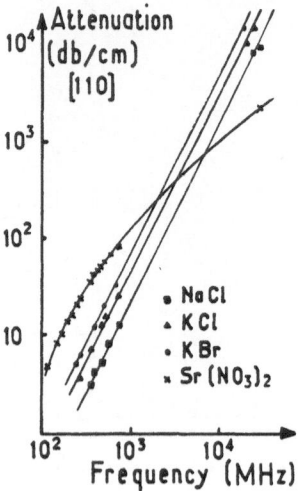

Figure 1

α(db/cm)	NaCl		KCl		KBr	
Ω(MHz)	$[100]$	$[110]$	$[100]$	$[110]$	$[100]$	$[110]$
200	1,4-1,4 (3)	1,1-1 (3)	2,6-2,8	2,2-2,6	3,5-3,5	3-3,2
700	15	11,7	30	25	40	35

Table I

can be characterized by Ω^{p} with $p<2$ for frequency as low as a few hundred Megahertz . This special behaviour can be combined with our previous study of the phase velocity dispersion of the elastic waves propagating in these compounds, to form a more complete picture of the acoustical behaviour (4) . In particular , we have shown that the velocity of the longitudinal waves propagating along the directions $[100]$ and $[110]$ in $Pb(NO_3)_2$ are greater when measured by BRILLOUIN scattering (20-30 GHz) than when measured by ultrasonic techniques (10-50 MHz) .

The whole of these results are consistent with the hypothesis of a change in the regime of propagation of the acoustic phonons as is predicted by many theoretical works (5)(6). In particular,the restriction of the first sound regime theory($\Omega \tau \ll 1$) is not satisfied in the whole range of investigated frequency . Nevertheless, other relaxation mechanisms that can be anticipated can be discarded (interaction with the dislocations, rotation of the NO_3^- groups...)(7).

It's worth noticing that in the hypothesis of a transition from

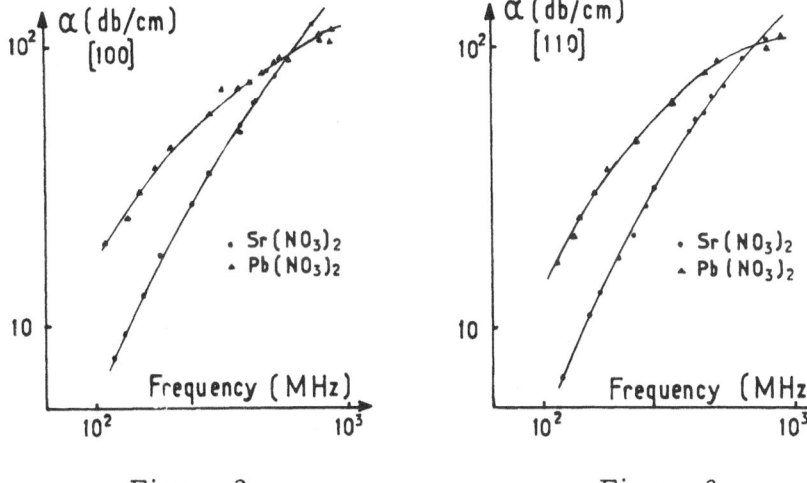

Figure 2   Figure 3

the first sound to the zero sound regime, the phonon lifetime τ which is to be considered is not the average value $\langle \tau \rangle$ obtained through thermal conductivity measurements .

This assertion is validated by the low value of $\langle \tau \rangle$ obtained recently (8) and reported in the same session .

We are beginning in our laboratory the determination of the third order elastic constants and these results could clarify the importance of the anharmonic interactions in these nitrates and could explain why the transition between the two regimes of propagation seems to occur at such low frequency .

REFERENCES

(1) - R. ADLER - I.E.E.E. Spectrum, 4, 42, (1967)

(2) - H.E. BOMMEL, K. DRANSFELD - Phys. Rev. 117 , 1245, (1960) .

(3) - L.G. MERKULOV, R.V. KOVALENOK, E.V. KONOVODCHEN-KO - Soviet Phys. Solid. State 11, 2241, (1970) .

(4) - F. MICHARD, A. ZAREMBOWITH, R. VACHER, L. BOYER-Phonons International Conf. Rennes, 321, (1971).

(5) - R.A. COWLEY - Proc. Phys. Soc. 90 , 1127, (1967)

(6) - H.J. MARIS - Physical Acoustics 8 , 279, (1971)

(7) - W.C. HAMILTON - Acta Cryst. 10 , 103, (1957)

(8) - A.M. de GOER, A. de COMBARIEU - Phonon scattering in solids. Nottingham University (1975) .

LOW TEMPERATURE THERMAL CONDUCTIVITY OF $(NO_3)_2Ba$, $(NO_3)_2Sr$ AND $(NO_3)_2Pb$ SINGLE CRYSTALS

A. de Combarieu and A.M. de Goër

Centre d'Etudes Nucléaires de Grenoble
Service des Basses Températures
BP 85 Centre de Tri, 38041 Grenoble-Cedex (France)

In view to investigate the evolution of phonon-phonon interaction in this serie of isomorphous compounds, where anharmonicity is the largest for the Pb nitrate [1], the thermal conductivity K of single crystals has been measured between 1.5 and 300 K (from about $\theta/60$ to $2\,\theta$).The crystallographic cell of these cubic crystals contains 36 atoms [2], so that several low-lying optic modes may contribute to the heat flow.

The samples(parallelepipeds about 3x3x15 mm) have been kindly supplied by M.Zarembovitch(Paris VI University). Measurements have been done by the classical method of stationary heat flow. Experimental problems arise from the large values of the thermal expansion coefficient [3]. The absolute incertitude on K is about 10 %. Experimental results are given on figures 1,2 and 3, and the curves have the typical shape for insulating crystals containing a large number of isotopes. Quantitative fits have been done in a simplified way : the Callaway's correction is omitted and the total relaxation time used is [3]: $\tau^{-1} = V/L + A\,\omega^4 + B\,\omega\,T^4$. Good fits have been obtained below 60 K (fig.1,2,3), with the "experimental values" of the parameters given in table 1(Debye temperatures and mean sound velocities, calculated from elastic constants [3], are also given).

TABLE 1

	θ (K)	V (cm.sec^{-1})	$(V/L)_{cal}$ (sec^{-1})	$(V/L)_{exp}$ (sec^{-1})	A_{calc} (sec^3)	A_{exp} (sec^3)	B_{exp} (K^{-4})
$(NO_3)_2$ Ba	112	$1.925.10^5$	$6.79.10^5$	$6.4.10^5$	$2.7.10^{-44}$	$1.7\ .10^{-44}$	$1.2.10^{-9}$
$(NO_3)_2$ Sr	142	$2.33.10^5$	$7.85.10^5$	$7.5.10^5$	$8.9.10^{-45}$	$11.9.10^{-44}$	$1.1.10^{-9}$
$(NO_3)_2$ Pb	107	$1.78.10^5$	$5.7.10^5$	6.10^5	$1.4.10^{-44}$	$1.1.10^{-44}$	$2.4.10^{-9}$

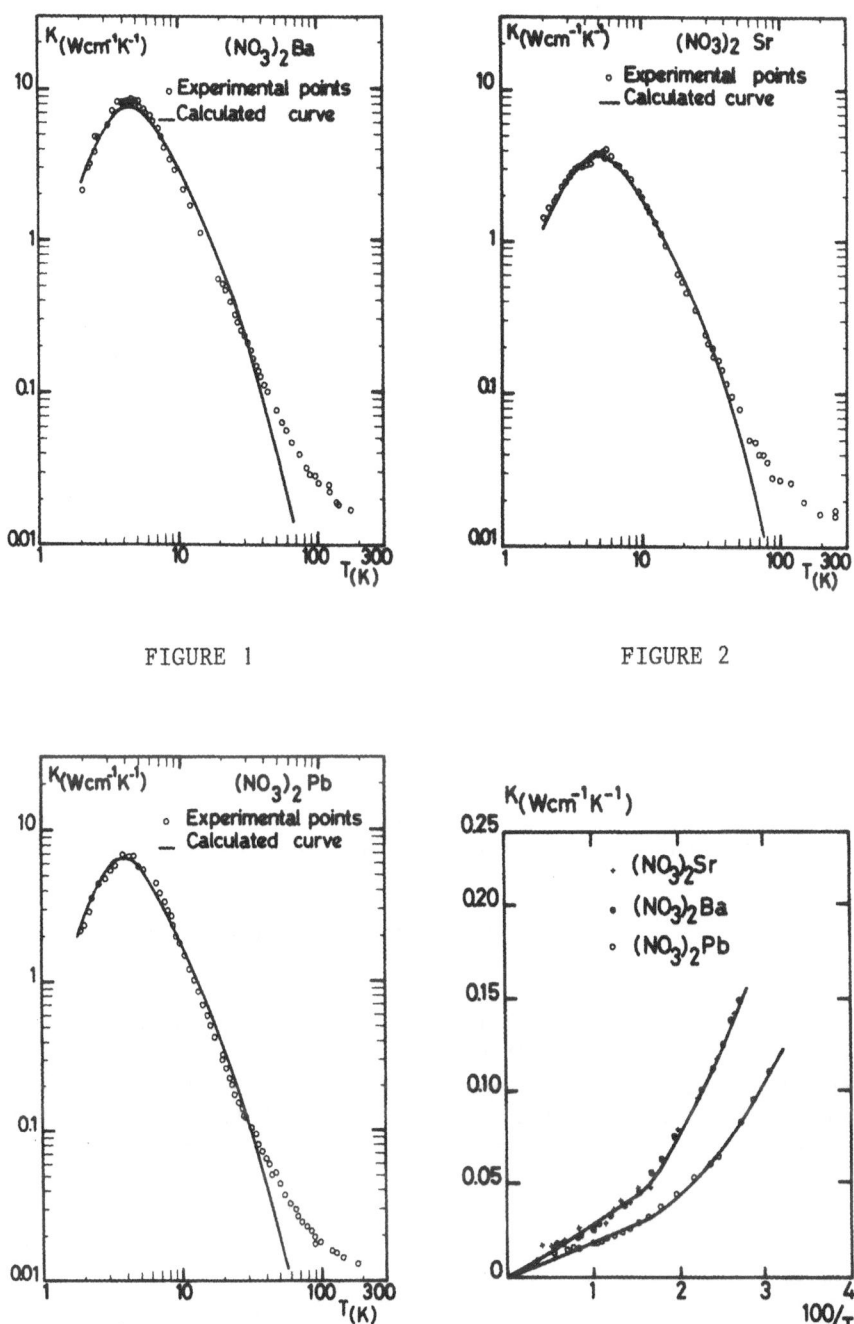

FIGURE 1

FIGURE 2

FIGURE 3

FIGURE 4

TABLE 2 (cgs units)

	M	V_O	γ	β
$(NO_3)_2$ Ba	261.3	$13.4.10^{-23}$	0.44	97
$(NO_3)_2$ Sr	211.6	$11.8.10^{-23}$	1.0	29.7
$(NO_3)_2$ Pb	331.2	$12.1.10^{-23}$	0.9	24.8

Firstly we note that the parameter B, measuring the strength of phonon-phonon interaction, is quite larger in the case of the lead nitrate : this confirms the greater anharmonicity of this crystal. Secondly the experimental values of the Casimir term and point defect scattering term compare favorably to the calculated ones (table 1). A_{calc}. corresponds to isotope scattering alone and the larger value of A_{exp} in the case of Sr nitrate can be due to chemical impurities. These results, and the fact that scattering by dislocations was undetectable, indicate the good overall quality of the crystals.

Finally, above nearly $\theta/2$ the product K.T is constant, as shown on figure 4. The constant is proportional to $\beta = M \theta^3 V_o^{1/3}/\gamma^2$ [4], which has been calculated (table 2) using the macroscopic Grüneisen constants γ, and the molecular values for M and V_o. The ratio of these β values for two nitrates can be compared to the experimental ones. This comparison shows [3] that the macroscopic value of γ is noticeably different from the γ corresponding to the modes which actually contribute to the heat flow, at least in the case of Ba nitrate.

REFERENCES

[1] MICHARD F. and ZAREMBOVITCH A. in PHONONS, edited by
    M.A. NUSIMOVICI p 321 (1971)
[2] MICHARD F. Thèse d'état, Paris (1973)
[3] de COMBARIEU A. and de GOER A.M.- note SBT 380/75 (1975)
[4] ZIMAN J.M. - Electrons and Phonons (Clarendon Press) p 296 (1960)

BALLISTIC PHONONS AND THE TRANSITION TO SECOND SOUND IN SOLID ^{3}HE

V Narayanamurti and R C Dynes

Bell Laboratories, Murray Hill, New Jersey 07974, U S A

We report a study of heat pulse propagation in solid ^{3}He as a function of temperature (between .05K and .5K) and molar volume (24.5 cm^3/mole to 23.7 cm^3/mole). We use the fast heat pulse technique[1] and present data which shows the transition from ballistic flow to second sound. We present evidence for unusual behaviour in the temperature dependence of the velocity and intensity of the slow transverse phonon (propagating close to [110]). The implication of these measurements for specific heat data[2] are discussed.

In figure 1 we have plotted the temperature dependence of the heat pulse velocities for a crystal with a molar volume of 23.72 cm^3/mole. At low temperatures ($\lesssim$ .1K) the limiting velocities of the longitudinal (L), fast transverse (FT), and slow transverse (ST) modes indicates an orientation close to [110] from Greywall's sound velocity data[3]. As the temperature is raised to 0.2K the ST mode velocity increases from 113 m/sec to 119.5 m/sec. It then decreases to a value of 101 m/sec as in the second sound region (0.3 to .34K). At higher temperatures diffusion is observed.

The velocity behaviour for the [110] ST phonon described above was also observed in a crystal of molar volume 24.45 cm^3/mole. No unusual behaviour was observed with crystals possessing a [100] orientation or in similar measurements on solid ^{4}He. A clue to this unusual behaviour is obtained by plotting the intensities of the three pulses (figure 2). In the region between 0.15 to 0.2K where the L and FT modes rapidly decay, the ST mode picks up in intensity. Above 0.22K, a rapid decay of the ST mode intensity also occurs consistent with observed velocity dispersion and a second maximum is reached at about 0.3K as the collective second sound mode forms.

Fig 1: Peak heat pulse velocities as a function of temperature in solid ^{3}He. Molar volume 23.72 cm^3/mole. Orientation [110].

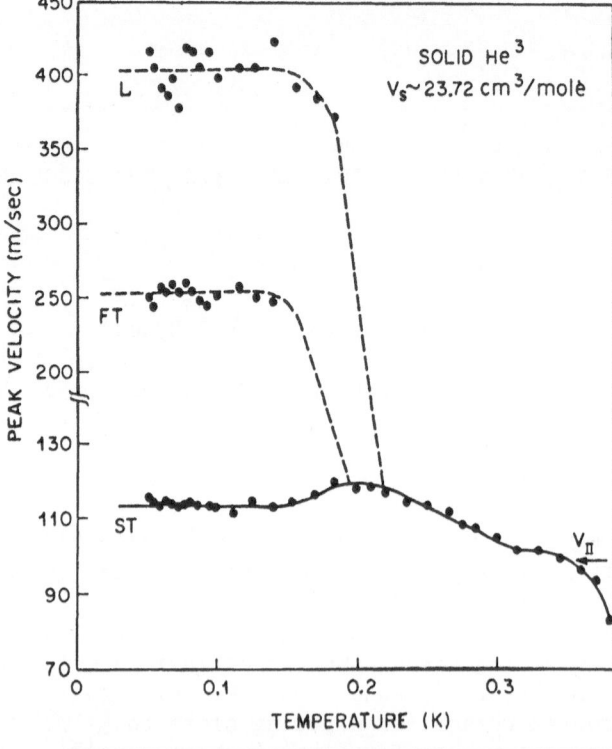

Fig 2: Peak intensity of heat pulses in solid ^{3}He as a function of temperature. Molar volume 23.72 cm^3/mole.

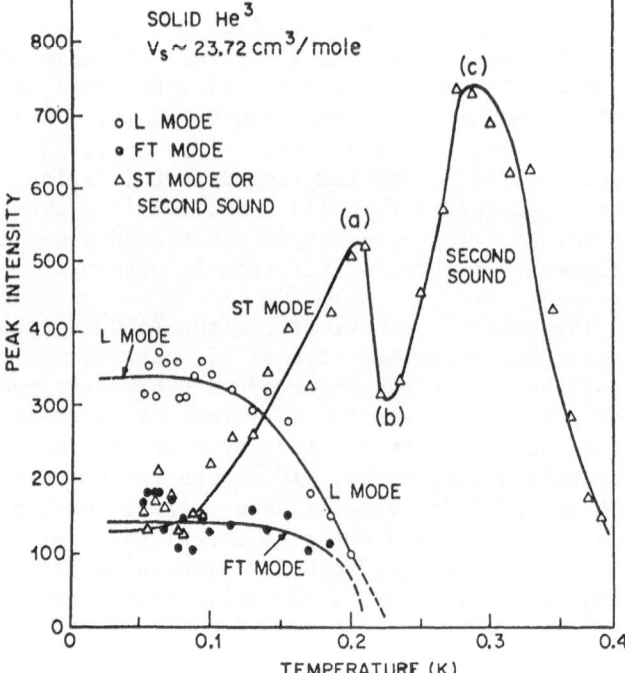

The velocity data of figure 1 are, therefore, interpreted as a mode pulling effect due to the higher probability for decay of the L and FT branches compared to the ST branch[4] by the 3 phonon process. At higher temperatures, the ST mode eventually also decays via higher order anharmonic N-processes and one observes second sound.

This interpretation is also consistent with calculations of the expected second sound velocity in an elastically anisotropic medium. Using Kwok's formula[5] and Greywall's elastic constants[3] we calculated the second sound velocity to be 96.3 m/sec for a molar volume of 23.72 cm^3/mole. This is very close to the measured velocity of 101 m/sec. The elastic Debye velocity divided by $\sqrt{3}$ has on the other hand a value of 135 m/sec.

Finally, the maximum in the ST velocity as a function of temperature implies that the density of states and hence the lattice specific heat, C_v, must show an unusual temperature dependence[2]. Note, however, that our data implies that the lattice specific heat (due to the phonons) should show the usual T^3 behaviour below .12K. Thus, the observed[2] departures from T^3 below .12K must be due to other causes, while that above could in part be due to the unusual temperature dependence of the slow mode.

REFERENCES

1. See for example R J von Gutfeld, in Physical Acoustics, edited by W P Mason (Academic, New York, 1968), Vol 5, p 233.
2. S H Castles and E D Adams, Phys Rev Lett 30, 1125 (1973).
3. D S Greywall, Phys Rev B11, 1070 (1975).
4. R Orbach and L A Vredevoe, Physics 1, 91 (1964).
5. P C Kwok, Physics 3, 221 (1967).

ON THE TEMPERATURE BEHAVIOR OF SECOND SOUND AND POISEUILLE FLOW IN SOLIDS

H. Beck[*]

IBM Zurich Research Laboratory

8803 Rüschlikon, Switzerland

In recent years second sound (SS) was observed in ^{3}He, ^{4}He, NaF and Bi [1] by means of heat-pulse techniques as well as in NaF by stimulated light scattering.[2] Moreover, the Poiseuille-flow (PF) regime of thermal conduction was identified in both heliums. These phonon-hydrodynamic phenomena[1] are usually treated on the basis of a Boltzmann equation for the density of phonons. A linearized collision operator L is used to describe normal (L_N) and Umklapp phonon interactions (L_U), as well as scattering from impurities and imperfections (L_I) and from boundaries (L_B). The transport equation is most easily solved by replacing the various parts of L by the inverse of mean relaxation times $\bar{\tau}_N$, $\bar{\tau}_U$, $\bar{\tau}_I$ and $\bar{\tau}_B$, respectively.

For an isotropic Debye model this approximation was used[3] to obtain quantitative information about heat pulse propagation in the temperature range covering ballistic, second-sound and diffusive regimes. In this simple framework SS appears as a pole $\Omega_{SS} = c_2(T) q - i \alpha_2 (q,T)$ of the energy density response function $\chi_E(q,\Omega)$. This collective excitation exists as soon as $\bar{\tau}_N(T)$ is small enough. The velocity $c_2(T)$ monotonically decreases with T. At the onset temperature T_0 of SS, c_2 lies half-way between the transverse sound velocity and the speed c_{II} of fully developed SS. [In very anisotropic materials, such as ^{3}He, it is conceivable that $c_2(T_0)$ is even higher than the lowest transverse velocity.] The temperature dependence of the $\bar{\tau}$'s in NaF was found by fitting the experimental[4] SS velocities. The fits for two NaF samples,[5] obtained with $\bar{\tau}_N = AT^3$, $\bar{\tau}_U = BT^4 \exp(-\theta/T)$, $\bar{\tau}_I = DT^4$ are shown in Fig. 1. Elementary arguments[1] would predict T^5 for $\bar{\tau}_N$, but powers higher than three are ruled out by considering the purest sample A (see the dashed curves

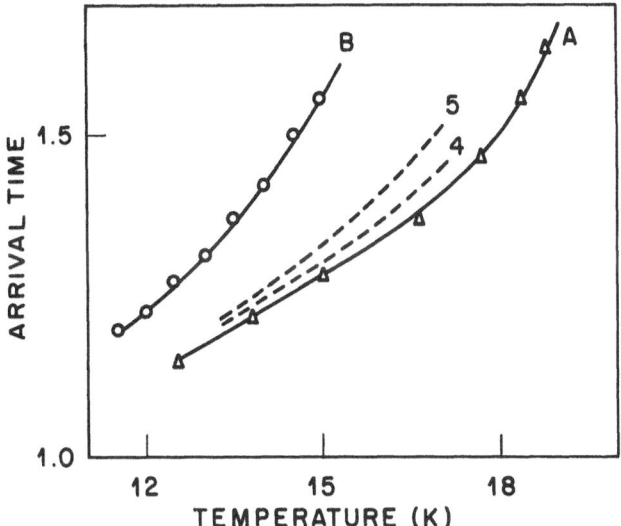

Fig. 1. Arrival time of the SS peak (normalized to the one of the transverse ballistic peak) as a function of T.   Full curve: theory, o and Δ : data from Ref. 4.   Dashed curves see text.

labelled by the corresponding exponents 4 and 5). Various explanations[5] for the deviation of $\bar{\tau}_N$ from T^5 have been proposed, e.g., wide-angle scattering or anharmonic interactions between the colliding phonons.

An important mathematical fact about L_N is the missing gap between the (degenerate) eigenvalue zero and the continuous spectrum. Taking this feature into account in a crude way by a wave-vector-dependent relaxation time[6] $\tau_N(k)$ with $\lim_{k \to 0} \tau_N^{-1}(k) = 0$ drastically alters the hydrodynamic equations. It may indeed lead to a temperature dependence of $\bar{\tau}_N$ weaker than T^5. Furthermore the radius dependence of thermal conductivity in the PF regime will be weaker than R^2, which also seems to be borne out by experiments on helium.

References

*
Postdoctoral Fellow from the Institute of Theoretical Physics of the University of Zurich, 8001 Zurich, Switzerland.

[1] For references see: H. Beck, P.F. Meier and A. Thellung, Phys. Stat. Sol. A 24, 11 (1974).

[2] D. Pohl, to be published.

[3]H. Beck and R. Beck, Phys. Rev. B $\underline{8}$, 1669 (1973).

[4]T.F. McNelly, Thesis, Laboratory of Atomic and Solid-State Physics, Cornell University, 1973, and Ref. 1.

[5]H. Beck, Z. Physik B $\underline{21}$, 209 (1975).

[6]H. Beck, Z. Physik B $\underline{20}$, 313 (1975).

THE PROPAGATION OF HIGH FREQUENCY THERMAL WAVES

IN DIELECTRIC CRYSTALS

I.F.I. Mikhail and S. Simons

Queen Mary College

Mile End Road, London, E.I.

1. INTRODUCTION

The present work is concerned with establishing the dispersion relation, that is the relation between frequency and wavenumber, for the propagation of high frequency thermal waves in dielectric crystals and its motivation is twofold: 1. To study the phenomenon of second sound in solids 2. To calculate the deviations from classical diffusive heat flow which occur at high frequencies in a more accurate manner than was done by Simons ('71).

To progress beyond the earlier work we shall take the following points into consideration: 1. Although second sound has been observed experimentally in several materials the experiments have so far only been performed with thermal pulses. It would clearly be desirable to do work with continuous waves for which the theoretical interpretation of experimentally measured quantities is more straightforward. 2. It is characteristic of most earlier theoretical treatments of the dispersion relation that a limited frequency range has generally been considered. 3. A more basic criticism of previous work is that it has consistently assumed a constant relaxation time τ independent of phonon wavenumber $\underline{k}$.

In view of the above considerations it was felt that a detailed calculation of the dispersion relation should be undertaken over the whole frequency range for which it exists, taking into account the proper k dependence of τ , and considering both resistive and normal phonon interaction processes.

2. DISPERSION RELATION

We begin by considering the time dependent Boltzmann equation, which in the case of small deviations from equilibrium can be expressed in the form

$$\partial \phi / \partial t = M\phi - v. \nabla \phi. \tag{1}$$

Here M is a linearized collision operator, and ϕ measures the deviation of the distribution function from the Base Einstein equilibrium distribution $F(E,T)$, where the usual notations are used. Since we are interested in a wave solution we assume $\phi = \phi(\underline{k}) \exp(i\omega t + i\underline{p}.\underline{r})$, where ω is the angular frequency and $\underline{p}$ is the wave number of the thermal wave. If a Callaway model for the collision operator is used then the dispersion relation can be shown to take the form

$$i < k^2 y^2 v \Delta \delta^{-1} > \quad < E^2 \eta (\omega + pvy) \delta^{-1} > \quad - < kyE v \eta \delta^{-1} >^2 = 0. \tag{2}$$

Here $\delta = \eta + i(\omega + pvy)$, $\Delta = \mu + i(\omega + pvy)$, v, μ and η are the reciprocals of the normal, resistive and total relaxation times, $y = \cos \theta$ where θ is the angle between $\underline{k}$ and $\underline{p}$ and the scalar product $<p \psi> = \int p \psi (\partial F / \partial E) d\underline{k}$.

The specification of $\eta(\underline{k}) (=\mu + v)$ depends, of course, on the type of phonon collision mechanisms which are assumed to operate. For dislocation and isotope scattering, $\mu(\underline{k}) \alpha k$ and k^4 respectively. For normal and umklapp processes we shall assume, as has been consistently done in earlier work on thermal conductivity, that the true form for $\eta(\underline{k})$ can be adequately represented by the power law which holds for small k. We therefore take $v \alpha k$, while in the case of umklapp processes the problem has been considered from first principles and it was found that $\mu \alpha k^3$.

Two complementary techniques are then utilized to solve equation (2), which can be summarized as follows:-
a) The series expansion method. In this method an analytical approach based on an infinite series expansion valid for small ω is used to find the solution. This expansion has in general a considerably more complicated structure than a simple power series and can involve fractional powers as well as logarithmic terms. In the following we summarize briefly the main results obtained by using this method.

1. In the case of diffusive flow it is found that the dispersion relation takes different forms according to the lowest power (n) of k in the expression for η. In the general case, however, when normal processes are taken into account the lowest power of k always arises from v and thus the dispersion relation takes the same form $(p^2 = \alpha\omega + \beta\omega^2 \ln a\omega + \gamma \omega^{5/2})$ for any combination of resistive processes.

2. In the case where $n \leqslant 2$ the first approximation to the solution is in complete agreement with the well known classical

result — $p^2 = -i\omega c/k$. If N processes could be completely neglected, n can be >2 and k then $\to \infty$ so that the classical solution breaks down. However, with the present approach a damped wave solution can be found exhibiting a rather involved $p^2 - \omega$ relationship.

3. In the case when $n \leqslant 2$ the first order deviations from the classical result differ in both the form and sign from those obtained by using a constant relaxation time (that is, $p^2 = \alpha\omega + \beta\omega^2$, Simons 1971). It is suggested that measurement of these deviations could be used to obtain the normal process relaxation time.

4. In the case of second sound it is shown that the equations used by earlier workers (Guyer and Krumhansl 1964 and 1966) assuming constant τ, remain true if a certain well defined mean relaxation time ($\bar{\tau}_N = < k^2 \tau_N > / < k^2 >$ and $\bar{\tau}_R = < k^2 > / < k^2 \tau_R^{-1} >$) is employed.

The average value of τ_N, however, has been defined wrongly in some earlier work, for example Jackson et al 1971 and Kimber et al 1973. The correct value of $\bar{\tau}_N$ will be higher than those given in these two references by about 20% and 35% respectively. This will consequently decrease the width of the window in which second sound can be observed by the same proportion.
   b) Iteration method. In this method equation (2) is first integrated with respect to y and the solution is then obtained numerically by using the Newton Raphson iteration technique. This procedure is effective for larger values of ω but is inferior to the series expansion approach for small ω.

The solutions obtained by using these two techniques were found to agree with each other to within 10^{-3}% at low frequencies. For higher frequencies, however, the effect of the neglected terms in the series solution becomes of importance and the agreement between the two solutions then gradually worsens.

With current techniques thermal wave frequencies $\sim 100KHZ$ can be generated and it is, therefore, of interest to make realistic estimates of the practicability of observing continuous wave (C.W.) second sound in specific materials. The results obtained above were therefore applied to different materials and these suggest that C.W. second sound is capable of being experimentally observed in solid He^4 at T = 0.6-0.7 k and in solid Ne at T = 2k. It is also found that it should be possible to observe the onset of second sound in very pure crystals of LiF and NaF at T=15-17k and T=10-11k respectively.

REFERENCES

Guyer R.A. and Krumhamsl J.A., 1964, Phys.Rev., 133, A1411-7.
Guyer R.A. and Krumhamsl J.A., 1966, Phys.Rev., 148, 778-88.
Jackson H.E. and Walker C.T., 1971, Phys. Rev., B3, 1428-39.
Kimber R.M. and Rogers S.J. 1973, J.Phys.C:Solid St.Phys.,6,2279-93.
Simons S, 1971, J.Phys.C: Solid St.Phys., 4, 2089-96.

LIGHT SCATTERING FROM SECOND SOUND IN NaF

Dieter W. Pohl

IBM Zurich Research Laboratory

8803 Rüschlikon, Switzerland

Second sound is wave-like rather than diffusive propagation of heat.[1] The existence of such a mode in a good crystal is to be expected from the close analogy between an ideal gas of atoms and the phonon "gas". Fluctuations in the particle density propagate as "sound" if the mean free paths of momentum-conserving (λ_N) and momentum-destroying collisions (λ_R) are sufficiently small and large, respectively, compared to the wavelength λ, i.e., if $\lambda_N < \lambda < \lambda_R$ (window condition).[1] At low temperatures and in sufficiently anharmonic crystals normal process may become frequent enough to satisfy the condition. The Rayleigh peak in the scattered spectrum which is a manifestation of entropy fluctuations then breaks up into a pair of second-sound peaks.[2] The displacement and width of the new peaks are a measure for the velocity and damping of the second-sound mode.

Second sound has so far been found in He,[3] NaF,[4,5] and Bi [6] only. Except for He with its unusual properties, SS was detected by means of heat pulses, a method that is applicable only if the damping length gets of the order of the crystal dimensions.[4-6] Light scattering allows for damping lengths of the order of one wavelength and therefore is considerably more sensitive. However, the classical techniques of light scattering run into difficulties because coupling coefficient, frequency offset, and scattering angle are extremely small.[7]

We therefore applied the technique of "forced Rayleigh scattering"[8] which allows the experimental limits to be pushed by a few orders of magnitude and enabled us to detect second sound in NaF by means of light scattering for the first time. An entropy

wave δT is generated with a pair of intense light beams (a,b) by
interference and absorption (Fig. 1). Its wavelength $\lambda = 2\pi/q$ and
frequency Ω can be adjusted by means of angular separation and
modulation of the pump beams. A third beam (i) with different frequency is scattered from δT. The scattered amplitude E_s is a
measure of the spectral response function $S(q,\Omega)$.

     The scattering in NaF was studied at modulation frequencies up
to 10 MHz (Fig. 2). Tunable 100% modulation of the exciting radiation is obtained by beating two feedback stabilized CO_2 lasers
with piezoelectric mirror mounts. The scattered light produced by
an argon laser beam is considerably weaker than the stray light (E_R)
propagating in the same direction (Fig. 1). The multiplier there-
fore receives the signal $S = |E_R|^2 + 2\,Re\,(E_R \cdot E_s{}^*)$. The recti-
fied output from the gate was integrated and recorded vs. frequency

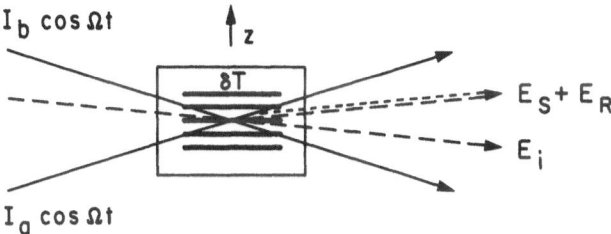

Fig. 1.   The principle of forced scattering.

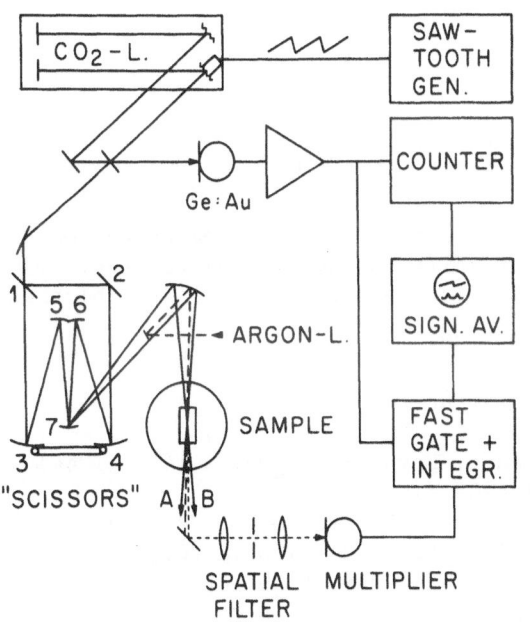

Fig. 2. Experimental set-
up for forced second-sound
scattering.

with a signal averager. For the present first measurements, λ =
295 μ was chosen. This wavelength is located well in between the
mean free path lengths for normal (N-) and resistive (umklapp-)
processes around 18 K [4,5] thus satisfying the window condition. A
weakly damped second-sound wave will be excited for $\Omega/2\pi = c_2/\lambda \sim$
6.3 MHz.

The preliminary results obtained so far fully support this
expectation (Fig. 3). At T > 17 K, a distinct peak at $\sim$ 6.5 MHz
rises out of the noise level. With T increasing above 20 K, the
signal gets strongly broader and weaker. Above 25 K the signal
extends smoothly over the whole frequency interval investigated
but starts to disappear in noise. The disappearance at T < 17 K
and T > 25 K is caused by vanishing dn/dT and increasing damping, respectively. The author is indebted to V. Irniger for technical assistance and to T.F. McNelly and H. Jackson for their
excellent NaF crystals.

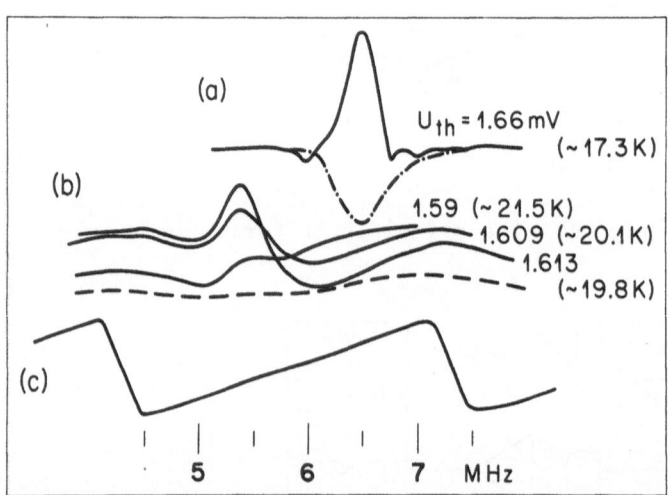

Fig. 3. Typical signals between 17 and 21 K. a) Demonstration of
phase reversal associate with a shift of the thermal grating. b)
Effect of increasing temperature (dashed curve: CO_2 lasers blocked).
c) Simultaneous record of the frequency sweep.

References

1. R.A. Guyer and J.A. Krummhansl, Phys. Rev. 133, A 1411 (1964).
2. A. Griffin, Phys. Lett. 17, 208 (1965).
3. C.C.Ackerman and W.C.Overton,Jr., Phys.Rev.Lett. 22, 764 (1969).
4. T.F. McNelly et al., Phys. Rev. Lett. 24, 100 (1970).
5. H.E. Jackson and C.T. Walker, Phys. Rev. B 3, 1428 (1971).
6. V. Narayanamurti and R.C. Dynes, Phys. Rev. Lett. 28, 1461 (1972).
7. D.W. Pohl and S.E. Schwarz, Phys. Rev. B 7, 2735 (1973).
8. D.W. Pohl,S.E.Schwarz and V.Irniger,Phys.Rev.Lett.31, 32 (1973).

DISCUSSION ON SECOND AND ZERO SOUND

Phonon Scattering in Liquid Helium   H J Maris   Page 77

J Jackle: Should one not expect a strong attenuation of those 'higher order second sound' modes, since in general well-defined, propagating collective phonon modes only exist if a sufficiently broad frequency window exists in the distribution of the eigenfrequencies of the phonon-collision operator?

H Maris: It seems that one does not need precisely a window but just that $\Omega\tau_i$ be > 1 for a number of ℓ.

H Kinder: In Sluckin and Bowley's theory of the attenuation length, there is still considerable attenuation above ε_c. Could you comment on that?

H Maris: My feeling is that at T = 0 there should be no attenuation at all above ε_c from any order phonon-phonon interactions.

****

Ballistic Phonons and the Transition to Second Sound in Solid ^{3}He
V Narayanamurti and R C Dynes   Page 93

H Namaizawa: Did you make your measurement at various pressures and if so what is the pressure dependence?

V Narayanamurti: In our present cell we have measured over the molar volume range of 23.8 cm^3/mole to 24.5 cm^3/mole. The velocities are, of course, different but the basic phenomena are the same.

H Maris: Another material, in which the Florida group have found a specific heat which is linear in T, is ^{4}He. We have measured the specific heat of solid ^{4}He and do not find a linear contribution. I do not know whether that implies it should be in ^{3}He or not.

On the Temperature Behaviour of Second Sound and Poiseuille Flow
in Solids   H Beck   Page 96

W C Overton Jr: I would like to ask about your experimental and
calculated arrival times. The usual procedure is to determine the
slope arrival time or the half amplitude arrival time from the observed signal leading edge. Could you comment on the manner in which
your arrival times were determined.

H Beck: They are all arrival times of the peaks of the pulses. It
turned out to be difficult to analyse leading edges or half heights
in a systematic way.

D Pohl: Jackson and Walker give a quantitative estimate on τ_N based
on their experimental data. They use a T^{-5} dependence for τ_N. Did
you compare your results with their data?

H Beck: No I did not. Mainly since in the theory described here,
the <u>magnitude</u> of τ_N, as opposed to its T-dependence, cannot be determined very accurately. Also the calculations (and the experiments
done on NaF) show that the 'lower end' of the window, i.e. $\Omega\tau_N \ll 1$,
can be replaced by $\ell\tau_N \lesssim 1$, which is less restrictive.

****

Multiphonon Absorption in Alkali Halides   D W Pohl and H Beck
                                                     Page 102

V Narayanamurti: How wide a frequency range can you scan to get
the frequency dependence?

D W Pohl: At present up to 20 MHz although the window for measurements of second sound shows we could go up to 100 MHz and that would
allow us to go up to 25K. It might also allow us to see second
sound in systems in which it is not observable by heat pulses.

H J Maris: I think one such material might be sapphire.

D W Pohl: I agree and the other alkali halides may be other candid-
ates.

PHONON SCATTERING IN AMORPHOUS SOLIDS*

Robert O. Pohl

Laboratory of Atomic and Solid State Physics, Cornell

University, Ithaca, New York   14853

The thermal conductivity of all amorphous dielectric solids shows a most remarkable uniformity which appears to be insensitive to the host and to chemical impurities. This is demonstrated in Fig. 1 in which the conductivities of all amorphous dielectric solids measured to-date are summarized (1,2). To within less than a factor of 10, the conductivity Λ depends solely on the temperature and not on the material, it has a plateau in the temperature region around 10K, and below 1K approaches a power low $\Lambda \propto T^{\delta}$ where δ falls into the range $1.8 < \delta < 2.0$. (Note in particular the recent data obtained by Lasjaunias et al (2) extending the temperature range in which this law appears to hold to 0.025K, with $\delta=1.95$). In this lecture, we want to review some of the studies undertaken to understand this remarkable fact.

In dielectric solids the heat flow is usually described with the aid of Debye plane wave longitudinal and transverse phonons travelling with the speeds of sound v_{ℓ} and v_t respectively. In amorphous SiO_2 and in a borosilicate glass, Brillouin light scattering experiments in the temperature range of liquid helium have given direct evidence for the existence of such thermally excited phonons (3). The phonons observed in this experiment were those that would be the dominant carriers of heat in the vicinity of 0.2K (phonon frequencies ~ 20 GH_z). It has also been possible to demonstrate a reduction of the thermal conductivity in very thin glass fibers with rough surfaces which has been in quantitative agreement with the Casimir scattering theory in a solid in which the heat is carried solely by Debye phonons (4). Similar results on borosilicate glass and on polycarbonate containing well-defined holes have given additional support to the picture that phonons are the sole carriers of heat in amorphous solids (5).

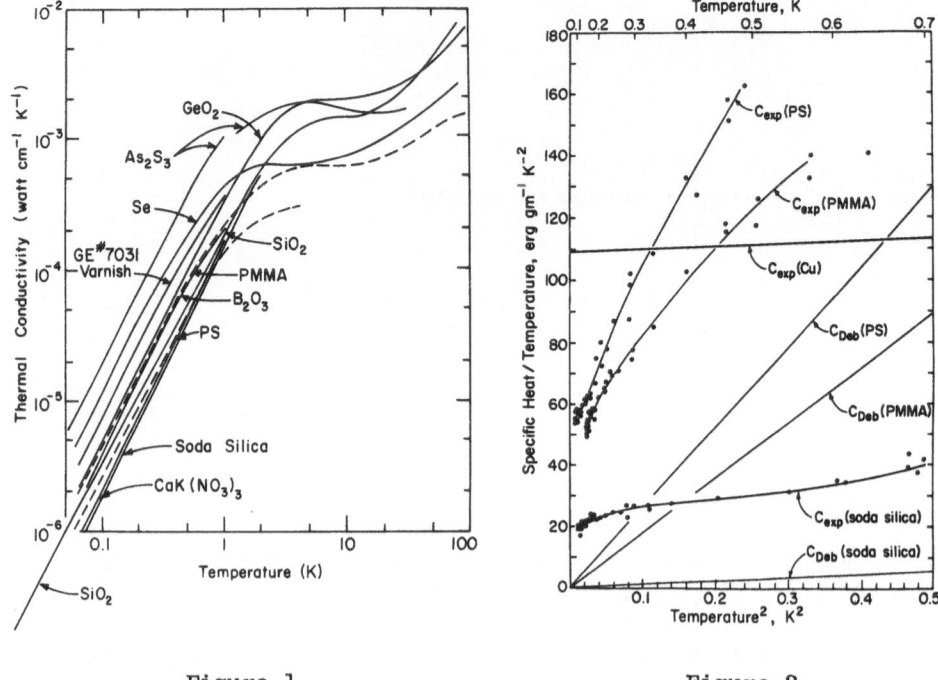

Figure 1   Figure 2

Figure 1 - Low temperature thermal conductivity of all amorphous
dielectric solids measured to-date, ref 1. Note the extension of
the SiO_2 data to 0.025 K, ref. 2.
Figure 2 - Specific heat of three non-crystalline solids plotted as
C/T vs. T^2. The specific heats predicted by the Debye model, the
lines marked C_{Deb}, are shown for each of the materials. The heat
capacity of Cu is shown for a comparison of the size of the excess
$(C_{exp} - C_{Deb})$.

From these experiments we may conclude that the understanding of
the thermal conductivity in amorphous solids requires an understanding of the scattering of Debye phonons, and is unlikely to be found
in the energy transfer through some hitherto unknown excitations in
these solids. The first measurements of the absorption of ultrasonic
waves with frequencies comparable to those responsible for the heat
transport near 1K yielded a distrubing and exciting result (6):
Phonons generated by Brillouin scattering of high power laser light
pulses were found to live several orders of magnitude longer than
expected on the basis of thermal conductivity experiments. (These
observations provided a major part of the stimulus to study the
phonon scattering in amorphous solids in detail.) Subsequent
measurements of the power dependence of the ultrasonic absorption
of longitudinal waves with standard ultrasonic techniques revealed
an increasing absorption with decreasing ultrasonic power (7).
Power-independent absorption was observed for powers typically less
than 10^{-7} Watt/cm^2. Phonon mean free paths observed at these low
power levels were in reasonable agreement with those deduced from

thermal conductivity. Further evidence that the heat in glasses is
carried by Debye phonons exclusively was recently provided through
measurements of the temperature dependence of the speed of longitudinal and of transverse ultrasound in the high power regime (8). From
these measurements the product of the density of states of the
scatterers, n_o, and the coupling energy M_ℓ and M_t, respectively,
could be determined, and from those quantities the thermal conductivity could be computed without adjustable parameters, only using
the assumption that $(n_o M_\ell)$ and $(n_o M_t)$ are independent of phonon
frequency. (We will see below, however, that we have no other, more
direct evidence for n_o being constant.) In the experiment, frequencies between 30 and 150 MHz were used; the conductivity, calculated
for T = 0.2K, where the dominant phonons have frequencies of order
20 GHz, agreed with the measured one to within 20%. Much has been
learned about the phonon scattering in glasses through ultrasonic
measurements, which we can mention only briefly: From the power
dependence of the absorption it follows that the scattering centers
should be highly anharmonic. This agrees with the theoretical model
that the centers are tunneling states of certain atoms or groups of
atoms with two or more almost equivalent equilibrium positions (9).
The ultrasonic attenuation has given evidence for both resonant
scattering and scattering by a relaxation process (10,11). We refer
in particular to the recent paper by Jäckle et al (10) which contains
an excellent review of the subject and also attempts to relate the
low temperature scattering states to those which have been known for
a long time from ultrasonic work at higher temperatures, and which'
had been interpreted through a classical, thermal relaxation process.
Finally, we should mention the direct determination of the lifetimes
of the scattering states (12) and of the elastic interaction between
such states (13), both performed with the aid of two ultrasonic
pulses. They have all been successfully interpreted within the
framework of the two-level (tunneling state)model (9). We mention
only one interesting discrepancy: The direct determination of the
excited state lifetime (12) yielded coupling energies to transverse
phonons M_t = 3eV, while measurements of ultrasonic absorption or of
temperature dependence of the speed of sound yielded approximately
ten times smaller values (8) (M_ℓ = 0.35eV, M_t = 0.24eV). For the
analysis of the second group of experiments, knowledge of n_o, the
density of states of the scatterers, had to be assumed. n_o was
derived from the low temperature linear specific heat anomaly,
apparently another characteristic property of glasses (1). In
order to bring these results into agreement with each other one
would have to assume that only 1% of n_o actually contributes to the
phonon scattering (since the phonon scattering rate is $\propto M^2$). In
this case, however, the phonon scattering states in these glasses
would have to be over 100 times stronger phonon scatterers than the
strongest phonon scatterers for low frequency phonons in crystalline
solids, namely tunneling defects like Li^+ or CN^- ions dissolved in
$KC\ell$ (14).

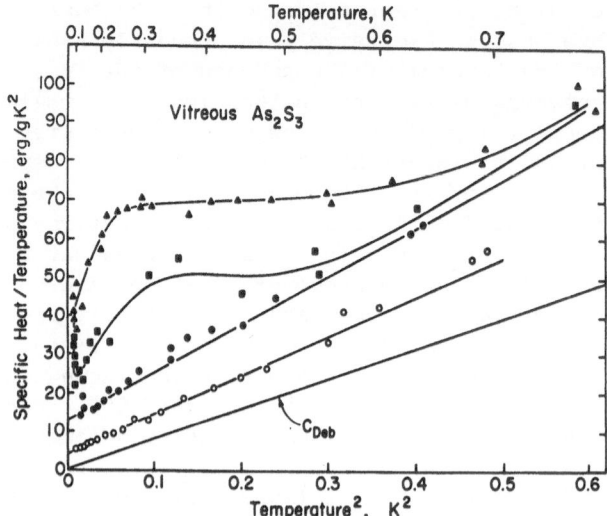

Figure 3 — Specific heat of four As_2S_3 samples plotted as C/T vs. T^2. The line marked C_{Deb} is the heat capacity predicted by the Debye model. The two upper curves are from samples which were produced at Cornell with successively more care to remove water of hydration. The next curve is from a sample which was supplied by A. J. Leadbetter. The lowest set of data is from a sample which was made by F. J. DiSalvo at Bell Labs. If one assumes that the differences between the upper three and lowest sets of data are due to the presence of two-level systems, one can calculate the density of these systems to be $68 \times 10^{16} cm^{-3}$ for the upper curve, 26×10^{16} cm^{-3} for the second one, and $\sim 10^{16} cm^{-3}$ for the third. Spark source mass spec analysis of the four measured samples showed the following impurity concentrations in the order given above: 1) 10^{19} to 10^{20} cm^{-3} Sb and 10^{17} to $10^{18} cm^{-3}$ Rb; 2) 10^{17} to $10^{18} cm^{-3}$ and 10^{16} to 10^{17} cm^{-3} Cd; 3) 10^{17} to $10^{18} cm^{-3}$ Ge; 4) nothing detectable; after (14).

Next, we want to review attempts to observe the phonon scattering states by other experiments. It is well known that amorphous dielectric solids also have an anomalous low temperature specific heat, which can be described by a polynomial expression $C_V = c_3 T + c_3 T^3$, where c_1 is of the order of 10-50 erg/(gram K^2), and c_3 is up to three times as large as the Debye term C_D based on the experimental sound velocity. Although the term linear in T usually dominates the specific heat at the lowest temperatures investigated ($\sim 0.1K$), the influence of the higher order term is still noticeable in most materials at these temperatures, which correspond to $\sim 10^{-3}$ of the Debye temperature. It appears natural to associate this excess specific heat with the scattering states. In the tunneling model (9), the density of tunneling states n_0 is usually computed from the linear specific heat term $c_1 T$ only; in this case, $n_0 =$ const. With this simplifying assumption, which appeared to introduce

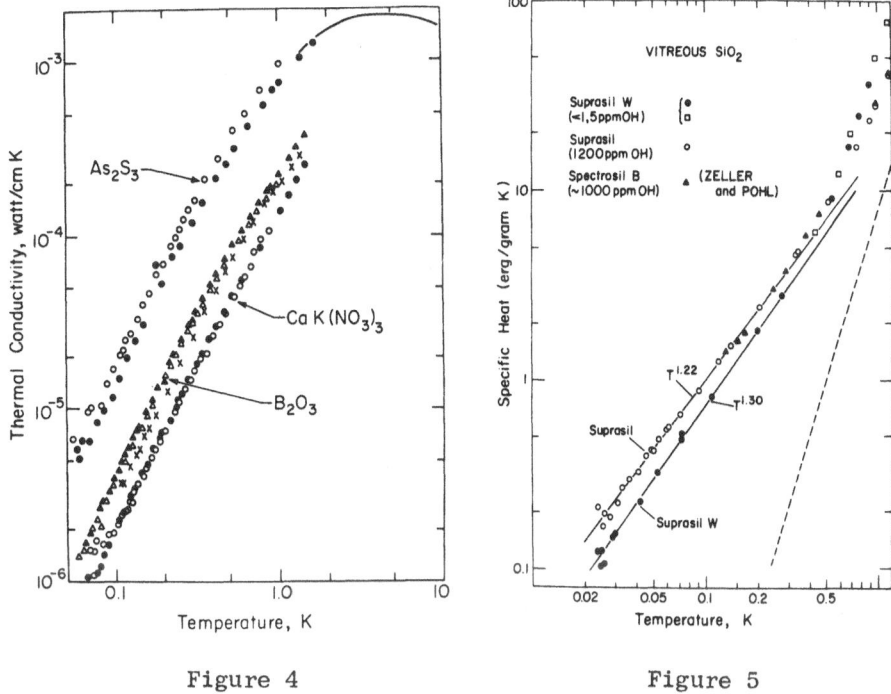

Figure 4

Figure 5

Figure 4 - Comparison of the thermal conductivity of "pure" and "impure" samples of As_2S_3, B_2O_3, and $CaK(NO_3)_3$. In all cases the data from the purer sample are solid points, and the data from the less pure sample are open ones. The solid line is from a measurement on the impure sample by A. J. Leadbetter. The discontinuity between the two measurements on As_2S_3 is probably due to a small error in determining, ℓ/A the geometry of the purer sample. If the open circles were shifted to line up with Leadbetter's measurements, one can see that there would be no difference between the conductivity of the two As_2S_3 samples measured at Cornell, just as for the B_2O_3 and $CaK(NO_3)_3$ samples. The x's are data on a B_2O_3 sample measured by Locatelli (Grenoble private communication, 1973) and are uniformly lower than our points by $\sim15\%$. The difference is probably also due to a small error in calculating ℓ/A. After (14).

Figure 5 - Specific heat of vitreous SiO_2 above 0.025K (2). Suprasil and Spectorsil B have large OH concentrations, but small metal ion concentrations. Suprasil has 130ppm chlorine and 100ppm fluorine. Suprasil W has low OH and metal ion concentrations, but 230oom chlorine and 290ppm fluorine (all numbers according to the manufacturers). Note that Spectrosil B and Suprasil have identical specific heats in the temperature range where both were measured (T > 0.1K). Note also that SiO_2 with low OH, but high (~50ppm) metal ion concentrations has been found to have lower specific heats that Spectrosil B above 0.5K, this difference, however, vanishes below that temperature (14). The dashed line marks the Debye specific heat of vitreous SiO_2, $C_{Deb}=8T^3$erg/(gramK^4), based on $v_t=3.75 \times 10^5$cm/sec and $v_\ell=5.80 \times 10^5$cm/sec.

a fairly insignificant error at temperatures below, say 0.5K (see
for instance, ref. (1), Fig. 9, i.e. for dominant phonon frequencies
$f < 50$ GHz, the T^2 temperature dependence of the thermal conductivity
could be easily explained (9). Recent, more careful measurements of
the low temperature specific heat, however, have demonstrated the
approximate nature of the simple polynomial (15). This is illustrated
in Fig. 2 for two polymers and the soda silica glass 5 SiO_2 Na_2O (14).
The temperature dependencies of the measured specific heats shows
enough variation from material to material to make one wonder why
the temperature dependences of their thermal conductivities show
that little variation. It seems obvious that, if the origin of
the phonon scatterings shows up in the specific heat at all, then
these states contribute only a part of the excess between the
measured specific heat and that calculated from the Debye model.
This conclusion has been confirmed by measurements on different
samples of the same material, as illustrated in Figs. 3 and 4 (ref.
14). With increasing sample purity and/or perfection, a Schottky
type anomaly disappears in As_2S_3, as does an anomaly beginning
below 0.1K, leaving an anomaly of the usual form of (closed
circles). Further care in producing the sample (chemical purity
and/or physical-perfection?), however, makes this anomaly shrink by
more than a factor of two (open circles). The thermal conductivity
of the latter two samples, however, differs by less than 20% (and
even this difference is believed to be the result of the poor
definition of the geometry of the sample (14). Similar results were
found for B_2O_3 and the ionic glass CaK $(NO_3)_3$ (14), (15). In these
glasses, a reduction of the anomalous specific heat resulted in
zero change of thermal conductivity, see Fig. 4. Consequently, the
entire removable portion of the specific heat anomaly apparently
does not scatter phonons. Whether all or at least part of the
excitations causing the excess specific heat in the sample with
the lowest specific heat scatter phonons, has to be left as an open
question. Recently, the specific heat of amorphous solids was
measured for the first time to temperatures well below 0.1K (2).
The results obtained on several samples of SiO_2 are shown in Fig. 5.
In addition to the sample dependence, the low temperature data also
show the limits to what is called the "linear specific heat anomaly",
as inferred from the data obtained above 0.1K (Fig. 5, see the data
by Zeller and Pohl). It is important to realize, though,that the
thermal conductivity of SiO_2 was found to be entirely sample independent (2). This result is indicated in Fig. 1.

One of the predictions of the tunneling model (9) was that a
wide range of relaxation times of excited tunneling states was to
be expected. So far as we can tell, this, in fact was the only
prediction specific to the assumption of tunneling as the cause of
the low energy scattering states. Goubau and Tait (16) tried to
study this through high speed specific heat measurements on thin
SiO_2 and PMMA samples. Their results, however, were negative:
Near 0.1K, no reduction of the specific heat during short measuring

times was observed, contrary to the prediction based on the tunneling model (9) (17). Mobile (tunneling) atomic or molecular units in glasses should be influenced by electric fields. A temperature dependent dielectric function ϵ has indeed been observed recently in vitreous silica between 0.3 and 5 K, at f = 1.1 GHz (18). Using the constant density of states n_o (in view of Fig. 5 clearly an approximation), these data can be interpreted with the assumption that each state has a dipole moment p_{el}.= 0.3 Debye = 0.063eÅ (uncorrected for local field). Far infrared absorption has also been observed in amorphous GeO_2, PMMA, and two samples of silica, differing by their water contents (19). Superimposed on a temperature independent absorption (60 GHz < f_{fir} < 150 GHz) a relatively small temperature dependent absorption was found, which could be explained as arising from two-level systems only. With the assumption of a oscillator strength of unity, this absorption corresponds to a density of states of absorbers of ~10^{-2} n_o (n_o from specific heat). The frequency range of the measurements is too narrow to tell whether this density is constant or frequency dependent. Finally, an attempt was made to influence the phonon scattering states with a large d.c. electric field (20). In an applied electric field of 10^5 Volt/cm, an insignificant decrease of the conductivity at 0.5 K of (0.4 $\pm$ 0.3)% was observed in silica (0.3 Debye x 10^5 volt/cm = k_B x 0.73 K). For the analysis of this result it is important to realize that a constant density of scattering states would be expected to yield zero change of the conductivity in an electric field. Assuming that the observed change (0.4%) was significant, Stephens calculated the dipole moment under the assumption that all states contributing to the specific heat (see eq. (1)) also scatter phonons, i.e. n = n_o + a ω^2, where a is a constant, and ω is the angular frequency of the excitations. This resulted in an upper limit of the dipole moment: p $\leq$ 0.13 Debye = 0.0027 eA. The analyses used to extract the dipole moments from these two experiments are obviously somewhat hampered by our ignorance of the proper density of states. Nonetheless, considering that atoms even in the covalently bonded silica carry charges on the order of one electronic charge, e, these measurements imply rather small separations of the potential minimma in the tunneling model, or, alternatively, lead to conclude that fairly large units of the glass are involved in the excitations.

In summary, although the phonon scattering responsible for the apparently universal low temperature thermal conductivity of glasses has been very thoroughly studied through ultrasonic techniques, no independent evidence for the centers causing the scattering has been found to-date. In particular, the frequently quoted low temperature linear specific heat anomaly must presently be viewed with caution. The density of states derived from it has to be regarded only as an upper limit of the scattering state.

REFERENCES

* Work supported by the U.S. Energy Research and Development Administration under contract AT(11-1) 3151, Technical Report No. COO-
3151-57 (unpublished).

1. R. B. Stephens, Phys. Rev. B8, 2896 (1973).
2. J. C. Lasjaunias, A. Ravex, M. Vandorpe, and S. Hunklinger,
    to be published.
3. W. F. Love, Phys. Rev. Lett. 31, 827 (1973).
4. R. O. Pohl, W. F. Love, and R. B. Stephens, Proc. of the 5th
    Intern. Conf. on Amorphous and Liquid Semiconductors, edited
    by J. Stuke and W. Brenig, Taylor and Francis, London 1974,
    page 1121.
5. M. P. Zaitlin and A. C. Anderson, Phys. Rev. Lett. 33, 1158
    (1974); M. P. Zaitlin and A. C. Anderson, and M. P. Zaitlin,
    L. M. Scherr, and A. C. Anderson, to be published.
6. W. Heinicke, G. Winterling, K. Dransfeld, J. Acoust. Soc. Am.
    49, 954 (1971).
7. W. Arnold, S. Hunklinger, S. Stein, and K. Dransfeld, J. Noncryst
    Solids 14 192 (1974), and B. Golding, J. E. Graebner, and
    B. I. Halperin, Phys. Rev. Lett. 30, 223 (1973).
8. S. Hunklinger and L. Piche, to be published.
9. W. A. Phillips, J. Low Temp. Phys. 7, 351 (1972), and P. W.
    Anderson, B. I. Halperin, and C. M. Varma, Phil. Mag. 25,
    1 (1972).
10. J. Jäckle, L. Piche, W. Arnold, and S. Hunklinger, to be
    published.
11. D. Ng and R. J. Sladek, same as ref. 4, page. 1173, and to be
    published.
12. W. Arnold and S. Hunklinger, Berh. Deutsche Physik. Ges. (VI)
    10, Münster 1975, page 645.
13. W. Arnold and S. Hunklinger, to be published.
14. See R. B. Stephens, Cornell Materials Science Center Report
    No. 2474, to be published.
15. J. C. Lasjaunias, D. Thoulouze, and F. Pernod, Solid State
    Comm. 14, 957 (1974).
16. W. M. Goubau and R. H. Tait, Phys. Rev. Lett. 34, 1220 (1975).
17. J. Jäckle, Z. Physik 257, 212 (1972).
18. S. Hunklinger, M.v. Schickfuss, W. Arnold, L. Piche, and
    K. Dransfeld, to be published.
19. K. K. Mon and A. J. Sievers, to be published.
20. R. B. Stephens, to be published.

LOW ENERGY PHONONS IN AMORPHOUS MATERIALS

Roger MAYNARD

Centre de Recherches sur les Très Basses Températures

CNRS, BP 166 Centre de Tri, 38042 Grenoble-Cédex, France

The Resonant interaction : All the properties[1] observed so far on glasses and polymers **are** conceptually strongly akin to the problem of the interaction between the electronic spins of paramagnetic impurities and phonons. More precisely the understanding of the low temperature properties requires the study of the propagation of phonons in a resonant medium characterized by a broad distribution of the energy splittings of the resonant spins. Therefore, the natural starting point is the theory of Jacobsen and Stevens[2] for the coupling of phonons and spins. Here, the spins are the configurational defects pictured by a double well potential proposed first by Anderson, Halperin, Varma[3] and Phillips[4]. Retaining only the double degrees of freedom, these defects can be represented by a fictive 1/2 spin $\vec{S}\alpha$ located at site $\vec{R}\alpha$, with an energy splitting $E\alpha$. The coupled Hamiltonian

$$\mathcal{H}_o = \sum_k \hbar\omega_k \, a_k^+ a_k + \sum_\alpha E\alpha S_\alpha^z \qquad (1)$$

where $\omega = vk$ is the frequency of a phonon k of a unique acoustic branch we consider for simplicity. The interaction Hamiltonian is a quadratic form of the spin and phonon variables :

$$\mathcal{H}_{int} = \sum_\alpha B_x^\alpha \, \eta(\vec{R}\alpha) \, S_\alpha^x \qquad \eta(\vec{r}) = \frac{1}{\sqrt{V}} \sum_k (\frac{\hbar\omega_k}{2\rho v^2})^{1/2} (a_k + a_{-k}^+) e^{i\vec{k}\vec{r}} \qquad (2)$$

ρ the mass density, $\eta(\vec{r})$ is the local strain operator (only the isotropic local dilatation is considered for simplicity) and S_α^x is the x-component of the Pauli matrix describing the spin flip process. B_x^α is the deformation energy measuring the strength of the flip-flop coupling. The calculation of the renormalization of the phonons to second order of the perturbation expansion in B_x^α is well known and leads to the coupled mode picture by writing the coupled equations of motion for the spins and the phonons. An alternative way[5] consists of analysing the propagation of a phonon in the resonant medium, described by the phonon propagator from r,t to r',t' :

$$D(\vec{r},t;\vec{r}',t') = - <T_{tt'} \; \eta(\vec{r},t) \; \eta(\vec{r}',t')> \qquad (3)$$

(The angular brackets denote the thermal average and $T_{tt'}$ is the temperature ordering operator). The poles of $D(\omega,k,k')$, the Fourier transform in time and space of this propagator, give the frequency shift and the life-time of the dressed phonon. From $\vec{r}$,t to $\vec{r}'$,t' the phonon can meet no spins, or can be scattered once, twice,... by the defects. All these events are contained in the following equation for the ω and k component of the propagator :

$$D(k,k')=D_k^0\delta_{k,k'} + \frac{1}{N}\sum_\alpha e^{-i(\vec{k}-\vec{k}')\vec{R}_\alpha} \; D_k^0 \; F_\alpha D_{k'}^0 \; + \; ... \qquad (4)$$

$$D_k^0 = \omega_k^2\left[2mv^2(\omega^2-\omega_k^2)\right]^{-1}. \qquad (m= \rho\frac{V}{N})$$

$$F_\alpha = \lim_{\delta\to0^+} (B_x^\alpha)^2 \; \frac{2E_\alpha}{(\omega+i\delta)^2-E_\alpha^2} \; \tanh(\frac{E_\alpha}{2k_BT}) \; + \; \text{terms in } (B_x^\alpha)^4 \; + \; ..(5)$$

The physical meaning of F_α is transparent : it represents the time correlation function of the spin S_α^x at the frequency ω of the incident phonon and the hyperbolic tangent reflects here the difference in population between the upper level and the ground state at the equilibrium and implies that the intensity of the wave is so small as to avoid any saturation of the two level systems. For one single spin $\vec{S}_\alpha$, it is obvious from (5) that D(k,k') is off diagonal in k space since translational invariance has been destroyed. However, for an arbitrary concentration of spins, the physical properties must be averaged over the various arrangements of the spins on the sites. Here, we are principally concerned with the the propagation of phonons for which we must take this ensemble average of the propagator : $\overline{D(k,k')}$. At site $\vec{R}_\alpha$, the probability of finding a spin with an energy splitting E and a coupling energy B_x is just proportional to the concentration $Vn(E,B_x)dB_xdE$ (n density per unit volume) of the spin packet $dEdB_x$ for a complete random distribution and the averaged vertex function $\overline{F_\alpha}$ is given by :

$$\overline{F}= \int Vn(E,B_x)dEdB_x \left[(B_x)^2 \; \frac{2E_\alpha}{(\omega+i\delta)^2-E_\alpha^2} \; \tanh(\frac{E_\alpha}{2k_BT})+\text{terms in } (B_x)^4\right] \qquad (6)$$

This new vertex does not depend on R_α and consequently introduces a translational invariance for the averaged medium for which $\overline{D_{k,k'}}=\overline{D_k}\delta_{k,k'}$. This is an important approximation in the calculation, equivalent to the assumption of a destructive interference for the scattering of the phonons off more than one spin at a time on the random sites of the spins. This result is then valid provided the mean free path (and not the wave length) is larger than the mean distance between the defects, a condition which seems largely fullfilled in glasses at low temperature. To the lowest order in B_x, one gets the coupled-mode approximation for the phonon propagator :

$$\overline{D_k}^{-1} = \frac{\rho v^2}{\omega_k^2} \left[\omega^2 - \omega_k^2 (1+ \frac{\overline{B_x^2}}{\rho v^2} \int_0^\infty n(E)dE \; \frac{2E}{(\omega+i\delta)^2-E^2} \; \tanh(\frac{E}{2k_BT})) \right]$$ (7)

where the independence of the random variables B_x and E has been assumed. This propagator contains both the attenuation of sound and the dispersion of the phonon k as the imaginary and real parts of the pole. The sound velocity can be written as :

$$v(T) = v \left[1+ \frac{\overline{B_x^2}}{\rho v^2} \; PP \int_0^\infty n(E)dE \; \frac{E}{\omega_k^2-E^2} \; \tanh(\frac{E}{2k_BT}) \right]$$ (8)

There is a very simple expression for v(T) for low frequency phonons $\hbar\omega \ll k_BT$, obtained by simplifying $\tanh(E/2k_BT)$ to $E/2k_BT$ for $E<2k_BT$ and to 1 for $E>2k_BT$. By assuming a constant energy density n(E)=n up to E_{MAX}, we get directly :

$$v(T) \simeq v \left[1- \frac{n\overline{B_x^2}}{\rho v^2} \; (1-\ln\frac{2k_BT}{E_{MAX}}) \right] \quad (\hbar\omega_k \ll k_BT)$$ (9)

This expression exhibits no frequency dependence and predicts a $\ln T$ increase with the temperature which has been observed[6] for the longitudinal and transverse waves. The $\ln T$ is important since it comes directly from the $\tanh(E/2k_BT)$ in the vertex (5) and therefore from the 2 degrees of freedom of the defects. Any resonant harmonic defects would exhibit a temperature independent F and so would fail to explain the sound velocity variation. When T→0, or more precisely $k_BT/\hbar\omega \to 0$, v(T) becomes temperature independent and reaches a value proportional to $\log(E_{MAX}/\hbar\omega)$. But a divergence subsists then when ω→0 in formal analogy with the Kondo problem[7]. The imaginary part of the pole is related to the lifetime of the phonons :

$$\Gamma_k = \pi n \; \frac{\overline{B_x^2}}{\rho v^2} \; \omega_k \; \tanh(\frac{\hbar\omega_k}{2k_BT})$$ (10)

This well known result gives a good description of the ultrasonic attenuation in the unsaturated regime[8]. For comparison with the spin-phonon problem one can remark that the coupled modes do. not give any absorption for an infinitively narrow level. Here, the finite lifetime comes mathematically from the continuous distribution of the energy E, but physically from the finite linewidth of the spin level which makes the distribution continuous (fig. 1). Saturation effects are also contained in (10), if the temperature of the spins becomes higher than the temperature of the phonons by increasing the imput power. The coupling energy is very large typically of the order of 0.3 eV (when n is taken from the specific heat data) instead of 0.1 eV for Cr^{2+} in Al_2O_3, one of the most strongly coupled magnetic impurities. Finally the dimensionless parameter $n\overline{B_x^2}/\rho v^2$ is of the order of 10^{-4}, a value which indicates that the contribution of the next order term in $\overline{B_x^4}$ is probably negligeable.

Propagation of light : The same analysis can be developped for the propagation of light in a resonant medium, where the dipole induced by the transitions between the two levels is $M_\alpha = <0|er_z^\alpha|1>$ where we choose the direction of the dipole moment operator to define the local z axis. By assuming that the wavelength of the photons is larger than the mean size of the dipole (dipole approximation), one may write directly the Hamiltonian of interaction between photons and spins (in cgs units) :

$$\mathcal{H}_{int} = \frac{i}{\sqrt{v}} \sum_{\alpha,k} \sqrt{\frac{2\pi\hbar\omega_k}{\epsilon_r}} \, M_\alpha S_x^\alpha (a_k + a_{-k}^+) e^{i\vec{k}\cdot\vec{r}_\alpha} \tag{11}$$

where now a_k^+ and a_k are creation or annihilation operators of a photon of energy $\hbar\omega_k$, ϵ_r the relative dielectric constant. By comparison with (2) one notices the analogy of the electric dipole $(2\pi/\epsilon_r)^{1/2} M_\alpha$ with the "elastic dipole" $B_x^\alpha (2\rho v^2)^{1/2}$, leading to the dispersion and attenuation of light in glasses. For example, the variation of the light velocity as a function of temperature can be written directly from (7) :

$$C(T) = C\left[1 - \frac{4\pi n \overline{M^2}}{\epsilon_r}(1 - \ln\frac{2k_B T}{E_{MAX}})\right] \tag{12}$$

This has been observed by SCHICKFUS et Al.[9]. It is tempting to take advantage of the strong similarity between sound velocity and light velocity to conclude that each two-level defect carries a dipole moment the mean value of which is of the order of 0.3 Debye in the quartz glass but larger in borosilicate glass.

Viscous regime for the phonons : At higher temperatures, (T>2K), the dispersion and attenuation of low frequency phonons departs strongly from the resonant law (9) and (10) : a hydrodynamic regime appears which is characterized by a relaxation time $\tau(E)$ of the two-level defect (spin-lattice relaxation time T_1) which has been measured directly by Arnold and Hunklinger[13] :

$$\tau^{-1} = (\frac{\overline{B_x^2}(long)}{v_\ell^5} + \frac{2\overline{B_x^2}(transv)}{v_t^5}) \frac{E^3}{2\pi\rho\hbar^4} \coth(\frac{E}{2k_B T}) \tag{13}$$

where ℓ and t refer to the longitudinal and transverse polarizations. The mechanism of this second viscosity is well known[10] :The ultrasound modulates the population of a spin packet of energy E through the deformation energy B_x, and the population returns to equilibrium with a time τ. This relaxation has been calculated in detail by Jäckle et al.[11] and can be derived from the assumption of the dependence of the stress tensor $\sigma(\eta,f)$ on the population f of the excited state of the two-level system and the local strain η. For one polarization, one gets :

$$\rho v^2 = \left.\frac{\partial\sigma}{\partial\eta}\right|_f + \left.\frac{\partial\sigma}{\partial f}\right|_\eta \frac{\partial f}{\partial\eta} \tag{14}$$

When the population f is modulated by ultrasound at frequency ω, $\partial f/\partial \eta = (\partial f/\partial E)B_x|1-i\omega\tau|^{-1}$ while $\partial\sigma/\partial f|\eta=n(E)B_y$ (the reciprocity relation (11)). By defining $\rho v_o^2=\partial\sigma/\partial\eta|f$, one can write by summing over the distribution n(E) :

$$\frac{\omega^2}{k^2} = v^2 = v_o^2 + \int_0^\infty \frac{n(E)\overline{B_y^2}}{\rho} (\frac{\partial f}{\partial E}) \frac{1}{1-i\omega\tau} dE \qquad (15)$$

The real and imaginary parts of k give the dispersion and the attenuation (which becomes unsaturable). It is remarkable that the same parameters n and B_x^2 are involved in the resonant and viscous regimes and explain fairly well the sound velocity and the attenuation up to 5 K. Furthermore the domains in temperature where each mechanism is dominant, are well separated which gives the curves a universal shape.

The interaction between defects : It has been suggested recently by JOFFRIN and LEVELUT[12] that the defects could interact via exchange of virtual phonons. At the same time, ARNOLD and HUNKLINGER [13] have measured the broadening of a saturated line and found that this width, proportional to $1/T_2$, is by two orders of magnitude larger than the natural linewidth of spin packet due to spin phonon T_1 of formula (14). Two mechanisms of interaction are possible from what we know : the elastic interaction through the elastic dipoles coupling and the electrostatic dipole-dipole interaction. We can take advantage of the r^{-3} variation with the distance for writing the mean absolute values of these interactions as : $W_{elastic}\simeq n_T B_x^2/\rho v^2$ and $W_{electrostatics}\simeq n_T M^2/\epsilon$ where n_T is the total density of the defects per cm^3. But n_T is an unknown parameter and we meet here the first difficulty. The ratio of these average energies which eliminates n_T is of the order of 10^2 in favour of the elastic coupling. However the situation is not so simple, since the elastic dipoles B_x are off diagonal (induced dipoles) which only couple the like "spins" (defects of same energy splitting) the concentration of which is many order of magnitude smaller than the total concentration. It has been estimated[12] that the permanent elastic dipoles B_z are only 10^{-2} B_x, leading to a mean elastic energy of 10 mK when all the atoms are considered as defects (for the same concentration the electrostatic interaction energy would be of the order of 1 K). The problem of a possible phase transition coming from the dipolar interaction is very analogous to the problem of the spin glasses in the sense that the interaction energy varying also in r^{-3} is a random variable due to the disorder of the configurations of the dipoles. But there is a strong difference originating in the distribution of zero field splitting E due probably to local constraint which inhibits probably the coopérative behavior. A very recent experiments[14] at 5 mK exhibits a strong sensitivity of the dielectric constant of BK7 with the electric field which reflects a strong correlated behavior of the dipoles (fig. 2).

The thermal conductivity : The model of the two-level defects offers the possibility to understand the thermal conductivity, not

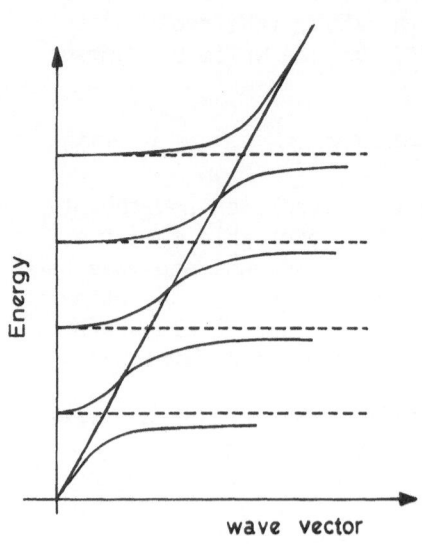

FIGURE 1

The multi-coupled modes

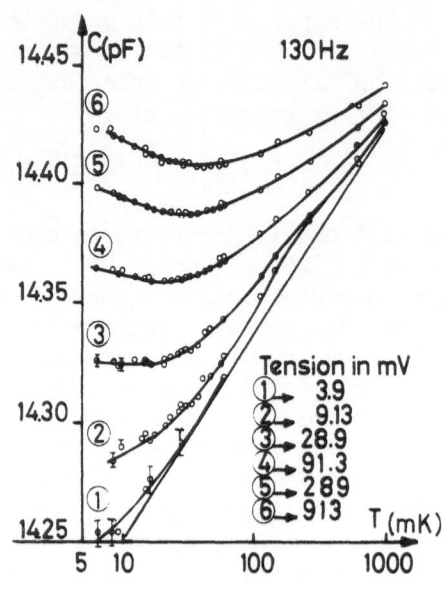

FIGURE 2

The capacitance of a borosilicate
glass BK7 as a function of the
temperature for various electrical
a.c. tension in mV (from
G. FROSSATI).

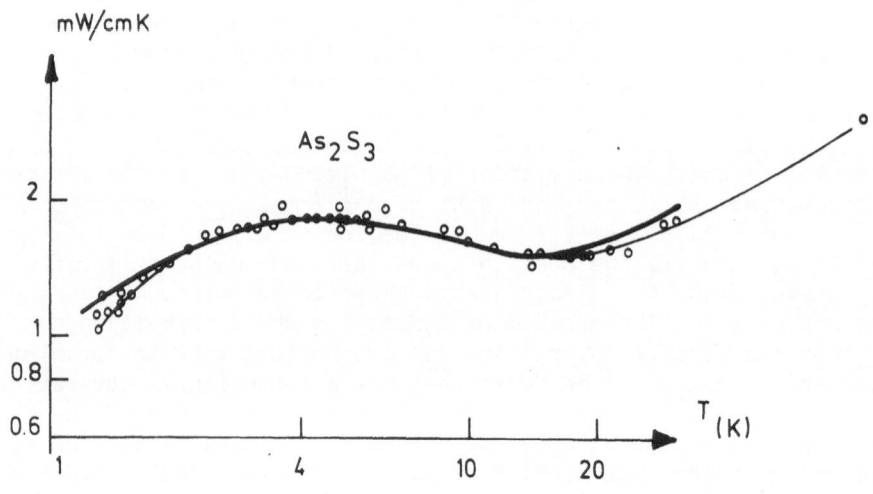

FIGURE 3

The thermal conductivity of As_2S_3 from ref.(16)
The solid curve represents a theoritical fit with a raman
process $\tau^{-1} \simeq \omega^4 T$.

only below 1 K where the resonant scattering explains fairly well
the T^2 regime but also near 10 K where the plateau occurs. Indeed
other theories, like the structural scattering of phonons[15]
might also explain the thermal conductivity of glasses but can't
describe the saturation or the temperature dependence of the sound
velocity. Some glasses, like As_2S_3[16] exhibit not only a plateau,
but a true minimum near 10 K. If one doesn't exclude this material
from the whole family of glasses, this circumstance, although weakly marked, could be important since most of the models proposed
so far are not capable to fit the curves . Actually, to get a minimum in thermal conductivity, the relaxation time must depend not
only on the frequency but on the temperature as well (and decrease
with increasing temperature). Then the tentative explanation in
terms of density fluctuations as moreover all the theories develop-
ped in the framework of the harmonic approximation, gives only
frequency dependent cross-sections. To enter the temperature in τ,
one must call for 1) anharmonic interaction (but the efficiency of
these processes estimated by MORGAN[15] seems to be too weak) 2)
relaxation of phonons on the two-level systems[17] (but the frequency of the thermal phonons is out of the range of validity for
this hydrodynamic regime) 3) the RAMAN interaction between systems
and phonons I propose to consider here. From the Hamiltonian

$$\mathcal{H}_R = \sum_\alpha \gamma_\alpha S_x^\alpha n(R_\alpha)n(R_\alpha) \qquad\qquad (16)$$

it is easy to write the relaxation time at the Born approximation.
However it is difficult to obtain a precise determination of τ
since the spectral densities at high energy are unknown. Nevertheless, for constant $n(E)$ and an acoustical spectral density, there
is a very simple scaling law in ω and T :

$$1/\tau(\lambda\omega,\lambda T) = \lambda^5 \, 1/\tau(\omega,T) \qquad\qquad (17)$$

from which one can take a typical term $\tau^{-1}\sim\omega^4 T$. From this, a fit
of the thermal conductivity of A_2S_3 follows and is illustrated in
fig. 3.

ACKNOWLEDGEMENTS
     I whish to thank Drs. ARNOLD, HUNKLINGER, LEVELUT, MATHO,
THOULOUZE, SCHICKFUS for fruitfull and stimulating discussions.

REFERENCES
(1) See for example the article from J.C. LASJAUNIAS, RAVEX and
    M. VANDORPE   this issue.
(2) E.H. JACOBSEN and K.W.H. STEVENS, Phys. Rev. 129, 2036 (1963).
(3) P.W. ANDERSON, B.I. HALPERIN and C.M. VARMA, Phil. Mag. 25 (1972).
(4) W.A. PHILIPS, J. Low Temp. Phys. 7, 351 (1972).
(5) G.A. TOOMBS and F.W. SHEARD, J. Phys. C 6, 1467 (1973).
(6) L. PICHE, R. MAYNARD, S. HUNKLINGER, and J. JÄCKLE, Phys. Rev.

Letters 32, 1426 (1974).
L. PICHE and S. HUNKLINGER (1975), this issue.

(7) K. MATHO and R. MAYNARD, to be published.

(8) S. HUNKLINGER, W. ARNOLD, S. STEIN, R. NAVA and K. DRANSFELD
Phys. Letters 42A, 253 (1972).
B. GOLDING, J.E. GRAEBNER, and B.I. HALPERIN, Phys. Rev.
Letters 30, 223 (1973).
W. ARNOLD, S. HUNKLINGER, S. STEIN and K. DRANSFELD, J. Non-
-Cryst. Solids 14, 192 (1974).

(9) M. VON SCHICKFUS, S. HUNKLINGER, and L. PICHE, to be published.

(10) L. LANDAU and E. LIFCHITZ, Mécanique des Fluides (1971),p.385.

(11) J. JÄCKLE, Z. Physik 257, 212 (1972).
J. JÄCKLE, L. PICHE, W. ARNOLD and S. HUNKLINGER (1975) to be
published.

(12) J. JOFFRIN and A. LEVELUT, J. de Physique 36 (1975).

(13) W. ARNOLD and S. HUNKLINGER, to be published.

(14) G. FROSSATI, private communication.

(15) G.J. MORGAN, J. Phys. C 2, 347 (1968).
G.J. MORGAN and D. SMITH, J. Phys. C 7, 649 (1974).

(16) A.J. LEADBETTER, A.J. JEAPES, WATERFIELD, unpublished work,
R.B. STEPHENS, Phys. Rev. B15 (1973).

(17) M.P. ZAITLIN and A.C. ANDERSON, to be published.

PROPAGATION OF ULTRASONIC LONGITUDINAL AND TRANSVERSE SOUND WAVES

IN GLASSES AT VERY LOW TEMPERATURES

L. PICHE and S. HUNKLINGER

Centre de Recherches sur les Très Basses Températures
and Max-Planck-Institut für Festkörperforschung
F 38042 Grenoble-Cedex, France

It was proposed /1/ that the thermal anomalies /2/ in glasses arise from a distribution of localised two-level systems (or very anharmonic oscillators) characteristic of the amorphous state. These systems are expected to scatter resonant phonons ; a first evidence is that the ultrasonic attenuation can be saturated /3/, thus indicating a saturation of two-level systems. We have pursued by measuring in extenso the acoustic properties of a typical glass (borosilicate BK 7) at temperatures between 0.28 K and 4 K, the data was obtained at $1\,mW/cm^2$ acoustic intensity, well above the saturation threshold of $10^{-7}\,W/cm^2$.

First we have measured the relative variation of the sound velocity $\Delta v/v = \left(v(T) - v(T_o)\right)/v(T)$ where $T_o = 0.3\,K$ is a reference temperature. Our results (Fig. 1) show that the general behaviour is the same for both polarizations : at the lowest temperatures, the sound velocity increases with temperature then becomes frequency dependent and decreases. The resonant interaction alone /4/ accounts for the behaviour at very low temperatures and gives new evidence for this process. The absorption and dispersion laws are given by /4,5/ :

$$\ell_{\alpha Res}^{-1} = \frac{n_o M_{\alpha}^2}{\rho v_{\alpha}^3}\,\pi\,\omega\,\tanh\left(\frac{\hbar\omega}{2k_B T}\right) \qquad (1)$$

$$\frac{\Delta v_{\alpha}}{v_{\alpha}}\bigg|_{Res} = \frac{n_o M_{\alpha}^2}{\rho v_{\alpha}^2}\,\ell n\left(\frac{T}{T_o}\right) \; ; \; \hbar\omega \ll k_B T \qquad (2)$$

where M_{α} is the coupling constant for the resonant interaction between a two-level system and a phonon of frequency ω and polarization α, and n_o the constant part of the density of states. For

123

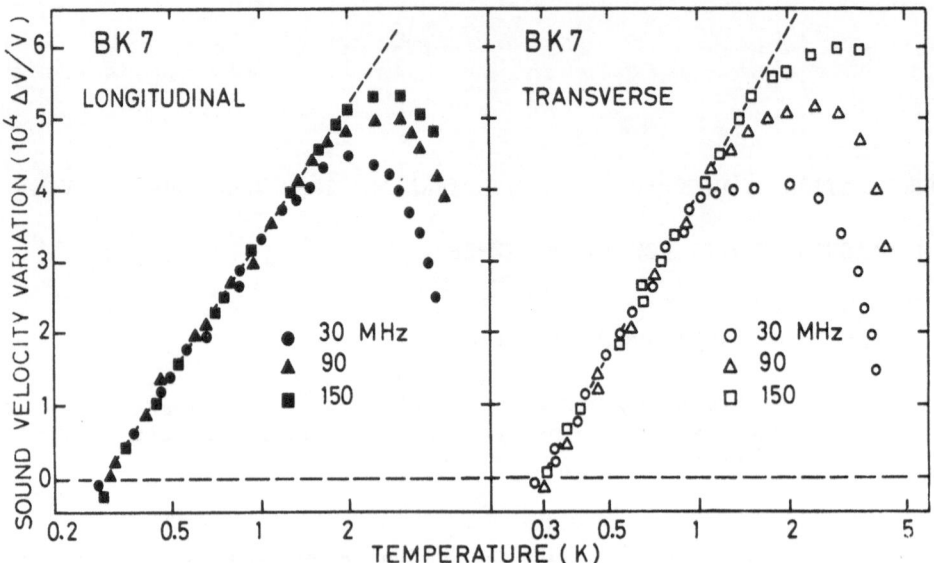

Fig. 1. Relative variation of sound velocity $\Delta v/v$ as a function of temperature. Straight dashed line, the contribution of the resonant interaction.

BK 7 the density is $\rho = 2.51 \text{ g/cm}^3$ and the sound velocities $v_\ell = 6.20 \times 10^5 \text{cm/sec.}$ and $v_t = 3.80 \times 10^5 \text{cm/sec.}$ From our data, we obtain $n_o M_\ell^2 = 2.6 \times 10^8 \text{erg/cm}^3$ and $n_o M_t^2 = 1.2 \times 10^8 \text{erg/cm}^3$; from these, we calculate the thermal conductivity K given by (1) and the kinetic formula : $K = 1.1 \times 10^{-5} \text{W/cm.K}$ at 0.2 K, in agreement with the measured value of $9 \times 10^{-6} \text{W/cm.K}$ /6/. Finally, using $n_o = 8.32 \text{erg}^{-1} \text{cm}^{-3}$, we find $M_\ell = 0.35 \text{ eV}$ and $M_t = 0.24$, in accordance with estimates from resonant absorption measurements /3/.

Above 1 K, the behaviour is governed by a relaxation process, which (in our case) determines the attenuation. The sound wave disturbs the thermal equilibrium of the two-level systems which relax via the emission or absorption of thermal phonons. The absorption due to this process is given as a function of the recombination rate $\tau(E)$ by /5/ :

$$\ell_{\alpha \text{Rel}}^{-1} = \frac{n D_\alpha^2}{\rho v_\alpha^3} \int \frac{df(E)}{dE} \frac{\omega^2 \tau(E)}{1 + \omega^2 \tau^2(E)} dE \qquad (3)$$

$$\tau^{-1}(E) = \left(\frac{M_\ell^2}{v_\ell^5} + \frac{2M_t^2}{v_t^5}\right) \frac{E^3}{2\pi\rho\hbar^4} \coth\left(\frac{E}{2k_B T}\right) \qquad (4)$$

where D_α is the shift in energy splitting E per unit strain, $f(E) = (\exp E/k_B T + 1)^{-1}$. At low temperatures, when $\omega\tau(E) \gg 1$, (3) reduces to $\ell_\alpha^{-1} \propto n_o D_\alpha^2 T^3$. Our results for the absorption are shown in Fig. 2.

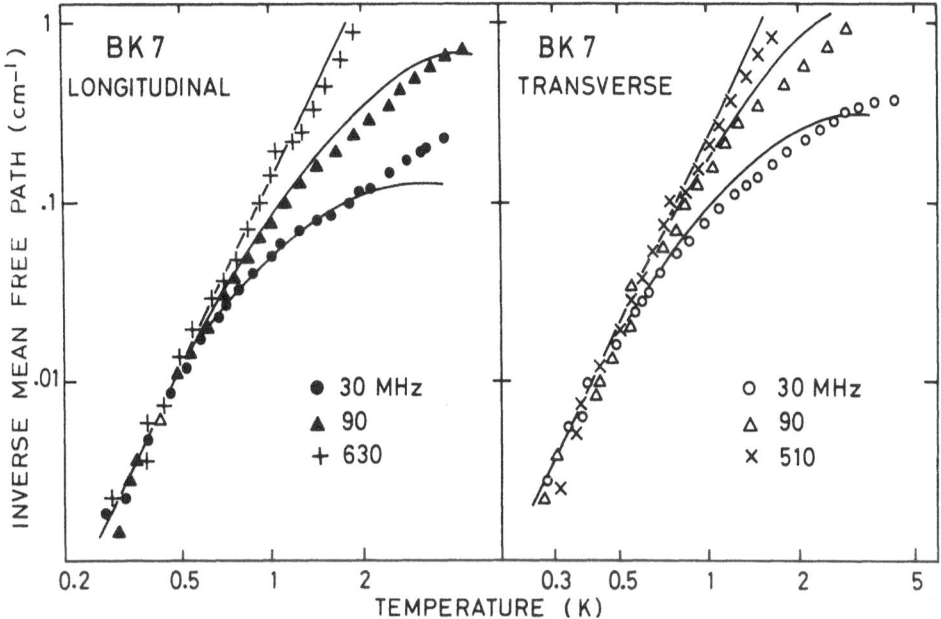

Fig. 2. Contribution of the relaxation process to the attenuation ; solid lines are theoretical prediction.

The behaviour is analogous for both polarizations, only the magnitude is higher in the case of transverse waves. At the lowest temperatures, the attenuation indeed obeys to an $\omega^{o}T^{3}$ law. The solid lines indicate the theoretical fit corresponding to $D_{\ell} = 0.9$ eV and $D_{t} = 0.6$ eV. From eq. 3, we can calculate through the Kramers-Kronig relation $(\Delta v_{\alpha}/v_{\alpha})$Rel. for the relaxation process. This contribution is opposed to that of the resonant process and leads to a decrease of the sound velocity. Using the values of D_{ℓ} and D_{t}, we are able to describe /4/ the high temperature variation of the sound velocity.

In conclusion, our results can be explained through the simple scheme of two-level systems. Using our results alone, we calculate the thermal conductivity and also describe the coupling of two level systems with longitudinal and transverse phonons.

1 P.W. Anderson, B.I. Halperin and C.M. Varma, Phil. Mag. 25, 1 (1972) ; W.A. Phillips, J. Low Temp. Phys. 7, 351 (1972).
2 R.C. Zeller and R.O. Pohl, Phys. Rev. B 4 , 2029 (1971) ; R.B. Stephens, Phys. Rev. B 8 , 2996 (1973).
3 S. Hunklinger, W. Arnold and Stain, Phys. Lett. 45 A, 311 (1973).
4 L. Piché, R. Maynard, S. Hunklinger and J. Jäckle, Phys. Rev. Lett. 32, 1426 (1974).
5 J. Jäckle, Z. Phys. 257, 212 (1972).
6 M.P. Zaitlin and A.C. Anderson, Phys. Rev. Lett. 33, 1158 (1974).

PHONONS IN SOME AMORPHOUS SOLIDS

M. Goda and S. Takeno[*])

Faculty of Engineering, Niigata University, Nagaoka,
Japan and[*]) Faculty of Industrial Arts, Kyoto Technical
University, Kyoto, Japan

ABSTRACT

An attempt is made for investigating phonons in topologically
disordered systems with configurational short-range order (SRO).
Though the system we investigate is the simplest one and the adopted
approximation is of low order, some general characters of the system
are obtained from the viewpoint that many quantities for phonons are
dominated by the SRO.

A THEORY OF PHONON

Assuming that some amorphous systems are made with (spherical)
pair- interaction potential, it is shown that the squared frequency
of the phonon $\omega^2(K)$ is given, within the framework of the
renormalized harmonic approximation, as a function of wave vector K

$$\omega^2(K)=(\rho/M)\int d\vec{R}\ g_2(R)\nabla\nabla V_{eff}(R)\left[1-\exp(i\vec{K}\cdot\vec{R})\right] \qquad (1)$$

Here, ρ is the number density of atoms in the system, M is the
atomic mass, $V_{eff}(R)$ is an effective pair-interaction potential
describing the interaction of each pair of atoms and $g_2(R)$ is the
pair-correlation function for the constituent atoms as a function
of distance R. When $V_{eff}(R)$ is replaced by V(R) the Eq.(1) is
essentially the same as that obtained by Hubbard and Beeby (]969).

Under some general assumptions which are natural for the system
with the SRO (for example that there exists the first peak in the
pair-correlation function), it is shown that the longitudinal phonon
dispersion relation of the system is of phonon-roton type as

observed in liquid He^{4}, and that the dispersion relations are
intimately connected with the SRO such as the pair-correlation
function. The energy $\Delta \varepsilon$ for the wave vector k_{0}, which corresponds
to the reciprocal lattice vector of the perfect crystal, decreases
when the degree of SRO increases and finally vanishes to be that
of the perfect crystal.

NUMERICAL RESULTS FOR HIGH FREQUENCY PHONONS IN SIMPLE LIQUIDS

Because the amorphous system we adopted here is rather an
idealized one and we do not know the detailed informations of $g_{2}(R)$
for the system,examinations of the Eq.(1) are made by calculating
the high-frequency phonons in simple liquids. In spite of the fact
that the fluidity of liquids is completely ignored in the Eq.(1) and
the adopted approximation is of the lowest order including only the
pair-corration function, the calculated dispersion relations for
Liq.Ar, Liq.He^{4} and Liq.Rb show rather good agreement with the
experimental results. It should be mentioned that the dispersion
relation for Liq.He^{4} can almost be realized also within the
framework of the theory when only quantum corrections are made in
the pair-potential. The experimental results of Henshaw(1960) and
Dietrich et al (1972) show that the dependence of the $\Delta \varepsilon$ to the
sharpness of the first peak of the $g_{2}(R)$ (when the pressure and the
temperature are varied) is at least qualitatively like the above
mentioned theoretical results.
Because lack of the fluidity causes longer lifetime of phonons,
we   have a confidence that the above mentioned characters are
general in topologically disordered systems with the SRO as far as
the wave number K is defined.

DISCUSSIONS ABOUT AMORPHOUS SOLIDS

With the above mentioned confidence we have a conjecture that
the general characters are seen also in some amorphous systems.
There seems to be some distance between the amorphous system and
the realistic one. The system is a "simple amorphous" system
corresponding to the simple liquid. The existence of the simple
amorphous system is not yet sure experimentally but the system at
least seems to realize some essential features of amorphous systems
as the simple liquid of liquids. To calculate the life time of the
phonons in the systems is an important problem. Some generalizations
of the theory such as to the cases of many atomic amorphous systems
are made in a quite straightforward manner but the description of
the amorphous systems with non isotropic pair-interaction potential
is yet unsuccessful.
Recently Zeller and Pohl(1971) reviewed the anomalous effects
of the specific heat and the thermal conductivity common in many

amorphous systems. We can understand some of these anomalous
effects by the existence of the "roton" part of the longitudinal
phonon. The suggested energy of the phonon for the roton part is
of the order of sevral degrees. The excess specific heat can be
understood by the existence of the excess state density due to the
phonons of the roton part in addition to that of the Debye phonon.
The anomalous thermal conductivity can also be understood by the
resonance scattering of the Debye phonons by the rather dirty
phonons of the roton part having zero group velocity.

Unfortunately the experimental informations are yet too little
to make the qualitative discussions of these problems.
Experimentally, the resemblance between the structure factors of
liquid state, of amorphous state and of policristalline state is
reported by several workers (such as Larsson 1968). Some phonons
are observed in amorphous GeO_2 (Leadbetter and Litchinsky 1970).
The detailed inelastic neutron scattering studies are requested
to check the above mentioned conjecture and to make the qualitative
discussions about these anomalous effects of amorphous solids.

REFERENCES

Dietrich O. W. et al 1972 Pys. Rev. A5, 1377
Goda M. 1972 Prog. Theor. Phys. 47, 1064
Henshaw D. G. 1960 Phys. Rev. 119, 14
Hubbard J. and Beeby J. L. 1969 Jour. Phys. C (Solid State Phys.)
    2, 556
Larsson K. E. 1968 Proc. Sympo. on Neutron Inelastic Scattering,
    Copenhagen, Vol. 1, 397
Leadbetter A. J. and Litchinsky D. 1970 Disc. Farad. Soc. 50, 62
Stephens R. B. 1973 Phys. Rev. B8, 2896
Takeno S. and Goda M. Prog. Theor. Phys. 45 (1971), 331;
    47 (1972), 790; 48 (1972), 724
Zeller R. C. and Pohl R. O. 1971 Phys. Rev. B4, 2029

OBSERVATION OF DIRECT INTERACTION BETWEEN LOW-ENERGY

EXCITATIONS IN GLASSES

W. Arnold, S. Hunklinger, and K. Dransfeld

Max-Planck-Institut für Festkörperforschung and Centre
de Recherches sur les Très Basses Températures
B.P.166, 38042 Grenoble Cedex, France

The striking anomalous thermal and acoustic properties of amorphous materials /1,2,3/ at low temperatures have given rise to the
hypothesis that two-level systems exist in these materials with an
almost constant density of states n_0 /4/. The resonant scattering
of phonons by these two-level systems depends on the acoustic intensity: The attenuation becomes smaller at higher powers when
both levels are equally populated (acoustic self-induced transparency or saturated attenuation). Consequently the attenuation increases with decreasing power and becomes independent below a critical intensity I_c /5/. For example $I_c = 10^{-7}$ W/cm^2 in Borosilicate glass BK 7 at 740 MHz and 0.5 K. The coupling between these
two-level systems and phonons is described by a deformation potential /6/, which is of the order of 1 eV. Such a large value obviously makes possible the existence of a strong direct elastic
coupling between these two-level systems and led to the prediction
/7/ of a broadened excited state.

We report here the first direct measurement of the width of the
upper state. We generated two acoustic pulses P_1 and P_2 (duration
1 μsec) travelling together through the sample of Borosilicate
glass BK 7. The pulse P_2 had an intensity of only 0.9×10^{-7} W/cm^2
which is just below I_c and its frequency was fixed at 740 MHz. On
the other hand pulse P_1 had variable frequency and much higher intensity than I_c. We measured the resonant of the very weak test
pulse P_2 as a function of the frequency of the saturating pulse P_1
As can be seen in Fig.I, a broad minimum in the attenuation occurs
- roughly 100 MHz wide -, which is much broader than the frequency
uncertainty of the 1 μsec pulse. Obviously pulse P_1 also populates
levels having an energy splitting quite different from its frequency. The observed broadening cannot be explained as a saturation

broadening /8/ because the observed linewidth was quite independen-
dent of the acoustic power level of the intense pulse (Fig.I).
The halfwidth of about 50 MHz can, however, be understood from the
strong coupling between two-level systems and phonons. Already a
small strain leads to a considerable shift of the energy splitting
for the two-level systems. The thermal excitation or recombination
of one system can cause a variation of the strain acting on adjacent two-level systems. Considering the two-level system as an
elastic dipole /9/, we can estimate the interaction energy between
two neighbouring two-level systems. The dipole moment is given by
$p = \sqrt{D^2/\rho v^2}$, where D is the coupling energy describing the variation of their level splitting in a strain field, ρ the density and
v the sound   velocity. For the interaction energy we can then
write $\Delta E = D^2/\rho v^2 r^3$, with r as the distance between the two-level
systems. The energy splitting of any given two-level system will
fluctuate quasi-statically in time due to thermal excitation and
recombination of the adjacent systems. In the low-temperature limit

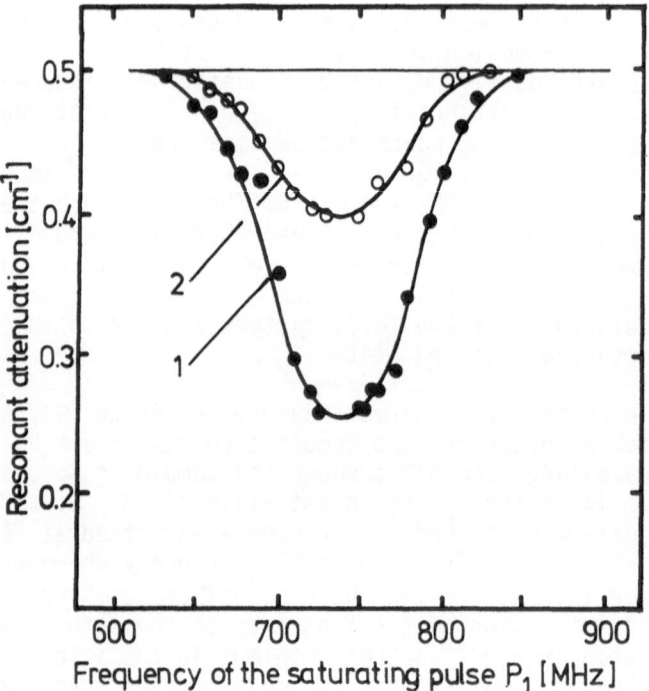

Fig.1: Resonant attenuation of the test pulse P_2 with frequency of
       738 MHz as a function of the frequency of the saturating
       pulse P_1 at 0.55 K. In Curve 1 the intensity of pulse P_1
       was 5×10^{-6} W/cm^2, in Curve 2 5×10^{-7} W/cm^2.

where the fluctuation rate is slow compared to $\Delta E/h$, the line
appears therefore homogeneously broadened by just ΔE.
A calculation taking into account the integration over all thermally excited adjacent two-level systems results in /10/:
$\Delta E \alpha n_0 D^2 kT/\rho v^2$. Inserting the value for $n_0 D^2$ derived from ultrasonic measurements /6/, we get for $\Delta E/h \simeq 60$ MHz at 0.5 K in good
agreement with the experimental value of 50 MHz. In a preliminary
experiment a similar broadening of the linewidth has been observed
in vitreous silica.

We want to acknowledge stimulating discussions with R.Maynard,
J.Jäckle, A.Levelut and R.Ranvaud.

References:

/1/ See for example R.O.Pohl, the present conference proceedings

/2/ S.Hunklinger, W.Arnold, S.Stein, R.Nava, and K.Dransfeld,
     Phys.Lett.42A, 253 (1972);
     B.Golding, J.E.Graebner, B.I.Halperin, and R.J.Schutz, Phys.
     Rev.Lett.30, 223 (1973)

/3/ L.Piché, R.Maynard, S.Hunklinger, and J.Jäckle, Phys.Rev.Lett.
     32, 1426 (1974)

/4/ P.W.Anderson, B.I.Halperin, and C.M.Varma, Phil.Mag.25, , 1972;
     W.A.Philips, J.Low Temp.Phys.7, 351, 1972

/5/ W.Arnold, S.Hunklinger, S.Stein, and K.Dransfeld, J.Non.Cryst.
     Sol.14, 192, 1974

/6/ J.Jäckle, L.Piché, W.Arnold, and S.Hunklinger to be published

/7/ J.Joffrin and A.Levelut, to be published

/8/ A.Abragam, The Principles of Nuclear Magnetism, p.47,
     Oxford University Press (1961)

/9/ A.S.Nowick and W.R.Heller, Adv.Phys.12, 251 (1963)

/10/ W.Arnold and S.Hunklinger, to be published in Sol.Stat.Comm.

*absorption

VIRTUAL PHONON EXCHANGE IN GLASSES

J Joffrin and A Levelut

Laboratoire d'Ultrasons[+], Universite P et M Curie

Tour 13, 4 Place Jussieu, 75230 Paris-Cedex 05, France

The phonon field has long been recognized to provide an efficient coupling of particles or excitations in crystals. The most famous example is the case of superconductivity where a pair of electrons is bound by virtual exchange of phonons. The same mechanism has also been invoked in the case of paramagnetic impurities in crystals (1). However because of the small value of the coupling constant of each spin with the lattice even for non Kramers ions, this mechanism yields a negligible contribution to the linewidth measured in E.P.R. (2). On the contrary, in the case of helium 3 or 4 impurities in helium 4 or 3 single crystals, the coupling between the impurities via the phonon field seems to be rather strong and probably contributes to the anomalous value of the diffusion coefficient of impurities, as observed in NMR (3). One might also ask whether the anomalous value at low temperatures of the dielectric constant of KCl crystals containing molecular impurities (OH⁻) is not due to the appearance of a collective effect ordered by the phonon field (4).

In view of these examples, the following question can be raised: how does the phonon field govern certain properties of amorphous materials and in particular, following the model of Anderson et al (5) and Phillips (6), the value of the T_2 relaxation time of the spin populations?

It is generally admitted that an amorphous material may be thought of as an assembly of defects with randomly distributed characteristics. Each of these defects has two conformation states separated by a potential barrier which can be crossed by tunnelling.

[+]Associated with the Centre National de la Recherche Scientifique.

Thus, each defect may be represented by an effective spin 1/2 in a
local magnetic field whose intensity and orientation varies from
site to site. The energy separation 2E of two states of the assembly extends over a large range of energy. It has been established
by thermal measurements at low temperatures that the distribution
N(E) of spins per unit energy varies very smoothly with E up to a
maximum value of the order of 1 Ev, giving a linear temperature
variation for the specific heat (7,8).

The other types of experiment which have been performed on
glasses are transport experiments: thermal conductivity (7,8) and
acoustic propagation (9). These experiments give information on the
dynamical behaviour of the spin system (relaxation of the population,
mean free path of the phonons, etc.) and all of them are related to
the change of properties of each spin as the strain associated with
the phonon varies. In fact, any strain of the environment of the
defect modifies the parameters of the double well. Consequently,
an elastic wave may induce resonant transitions between the spin
levels provided that the frequency of the phonons is on speaking
terms with the energy 2E of the spin. Following this simple model,
the dynamics of the spin system is conveniently described by the
same concepts used in a paramagnetic system: spin population, longitudinal (T_1) and transverse (T_2) relaxation times, line saturation,
self induced transparency for the propagation of an acoustic wave in
a resonant medium, etc. However, until now the results of experiments in amorphous materials have been interpreted with the use of
the *a priori* relation $2T_1 = T_2$ (9,10). Such an approach is attractive
because magnetic dipolar interactions or exchange interactions do
not occur here for fictitious spins. However, we should like to investigate the following problem: due to the large spin phonon coupling observed in most of the glasses, what is the influence of the
phonon field on the static properties of the spins? In particular,
what is the indirect coupling induced between two spins and what are
its consequences? In fact we show that the indirect interaction is
easily represented by an exchange constant between pairs of spins.
We first calculate the order of magnitude of this interaction. Once
it has been estimated, we calculate more carefully the different relaxation times. A technique developed by Van Vleck thirty years ago
for a mixture of spins (11) will be used. Because T_2 appears to be
smaller than T_1, we are led to recalculate the behaviour of an acoustical wave in such a resonant medium. Finally currently available
results whose interpretation depended until now on the hypothesis T_1
$\approx T_2$ will be reinterpreted within the framework of our theory. In
this way, some apparently contradictory data can be reconciled.

A few elements of the theory developed here are not entirely
new, but they are included in order to present a self contained paper
from the starting equations to the numerical estimation of the parameters introduced in the model. The difficult question of the anomalous propagation of a wave in a resonant medium when the stationary

state is not reached is beyond the scope of this article.   This is
the specific problem of self induced transparency.

Further details of this work are given elsewhere (12).

REFERENCES

(1) D McMahon and R Silsbee, Phys Rev A135 (1964) 91.
(2) R Orbach and M Tachiki, Phys Rev 158 (1967) 524.
(3) W Huang, H Golberg, M T Takemori and R A Guyer, Phys Rev Lett
     33 (1974) 283.
(4) F Luty, J Physique Colloques 28 (1967) C4-120.
(5) P W Anderson, B Halperin and C Varma, Phil Mag 25 (1972) 1.
(6) W A Phillips, J Low Temp Phys 7, (1972) 351.
(7) R C Zeller and R O Pohl, Phys Rev B4 (1971) 2029.
(8) J C Lasjaunias and R Maynard, J Non Cryst Solids 6 (1971) 101.
(9) W Arnold, S Hunklinger, S Stein and K Dransfeld, J Non Cryst
     Solids 14 (1974) 192.
(10) B Golding, J Graebner, B Halperin and R Schutz, Phys Rev Lett
     30 (1973) 223.
(11) J H Van Vleck, Phys Rev 74 (1948) 1168.
(12) J Joffrin and A Levelut, Journal de Physique 36 (1975) 811.

INFLUENCE OF PHASE SEPARATION ON THE THERMAL CONDUCTIVITY OF POTASSIUM BOROSILICATE GLASSES BETWEEN 1 K AND 10 K

F. Canal, P. Carrara, J.P. Redoules, and M.C. Schmidt

Laboratoire de Physique des Solides, associé au CNRS

Université Paul Sabatier, 31077 Toulouse Cedex (France)

INTRODUCTION

Phase separation has long been recognized in glasses and often develops as a result of thermal treatment (1). Since Kapitza's first report, thermal resistance has also been observed in boundaries between different solid phases (2-4). The purpose of the present study is to investigate the influence of particule sizes and boundaries on the thermal conductivity at low temperatures (1K-10K) of five samples of a phase separated potassium borosilicate glass having undergone different lengths of heat treatment.

EXPERIMENTAL PROCEDURE

The molar composition of our glass was : $50SiO_2-41.7B_2O_3-8.3K_2O$. Five rectangular samples of equal shape and size (dimensions 8x1x0,5 cm) were cut out of this glass and treated at a temperature of 565°C for 0h, 1/2h, 2h, 8h and 24h respectively. Later, they were reheated at 450°C for an hour. The average dimension of the growing particules of the phase poor in silicium was measured by transmission electron microscopy (480Å, 600Å, 1000Å, 1500Å and 2200Å respectively)

RESULTS AND DISCUSSION

The thermal conductivity graphs of the five samples are given in Fig.1. The values for the untreated sample are very close to those found by Slack and Fisher for amorphous SiO_2 (5-6). All the samples present between 6K and 10K the usual plateau of amorphous materials. A relevant characteristic of the curves is the dependence of the thermal conductivity below 6K on the duration of the

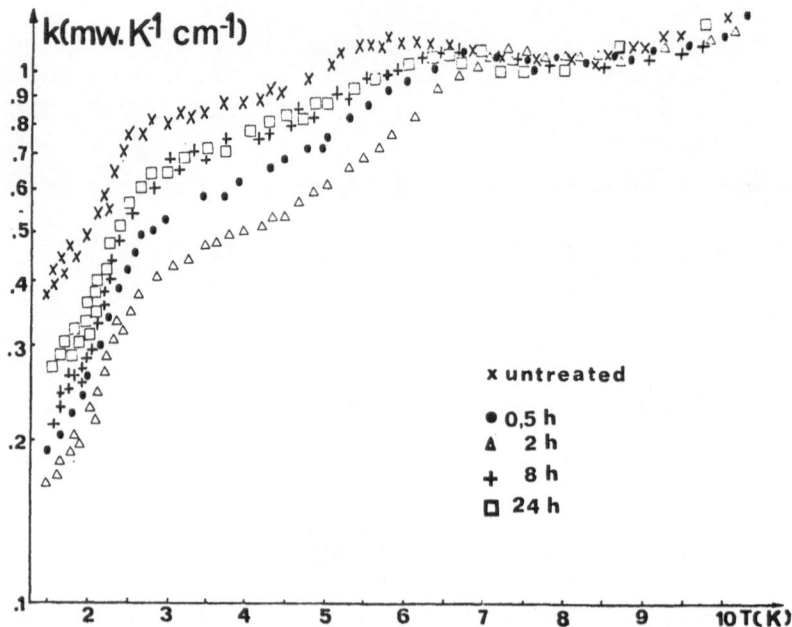

Fig.1 Thermal conductivities of the potassium borosilicate samples.

heat treatment : for the conductivity corresponding to a treatment
lasting approximately 75 minutes a minimum is observed. Furthermore
the time evolution of the average dimension of the growing particles
follows a $t^{1/3}$ law typical of an Ostwald ripening process. So the
minimum of the conductivity corresponds to a mean dimension of 900Å
(Fig.2). In this report, attention is focused on the thermal treatment dependance of the conductivity.

    In order to explain the minimum, a simple model where the pre-
cipitated phase of conductivity k_A is composed of cubes with side
β regularly distributed within a matrix of conductivity k_B, is
considered. Taking into account the Kapitza resistance at the interface, a simple calculation gives then a good approximation of the
thermal conductivity dependance as a function of the mean dimension β
of the grains :

$$\frac{k}{k_B} = 1 - \frac{\lambda y^3}{1+\lambda y} \; ; \text{ where } : \lambda = \frac{k_B}{k_A}(1 + 2\frac{r_K\,k_A}{\beta}) - 1$$

$$y = (\frac{\text{volume of the precipitate}}{\text{total volume}})^{1/3} = 0.84; \; r_K = \text{Kapitza resistance.}$$

    The minimum of the conductivity may be due to a change in
Kapitza resistance during the thermal treatment that modifies the
structure of the grain boundaries. Thermal treatment leads to a
sharpening of the interface between both phases, so, Kapitza resistance increases rapidly in the early stages of the thermal treatment

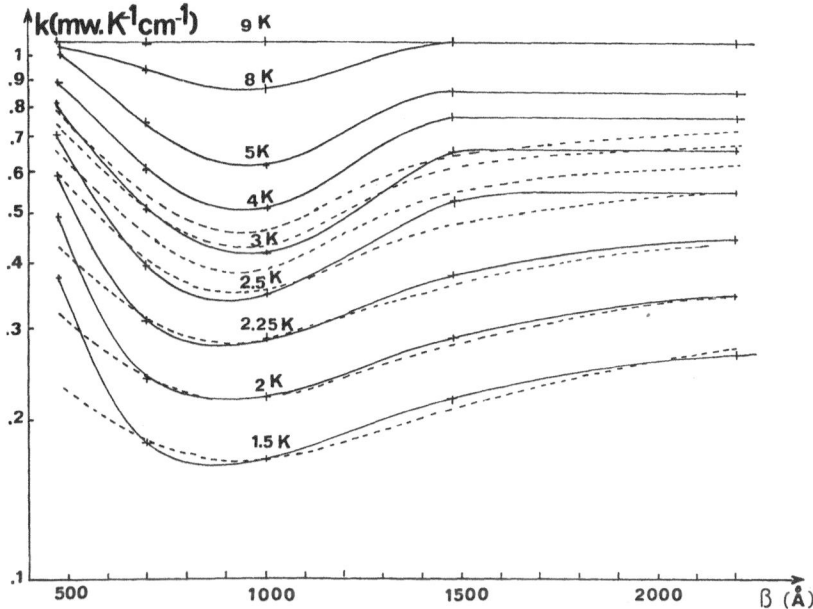

Fig.2. Variation of k as a function of the grains mean dimension :
- Experimental --- Theoretical for 1.5K, 2K, 2.25K, 2.5K, 3K, 4K, 5K.

from a very weak initial value for the untreated sample. The width
of the interfaces is related to the grains average dimension, so
that Kapitza resistance is taken to be of the form :

$$r_K(\beta) = r_K(\infty) \left[1 - \exp(\frac{-\beta}{\beta_o}) \right]$$

The parameters β_o and $r_K(\infty)$ are calculated in order to fit the
experimental results(Fig.2) :

$$\beta_o = 1300\overset{o}{A} \quad ; \quad r_K(\infty) = 2 \times 10^{-2} \ cm^2 \ KW^{-1} \ .$$

REFERENCES

1.J.Zarzycki, Discuss. Faraday Soc. 50, 122 (1970)
2.C. Schmidt, Kernforschungszentrum Karlsruhe. Report n°2030, 1974
3.R.E. Peterson and A.C. Anderson, J. Low.Temp. Phys. 11, 639, 1973
4.L.J. Challis and J.D.N. Cheeke, Phys.Lett. 5, 305, 1963
5.G.A. Slack, Cryogenics, 389, oct. 1969
6.R.A. Fisher, G.E. Brodale, E.W. Hornung and W.F. Giauque, Rev.Sci.
   Instr. 40, 365, 1969

ACKNOWLEDGMENTS

We wish to thank J Zarzycki, Larche and Pernot (Laboratoire des
Verres, Montpellier) for providing the samples. We are also grateful
to R Maynard (CRTBT Grenoble) for his valuable suggestions.

INFLUENCE OF OH CONTENT ON THE SPECIFIC HEAT AND THERMAL

CONDUCTIVITY OF VITREOUS SILICA BELOW 1 KELVIN

J.C. Lasjaunias, A. Ravex, D. Thoulouze and M. Vandorpe

Centre de Recherches sur les Très Basses Températures

CNRS, BP 166, Centre de Tri, 38042 Grenoble-Cédex, France

Previous measurements of the specific heat of vitreous B_2O_3 down to 60 mK indicated a non linear temperature dependence, as $T^{1.45}$, which had been tentatively interpreted by the existence of a gap in the density of states at very low energy[1]. In order to check this idea, we tried to extend the measuremnts at lower temperatures and to compare them to heat conductivities. We began by SiO_2 which has been yet extensively studied; it can be obtained with different concentrations of OH impurities, which influence on specific heat and heat conductivity is interesting to know. We used Suprasil samples with two OH contents : 1200 and less than 1.5 ppm (hereafter called Suprasil W).

The specific heat results are shown in fig. 1. The anomaly previously observed down to 0.1 K on Spectrosil B[2] extends down to 25 mK for both samples, its value reaching 10^3 times the elastic component at this temperature. Below 0.5 K, the presence of OH groups increases the amplitude of the specific heat by about 50 % at 25 mK. In a temperature range of one order of magnitude, this anomaly can be fitted by simple power laws : $T^{1.30}$ for Suprasil W and $T^{1.22}$ for Suprasil. Then for Suprasil W, the corresponding density of states of the low energy excitations, supposing that they are two-level systems, can be written as $E^{0.30}$. However, it is possible to describe this density of states by different laws, and particularly [3]

$$n(E) = \bar{P} \log \frac{E}{E_o} . \tag{1}$$

In the framework of the tunneling model, this law expresses the existence of a gap E_o obtained if the values of the potential barriers are limited. A very good agreement is obtained for Suprasil W with $\bar{P} \simeq 0,3.10^{32}$ erg^{-1} cm^{-3} and $E_o \simeq 2mK$. For the

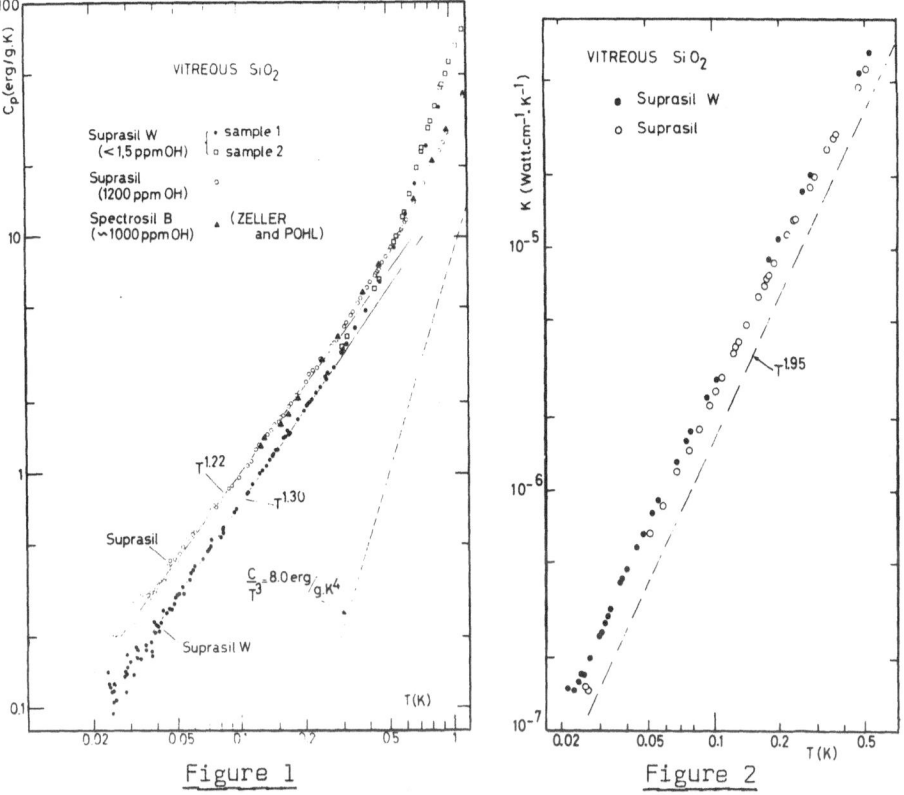

Figure 1

The specific heat of vitreous Silica with different OH contents. The dashed line represents the acoustic component calculated from the mean density (2.20 g/cm^3) and the sound velocities in liquid helium range (v_ℓ = 5.80 × 10^5 cm/s and v_t = 3.75 × 10^5 cm/s).

Figure 2

The thermal conductivity measured on the same samples of Suprasil as used for the specific heat measurements. The dashed line shows a $T^{1.95}$ variation, shifted from the experimental points for the clarity.

other Suprasil glass, the additional energy density due to the OH groups appears to be constant in energy, thus leading to an additional linear term of specific heat. The total specific heat then exhibits a variation closer to the T law.

In contrast to the specific heat, the presence of OH groups scarcely affects the thermal conductivity (fig. 2.) The temperature dependence is the same for both samples, close to $T^{1.95}$ below 0.5 K, and the amplitudes only differ by 10 to 15 % in the whole temperature range. If the excitations responsible for the specific heat anomaly are also the dominant scattering centers for the phonons (with a coupling constant M), the phonon lifetime is inversely proportional to $n(E) \times M^2$. Taking M as a constant[4], it is not possible to take into account the heat conductivity results with the values of the density of states deduced from the specific heat, which provide a $T^{1.74}$ variation. However, with the supplementary hypothesis of the tunneling character of the excitations, the density of states does not appear in the same analytic form in both expressions, and this leads to a temperature variation much closer from the experimental one.

To explain the difference between the two types of Silica we need to suppose that the supplementary excitations brought in by the additional impurities are more weakly coupled to the phonons than those characteristic of the pure vitreous structure, and so have very little influence on transport properties as thermal conductivity, velocity of sound,...[5].

We want to thank Prof. R. Maynard for many helpful discussions and B. Picot for this experimental assistance.

References

(1) J.C.Lasjaunias, D. Thoulouze and F. Pernot, Sol. State Comm. 14, 957 (1974).
(2) R.C. Zeller and R.O. Pohl, Phys. Rev. B, 4, 2029 (1971).
(3) R. Maynard, private communication.
(4) L. Piché, R. Maynard, S. Hunklinger and J. Jäckle, Phys. Rev. Lett. 32, 1426 (1974)
(5) S. Hunklinger, L. Piché, J.C. Lasjaunias and K. Dransfeld - to be published.

THE THERMAL CONDUCTIVITY AND DIFFUSIVITY OF EPOXY-RESINS FROM 2 TO 80 K

S. Kelham and H. M. Rosenberg

The Clarendon Laboratory, University of Oxford

Oxford, OX1 3PU, U.K.

Abstract

The anomalous behaviour of the thermal conductivity and the specific heat of glassy materials at low temperatures is being studied using an epoxy-resin, Epikote 828, as the sample material. Variations in the curing cycle can be used to change the degree of disorder and cross-linking in the resin and these effects alter the thermal conductivity and diffusivity.

Introduction

It is well known that the thermal conductivity of all types of glassy materials shows anomalous behaviour at low temperatures. The most characteristic feature of the general behaviour is that the conductivity displays a fairly constant plateau region in the temperature range around 10 K (e.g. see refs. 1-3), whereas one would expect that if the phonon mean free path is very short, which it will be in a disordered material, then the conductivity should show a monotonic decrease as the temperature is reduced.

In order to investigate this phenomenon it would be desirable to be able to change the physical characteristics of the glass and to observe the effect that this would have on the conductivity behaviour, but this, in the case of ordinary glasses, is not really possible without changing their chemical constitution.

Specimens and experiments

The present experiments have been initiated in order to over-

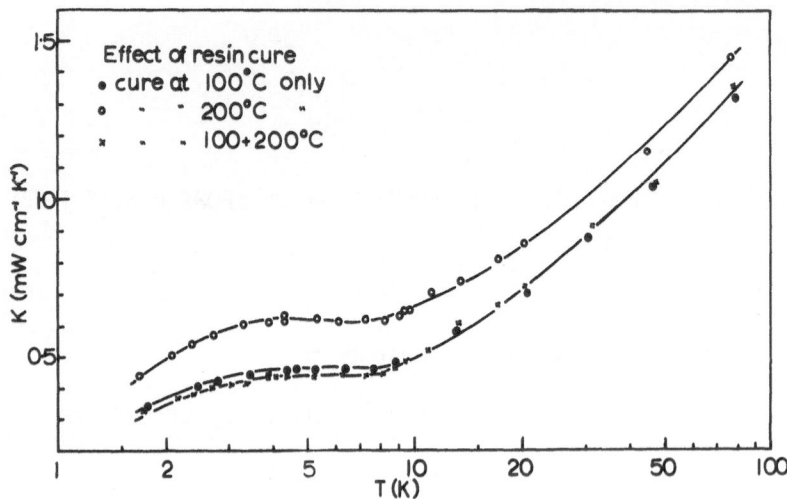

The thermal conductivity of Epikote 828 as a function of temperature. This illustrates the manner in which a change in the curing
cycle affects the conductivity.

come this problem. The glassy material used is an epoxy resin,
Epikote 828, and its properties can be changed by altering the curing
conditions. The resin is used in conjunction with Epikure NMA
hardener and BDMA accelerator in the proportions 100:90:0.5 by weight
respectively. The normal curing cycle of the resin consists of a
precure at $100^{\circ}C$ for 2 hours followed by a cure at $200^{\circ}C$ for 4 hours.
During the precure polymerization occurs and during the final cure
cross-linking between neighbouring polymer chains is the main effect,
although of course, both mechanisms will to a certain extent occur
during each of the curing processes.

The thermal conductivity of specimens which have been subjected
to different curing cycles has been measured in the range 2–80 K
using the conventional Searle's bar technique. In addition the
thermal diffusivity has been measured by applying a sinusoidal heat
input to the specimen and then measuring the velocity of the thermal
waves and also the relative amplitude of the waves at two points
along the specimen.

Results and discussion

Some preliminary results of the thermal conductivity experiments
are shown in the figure. It will be noted that all specimens exhibit
the plateau region but the conductivity is appreciably higher for
those specimens which were cured only at $200^{\circ}C$. Since the polymer

chains would be less developed in these samples it would appear that polymerization is the major mechanism for depressing the conductivity and that cross-linking has very little influence on the thermal properties. Preliminary measurements of the thermal diffusivity, however, indicate that above the thermal conductivity plateau the diffusivity is <u>independent</u> of the curing cycle, i.e. at a given temperature the thermal conductivity of all specimens is directly proportional to their specific heat. This would suggest that in this temperature range the differences between the thermal conductivities of the various samples are due only to the differences in their specific heats, i.e. to the phonon spectrum, and <u>not</u> to any phonon mean free path effects.

In the region of the thermal conductivity plateau and below, however, the diffusivity of the samples is <u>not</u> the same and hence the effect of different mean free paths in the various specimens must be taken into account.

These experiments are still in progress.

<u>References</u>

1.   R. Berman, Proc. Roy. Soc. <u>A208</u>, 90 (1951).
2.   W. Reece, J. Appl. Phys. <u>37</u>, 3227 (1966).
3.   R. C. Zeller and R. O. Pohl, Phys. Rev. <u>B4</u>, 2029 (1971).

VERY LOW FREQUENCY RAMAN SCATTERING OF GLASSES

G. Winterling

Max-Planck-Institut für Festkörperforschung

Stuttgart, Federal Republic of Germany

Dielectric glasses exhibit at helium temperatures anomalous thermal properties. Fulde and Wagner (1) interpreted the excess specific heat as arising from the low-frequency tails of the spectral functions of high momentum acoustic phonons. Anderson, Haperin and Varma (2), on the other hand, accounted for the thermal anomalies by assuming a distribution of structural defects giving rise to very low frequency tunneling modes. The present experiments were mainly stimulated by the question whether the defects would give a detectable contribution to the spectrum of the scattered light.

In an amorphous solid, all vibrational excitations, in principle, can take part in light scattering due to the intrinsic disorder. The scattered light intensity is generally (3) written as:

$$I(\omega, T) = \sum_b c_b(\omega) \rho_b \left[n(\omega, T) + \left(\begin{smallmatrix} 1 \\ o \end{smallmatrix} \right) \right] \omega^{-1}$$

where ρ_b is the density of states in band b, $n(\omega,T)$ is the occupation number of a harmonic oscillator and $\omega/2\pi$ is the frequency shift of the scattered light. $c_b(\omega)$ refers to the coupling constant; the 1 and o in the brackets refer to the Stokes and Antistokes scattering, respectively. Because of the thermal anomalies our interest here is mainly with the very-low frequency scattering.

Martin and Brenig (4) studying theoretically this frequency region describe the amorphous solid with a continuum model. Correspondingly, the scattering at small ω is expected to arise from modes having sound wave character; the coupling constant c_b of these modes is predicted to vary with ω^2 for small ω and the spectrum of the scattered light is expected to show besides the Brilliouin doublets a weak "amorphous" background which should vary at small ω and at higher temperatures T with the density of states, i.e., ω^2. In addition, the temperature dependence of the scattering should correspond to one-phonon processes.

We measured the low frequency Raman spectra of several glasses as close as 3cm^{-1} (~0.36meV) to the exciting line. So far as we know previous Raman data on glasses have been only available for $\omega > 15\text{cm}^{-1}$. Figure 1 shows the depolarized spectrum of the borosilicate glass BK7 recorded at 90° and using 647nm light; the Brillouin lines (here scattering from transverse sound waves) located at $|\omega| < 0.5\text{cm}^{-1}$ are not resolved because of the larger instrumental width of $\sim 1.5\text{cm}^{-1}$. In contrast to the above prediction, the measured scattered intensity I remains finite for $\omega \rightarrow 0$. A similar behavior was found in the spectra of vitreous silica (see Fig. 2) and other glasses.

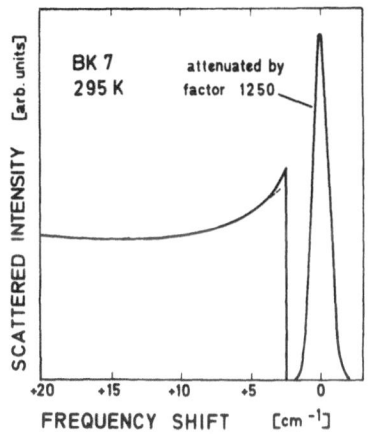

FIG. 1: The Raman spectrum of the glass BK7 at very small frequency shifts. The dashed curve represents the spectrum obtained after subtraction of the instrumental tail of the strong elastic scattering.

The comparison with the above theory shows that there is an excess scattered intensity ΔI at small ω. ΔI was found to be characteristic of the glassy state (5). It is depolarized with the same ratio D as the low frequency scattering ($\omega < 50\text{cm}^{-1}$); this ratio differs significantly from the D of of the Brillouin scattering. The integrated intensity of ΔI in Suprasil W1 was determined to be about 100 times weaker than the integrated intensity of one polarized Brillouin line (5). The temperature dependence of I in vitreous silica was measured to agree with the theoretical expectation of one-phonon processes (3,4) for $\omega > 20\text{cm}^{-1}$, however, for $\omega < 20\text{cm}^{-1}$ a stronger temperature dependence was observed (see Fig. 2).

ΔI can be interpreted with respect to disorder induced scattering from damped high-frequency sound waves. The above equation is based on the assumption of excitations narrow in the frequency domain. In glasses, however, the higher momentum sound waves are strongly damped; correspondingly, their spectral function can exhibit a low frequency tail. Because of the high density of states with large momentum these tails can give rise to an excess scattered intensity at small ω. We analyzed our data (for details, see

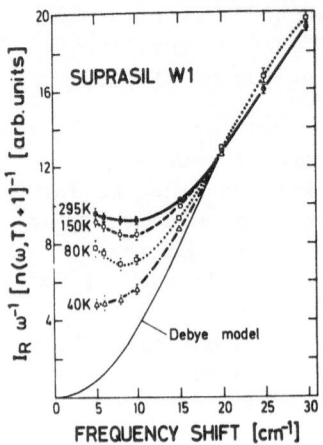

FIG. 2: Raman data (in-
strumental width 1.9cm^{-1})
at various temperatures
normalized at ω=-420cm^{-1}
according to one-phonon
processes. The theoreti-
cal expectation based on
a Debye model is included
for comparison.

Ref. 5) by describing the damping of
the phonons as in Ref. 1 with a structural relaxation which, at lower temperatures, gives rise to an additional but weak central component.
From the magnitude and the spectral
shape of ΔI values for the relaxation strength and mean relaxation
rate, respectively, have been derived
which are consistent, at least for
T>80K, with the values obtained from
ultrasonic experiments.

In a microscopic description,
the structural relaxation arises
from the mechanical coupling of the
sound waves to the structural defects
being present in the glass (2).
Correspondingly, the defects can be
seen indirectly in light scattering
via their mechanical coupling to the
sound waves.

In principle, however, it is possible
that the structural defects can also
directly couple to light if their
effective polarizability is changing
during a transition between two
states; it is conceivable that polarizability tensor of a defect changes
its spatial orientation during a
transition. This possibility is similar to the (depolarized) anisotropy scattering of a gas being composed of optically anisotrop
molecules. Using a gas model one can understand both the spectral
shape and the magnitude of ΔI (5). On the other hand, however, it
is difficult to account for the observed material dependence of
the depolarization ratio D.

REFERENCES:
(1) P. Fulde and H. Wagner, Phys. Rev. Lett. 27, 1280 (1971).
(2) P.W. Anderson, B.I. Halperin and C.M. Varma, Phil. Mag. 25, 1
    (1972).
(3) R. Shuker and R.W. Gammon, Phys. Rev. Lett. 25, 222 (1970).
(4) A. Martin and W. Brenig, Physica Status Solidi B64, 163 (1974).
(5) G. Winterling, Phys. Rev. B, August 1975.

BRILLOUIN SCATTERING STUDY OF THERMAL PHONONS IN CRISTAL QUARTZ,

FUSED QUARTZ AND BOROSILICATE GLASS AT LOW TEMPERATURE

R. Vacher and J. Pelous
Université des Sciences et Techniques du Languedoc
Laboratoire de Spectrometrie Rayleigh-Brillouin
Place E. Bataillon
34060 Montpellier-Cédex - France

The interest in the study of acoustic phonons in glass at very low temperatures originates from the numerous anomalies observed in the physical properties of amorphous media at these temperatures.

We present here the results obtained from spontaneous Brillouin scattering for fused quartz (Puropsil), borosilicate glass (composition : SiO_2,40 ; B_2O_3,50 ; K_2O,10 in mole percent), as well as in crystal quartz used as a test material. The frequencies of hypersonic waves responsible for the scattering are in the range from 30 to 40 GHz.

The experimental conditions have been described elsewhere. The wavelength of the incident light was $\lambda = 4880$ Å. A double-passed plane Fabry-Perot was used as a monochromator ; the resolving unit was a spherical Fabry-Perot of 50, 100, or 250 mm thickness. With this instrument, a resolving power of 10^8 was achieved, with a contrast of 10^7.

From the dynamical theory of crystal lattices and the quantum theory of Brillouin scattering, it is known that, in crystals, the Brillouin stokes (I_S) and antistokes (I_{AS}) are proportional to <n> + 1 and <n> respectively, where n is the Bose's occupation number. The intensity measured in function of temperature is in agreement with the above theoretical result for the three materials. These measurements extend the previous results given by Love[2]. At low temperatures, differences between stokes and antistokes intensities have been noted, which agree with the calculated value : $I_{AS}/I_S = \exp -(h\nu/kT)$. These results show that phonons in glasses are well-defined excitations.

A minimum in the velocity is observed for fused quartz near 70K. By comparing this result to ultrasonic measurements, the position of the minimum is found independent on the frequency : this observation eliminates the surprising shift to low temperatures

with increasing frequency, noted by Krause[3] from ultrasonic and
stimulated Brillouin scattering (SBS) measurements. The minimum in
the velocity is noted in borosilicate glass near 120K which is the
temperature of the absorption peak in this glass. The relative accuracy of our measurements (10^{-3}) was not high enough to allow
the observation of the velocity maximum observed by Piché et al[4]
at very low temperatures : the velocity seems to be approximately
constant below 5 K.

The measurements of the attenuation in the range 4.2 to 300 K
for the three materials are given in fig. 1. The curves for the
two glasses are similar in the large ; the peak for borosilicate is
broader than that for fused quartz. In the previous SBS measurements[5], the attenuation was found roughly proportional to T^4 below
10 K. In our measurements, the linewidth to be measured varies
from 1 to 10 MHz. The instrumental linewidth is about 10 MHz ; the
broadening due to the aperture of the beams is in the range from 2
to 5 MHz ; the effect of finite volume dimensions can cause a broadening of about 1 MHz. It is therefore difficult to study the temperature dependence with a sufficient accuracy. After eliminating
the above parasitic broadenings, and substracting a residual attenuation of 1 MHz attributed to geometrical scattering, the attenuation is found roughly proportional to T^2 between 4.2 and 15 K,
which is not in agreement with the SBS results , on the other hand,
by assuming that the attenuation is approximately independent of

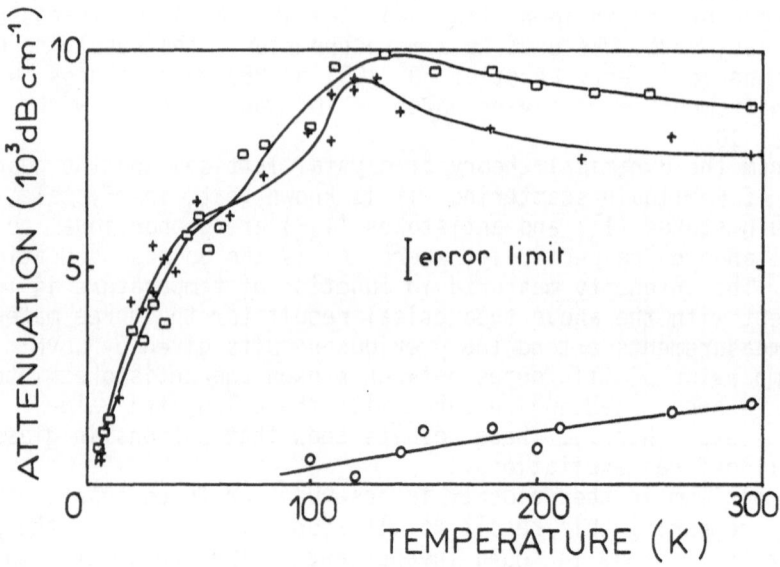

Fig 1:Attenuation vs temperature.

+ fused quartz,o cristal quartz,□ borosilicate glass .

frequency in this temperature range, our results agree with that
given by Arnold *et al*[6] for 1 GHz.

The "knee" noted near 30 K had to be compared to the slight
secondary maximum noted by Jones[7] near 5 K for frequencies of a-
bout 1 GHz. The dependence with temperature and frequency of this
"knee" agrees with the activation energy of 60 cal.mol^{-1} noted
previously[8].

In the resonant tunelling model theory given by Jackle[9], the
sound attenuation is found proportional to ω^2/T for the resonant
absorption process at very low temperatures ; for the relaxational
mechanism, occuring at higher temperatures, the attenuation is pro-
portional to T^3 when $\omega\tau_m \gg 1$ and to ω when $\omega\tau_m \ll 1$; τ_m is the
minimum value in the distribution of relaxation times of the two-
level systems. In our experiments, the relation $\omega\tau_m \gg 1$ is verified at low temperatures, while $\omega\tau_m \simeq 1$ at higher temperature. The
T^2 dependence indicated above and the "knee" observed can be interpreted tentatively as the transition between the two limiting
regimes ; such an interpretation has been proposed by Ng and
Sladek[10] for amorphous As_2S_3. Our results cannot be accounted for
through the theory of the scattering of elastic waves by density
fluctuations[11] ; the contribution of this process, obviously independent of temperature, is lower than 2 MHz for our frequencies.

1. R. Vacher, Thesis, Montpellier (unpublished)
2. W.F. Love, Phys. Rev. Lett., 31, 822 (1973).
3. J.T. Krause, Phys. Lett. 43A, 325 (1973).
4. L. Piché, R. Maynard, S. Hunklinger and J. Jäckle, Phys. Rev.
   Lett., 32, 1426 (1974).
5. W. Hecnicke, G. Winterling and K. Dransfeld, J. Acoust. Soc.
   Am., 49, 954 (1971).
6. W. Arnold, S. Hunklinger, S. Stein and K. Dransfeld, J. of Non
   Cryst. Solids, 14, 192 (1974).
7. C.K. Jones, P.G. Klemens and J.A. Rayne, Phys. Lett. 8, 31
   (1964).
8. J.T. Krause, J. Am. Ceram. Soc., 47, 103 (1964).
9. J. Jackle, Z. Phys., 257, 212 (1972).
10. D. Ng and R.J. Sladek, Phys. Rev., B.11, 4017 (1975).
11. D. Walton, Solid. Stat. Com., 14, 335 (1974).

DISCUSSION ON AMORPHOUS SYSTEMS

Phonon Scattering in Amorphous Systems   R O Pohl   Page 107

H Weinstock: What happens in the silica during irradiation with
fast neutrons?

R O Pohl: It is believed that they cause local melting and subsequent crystallization. This change in the local disorder results in
a decrease of the phonon scattering, as observed in Assfalg's measurements. The explanation is due to Klemens.

R Berman: What is the situation regarding the T^3 specific heat
which is several times greater than the Debye specific heat?

R O Pohl: This part of the excess specific heat is usually also
ascribed to the same kind of excitations which cause the linear anomaly. As you could see, this T^3 part decreases together with the
linear part; so far, however, we have not been able to reduce either
of them to zero.

D W Pohl: T^2 heat conductivity could be explained by a constant
density of states for the very low lying energy states. It is difficult for me to imagine that the random arrangement of molecules
producing those tunnel states should result in a constant density
of states.

R O Pohl: One assumes these states to arise from a broad distribution of internal stresses whose probability distribution is nearly
constant around zero internal stress. From this follows a constant
density of states rather easily, as explained in ref 9. The more
serious question is why the density of states is so similar in all
glasses. For this, no physical explanation has been put forth to
date. As I tried to show in this talk, however, part of the anomaly
is almost certainly not caused by the same class of defects which
provides the phonon scatterers.

K Dransfeld: Does not the decreased thermal conductivity in vycor
near the plateau provide strong evidence that the heat in glasses
is carried by plane wave phonons and not by some novel excitations
as sometimes speculated?

R O Pohl: If we believe that these novel excitations could not see
the voids, this conclusion would be correct. However, I believe
that the observation of Casimir boundary scattering provides more
direct evidence that the heat is carried predominantly by phonons.

****

Phonons in Amorphous Materials   R Maynard   Page 115

R O Pohl: I think one must emphasise that the value of the coupling
constant of 0.3 eV that you mentioned is determined using the density
of states obtained from the specific heats. This may not be reliable
and the value of 0.3 eV really should be considered as a lower limit.

R Maynard: I agree and as we know the value that comes rather directly from Arnold's two pulse experiment is $\sim$ 3 eV.

G Winterling: Did you take into account the dispersion of the velocity when describing the data on the thermal conductivity in the
neighbourhood of the 'plateau'?

R Maynard: No, the model only takes into account an average Debye
velocity;   the temperature dependence is only accounted for by the
temperature and frequency dependence of the mean free path.

****

Phonons in Some Amorphous Solids   M Goda and S Takeno   Page 126

J Jackle: What can you say about the lifetime broadening of the
roton excitations?

M Goda: The lifetime can in principle be calculated from the 3 body
correlation function but there is no data for this.

H Namaizawa: In your basic equation for the phonon frequency you
have two quantities: one is the pair correlation function $g_2(R)$ and
the other is the effective pair potential $V_{eff}(R)$. In the case of
He^4, for example, you included in V_{eff} the short-range correlation
or the hard-core effect. Since you simply inserted the experimental
value for $g_2(R)$, you also included the hard-core effect in $g_2(R)$.
Don't you commit a kind of double counting of the hard-core effect
in your approach? And can you correctly include zero-point fluctuations?

M Goda: Equation (1) represents the hard-core effect correctly and
it is appropriate to use the experimental values for $g_2(R)$ in this
equation. With regard to the zero-point fluctuations, there is some
inadequacy in using the experimental values for $g_2(R)$ in equation
(1).

Virtual Phonon Exchange in Glasses   J Joffrin and A Levelut
                                                      Page 132

H Kinder: Arnold et al observe that the width of the saturated
band is proportional to T. Can you explain this with your theory?

J Joffrin: The theory previously presented was the 'high temperature limit' version. It is easy to modify it in order to take
the temperature variation into account.

K Dransfeld: You are getting a critical frequency both theoretically and experimentally of about 100 MHz. Will the glass experience a phase transition at a temperature equivalent to this
frequency?

J Joffrin: The system is complicated but has some similarity to
the spin glass system and using the solution for this due to
Anderson and using the value of T_2 it would appear that there
should be an elastic transition at a few mk. This should be
observable in the elastic constants.

K Dransfeld: Would this also lead to an additional specific heat
above this transition - perhaps rather smeared out?

J Joffrin: There should be no anomaly if we argue again from the
spin glass problem where Anderson showed that there is no critical
specific heat but only a change in slope.

****

Influence of Phase Separation on the Thermal Conductivity of
Potassium Borosilicate Glasses Between 1K and 10K   F Canal,
M C Schmidt, J P Redoules and P Carrara   Page 135

D Greig: What is the explanation for the change in slope of the
thermal conductivity at 3K?

F Canal: We still don't have a satisfactory explanation for this.

****

Influence of OH Content on the Specific Heat and Thermal
Conductivity of Vitreous Silica Below 1 Kelvin   J C Lasjaunias,
A Ravax, D Thoulouze and M Vandorpe   Page 138

W A Phillips: Over how wide a temperature range did the logarithmic
dependence of n(E) extend?

J C Lasjaunias: From the lowest temperatures up to 0.3K. Above
this temperature there may be other excitations other than 2-level
systems.

****

Very Low Frequency Raman Scattering of Glasses   G Winterling
                                                     Page 144

W A Phillips: How do you know that the coupling constant goes as ω^2 at low frequencies and not for example as ω?

G Winterling: We expect this from a Debye- like model and a variation as ω would not account for my data. There is also independent evidence on the frequency dependence from Raman scattering in amorphous silicon.

****

Brillouin Scattering Study of Thermal Phonons in Crystal Quartz, Fused Quartz and Borosilicate Glass at Low Temperature
R Vacher and J Pelous   Page 147

G Winterling: Did you observe a decrease in the velocity below about 5K that would correspond to the decrease seen at low frequencies?

R Vacher: We hope to do this experiment very shortly, and I think the accuracy of our technique is sufficient to observe the size of change expected.

SPIN-PHONON INTERACTIONS

F W Sheard

Department of Physics, University of Nottingham

University Park, Nottingham NG7 2RD, U K

The scattering of light by an atomic system was considered in the early days of quantum theory and is described by the Kramers-Heisenberg scattering formula[1]. This may be derived by straight-forward perturbation theory[2] though certain refinements are necessary to include level shifts[3]. For a two-level atom the scattering is elastic. The atom, initially in the ground state absorbs a photon k, makes a virtual transition to an excited state, then emits a photon k' returning to its initial state. Energy conservation requires equality of the frequencies $\omega_k = \omega_k'$. The frequency dependence of the scattering cross-section σ_k divides into three regions. For low frequencies $\omega_k \ll \omega_0$, where $\hbar\omega_0$ is the separation of the atomic levels, we have Rayleigh scattering and $\sigma_k \sim \omega_k^4$. For high frequencies $\omega_k \gg \omega_0$, the electrons in the atom behave as if essentially free and we have frequency-independent Thomson scattering. For $\omega_k \sim \omega_0$ the cross-section exhibits a sharp peak which is the region of resonance fluorescence.

Phonon scattering by a localised impurity with discrete energy levels may be treated in the same way provided one has a simple model of the impurity-lattice coupling. This approach was first used by Griffin and Carruthers[4] for phonon scattering by donor levels in semiconductors and a similar treatment was suggested by Seiden[5] for magnetic ions. Despite the similarities between the phonon and light scattering problems there have been many subsequent treatments of spin-phonon scattering using both perturbation theory[6-9] and more sophisticated Green-function techniques[7,10-17]. These have arisen at least partly in order to account for the different features of spin-phonon scattering which are not all included in the Kramers-Heisenberg calculation of photon scattering by an isolated atom. Thus we may wish to take account of the following:

(a) Thermal populations - the level splittings $\hbar\omega_0 \sim$ 1-100 cm^{-1} are frequently $\sim k_B T$, where T is the temperature of the host crystal.
(b) Interactions between magnetic ions (spin-spin interactions) - lead to level broadening and ultimately to cooperative effects[18].
(c) Velocity dispersion - resonant scattering must lead to dispersion which influences ultrasonic propagation and thermal conduction.
(d) Finite ionic concentration - interference between the scattering by different spins leads to coherent mixing of the phonon and spin excitations described by coupled spin-phonon modes.
(e) Strong ion-lattice coupling - this modifies the bare single-ion states to give coupled ion-lattice or vibronic states (Jahn-Teller effect) from which phonons are scattered[19,20].

   Experimental studies are made principally by APR and thermal conductivity measurements. But APR has been restricted to relatively small splittings $\hbar\omega_0 \sim$ 1 cm^{-1} when the resonant lineshape is dominated by inhomogeneous strain broadening for strong spin-phonon coupling or spin-spin interactions for weak coupling. Thus the frequency dependence of the scattering has not been investigated by this means but high-frequency tunnel-junction phonon generators promise progress in this direction.

   In thermal conduction a much larger energy range $\hbar\omega_0 \sim$ 1-100 cm^{-1} may be studied but at the expense of using a broad-band thermal source. At first sight it might seem that such experiments would be insensitive to the details of the frequency dependence of the phonon scattering and that only the integrated lineshape would be important. But this is in general not the case since the thermal conductivity is a measure of the phonon relaxation time τ rather than the scattering rate τ^{-1}. Thus, when phonons are scattered by spins with rate τ_{sp}^{-1} and the background scattering rate due to other processes is τ_B^{-1} we have $\tau^{-1} = \tau_{sp}^{-1} + \tau_B^{-1}$ or $\tau = \tau_B/\{1 + (\tau_B/\tau_{sp})\}$. If we attempt to ignore the details of the resonant lineshape by simply regarding it as a delta function $\tau_{sp}^{-1} \sim \delta(\omega_q - \omega_0)$, then $\tau(\omega_q) = \tau_B$ everywhere except at the single point $\omega_q = \omega_0$. But this one point will not affect the value of the kinetic integral for the conductivity which is then completely unaffected by spin-phonon scattering. The point is that if over a certain frequency band the spin-phonon scattering is much greater than the background scattering, then this band of phonons is effectively removed from the conduction process. The band may be sufficiently wide that the frequency dependence of the resonant relaxation time may only be needed in the wings of the resonance line. Only if the peak scattering rate $\tau_{sp}^{-1}(\omega_0) \ll \tau_B^{-1}$ (which in practice means a very broad lineshape or very weak spin-phonon scattering) will the conductivity depend simply on the integrated lineshape.

PERTURBATION-THEORY CALCULATIONS

For a very dilute concentration of magnetic ions between which interactions are negligible, we may use the same approach as in the Kramers-Heisenberg theory of light scattering by an isolated atom. A simple model of a two-level system, which was first introduced by Jacobsen and Stevens[21] and has been much used since, is defined by the Hamiltonian (taking $\hbar = 1$)

$$\mathcal{H} = \sum_{qp} \omega_{qp} a_{qp} a_{qp}^{\dagger} + \omega_o S_{nz} + \mathcal{H}_{sp}$$

$$\mathcal{H}_{sp} = N^{-\frac{1}{2}} \sum_{qp} (A_{qp} S_{n+} + A_{qp}^* S_{n-})(a_{qp} + a_{-qp}^{\dagger}) \exp(iq.R_n)$$

The $S_{n\alpha}$ are spin-$\frac{1}{2}$ operators at site R_n and a_{qp}, $a_{qp}^{\dagger}$ are phonon operators for wave vector q and polarization index p; A_{qp} is the coupling parameter and N the number of atoms. We note that in first-order perturbation theory only energy conserving absorption and emission processes are allowed and the transition rate $\sim \delta(\omega_{qp} - \omega_o)$. But this lineshape is inadequate to account for the effect on thermal conductivity so we must proceed further. A common procedure is then to include lifetime broadening of the spin-levels by replacing the delta function by a Lorentzian lineshape but this is going beyond first-order theory.

We therefore continue to second order where a phonon q may be elastically scattered into modes q' via intermediate states in which the spin has flipped (fig 1(a)). For a spin initially in the lower state the second-order matrix element is of the form[8,16]

$$M_-(qp, q'p') \sim \frac{A_{qp} A_{q'p'}^*}{\omega_{qp} - \omega_o} - \frac{A_{-qp}^* A_{-q'p'}}{\omega_{qp} + \omega_o} \tag{1}$$

If initially in the upper state the matrix element is $M_+ = -M_-$. For a thermal distribution of initial states we must weight the corresponding transition probabilities with the population factors p_- and p_+, where $p_- - p_+ = \tanh\frac{1}{2}\beta\omega_o$ ($\beta = 1/k_B T$). The transition rate for scattering from q to q' is thus

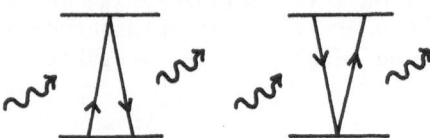

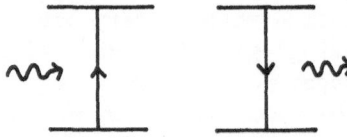

Fig 1(a) Second-order elastic   Fig 1(b) Direct absorption-emission
scattering   processes

$$W(qp,q'p') = 2\pi \ (p_-|M_-|^2 + p_+|M_+|^2) \ \delta(\omega_{qp} - \omega_{q'p'})$$

$$= 2\pi \ (p_- + p_+)|M_-|^2 \ \delta(\omega_{qp} - \omega_{q'p'})$$

and is temperature independent for a two-level system since $p_- + p_+$ = 1. Of course in a multilevel ion the sum of the populations of the two resonating levels is temperature dependent.

Calculation of the phonon relaxation time requires explicit knowledge of the form of the coupling. In the average coupling model[12] $A_{qp} = \varepsilon(\omega_o\omega_{qp}/12)^{\frac{1}{2}}$, where ε is dimensionless and the ω_{qp}^2 dependence follows since the coupling depends on the local lattice strain. For N_s independent spins giving a fractional concentration $c = N_s/N$ the phonon relaxation time for elastic scattering is given by[8,16]

$$\frac{1}{\tau_{qp}^{elast}} = \frac{\pi c \varepsilon^4 \omega_o^4 \omega_{qp}^4}{2\omega_D^3(\omega_{qp}^2 - \omega_o^2)^2} \tag{2}$$

assuming a Debye lattice spectrum. This simple formula shows all the features described previously for the scattering of light. Owing to neglect of lifetime effects, τ^{-1} diverges as $\omega_{qp} \to \omega_o$ but this is not necessarily serious in applications to thermal conduction since the phonons close to resonance contribute least to the heat current.

Although not apparently unduly restrictive the average coupling model has a particular consequence for the frequency dependence above resonance. If A_{qp} is not purely real then the terms in the matrix element (1) combine in different ways and give a high-frequency ω_{qp}^2 dependence. This occurs for the isotropic coupling model[22] in which the spin phonon interaction $\sim A_q(e_{qp} \cdot S_n)$, where e_{qp} is the phonon polarization vector, giving a complex $A_{qp} = \frac{1}{2}(e_{qp}^x \pm ie_{qp}^y)A_q$. The relaxation time is then of the form[8]

$$(\tau_{qp}^{elast})^{-1} \sim \omega_{qp}^4 \ (\omega_o^2 + \omega_{qp}^2)/(\omega_{qp}^2 - \omega_o^2)^2$$

Since the dependence of the coupling on phonon polarizations is usually complicated[23] the latter form may well be more generally valid than eq (2). A high-frequency ω_{qp}^2 law can also result from level degeneracy[24].

For non-dilute concentrations it is necessary to take account of spin-spin interactions. We shall not discuss cooperative phenomena but only the effects of level broadening. First-order absorpt ion-emission processes (fig 1(b)) can now affect the thermal conductivity. Calculating the phonon relaxation given by the difference in the rates for resonant absorption and emission of single phonons, the so-called direct process[6], gives

$$\frac{1}{\tau_{qp}^{direct}} = 2\pi N^{-1}|A_{qp}|^2(p_- - p_+)\frac{(\Gamma_{ss}/\pi)}{(\omega_{qp}-\omega_o)^2+\Gamma_{ss}^2} \tag{3}$$

We have taken a Lorentzian lineshape with spin-spin width Γ_{ss} though Gaussian lineshapes have also been used in data analysis.

It is interesting to observe that if we regard the level broad-ening as arising from spin-lattice relaxation then τ^{direct} becomes equal, close to resonance, to τ^{elast}[8]. In particular the replace-ment $\Gamma_{ss} \to \Gamma_{sL} = 1/2\tau_{sL}$ in eq (3), where τ_{sL} is the spin-lattice relaxation time, removes the temperature dependence (provided Γ_{sL}^2 is neglected in the denominator) since $\Gamma_{sL} \sim \coth\frac{1}{2}\beta\omega_o$ whilst $p_- - p_+$ $= \tanh\frac{1}{2}\beta\omega_o$. But this does not imply that the direct-process damping and the elastic scattering are equivalent mechanisms. In the elast-ic scattering the absorption and emission proceed via an intermedi-ate virtual state. The wavefunction for the system develops coher-ently with time and it is not possible to split up the process into independent absorption and emission events[3]. In the one-phonon dir-ect process this coherence is interrupted by the spin-spin interact-ions and the absorption and emission can be treated as real indep-endent events. Similar apparent equivalences have been noted in the literature on light scattering and the relation between resonance Raman scattering and hot luminescence has been the subject of some discussion[25,26].

PARAMAGNETIC COUPLED SPIN-PHONON MODES

So far we have neglected the spatial coherences which arise in the interaction of the phonons with different ions. These give rise to a mixing of the phonon and spin excitations which are then descr-ibed by a coupled-mode spectrum as shown in fig 2. This was first obtained by Jacobsen and Stevens[21] by linearising the semiclassical equations of motion. Taking account of nonlinearities leads to a damping of the excitations whose lifetimes were subsequently derived by a variety of Green-function methods[10-16]. The original motivation for using finite-temperature Green-function techniques was to inves-tigate the thermodynamic effects of spin-phonon coupling. Difficul-ties were experienced in obtaining consistent results[13,27] and cal-culations of the damping were also not all in agreement[16]. However for low concentrations the damping on the phonon-like parts of the coupled-mode spectrum must reduce to that given by the perturbation formula (3) for elastic scattering. This consistency requirement was neglected in most treatments and could not in any case be applied to those calculations[10,15] restricted to unit concentration c = 1. Here we give a brief summary of the coupled-mode theory emphasising the physical meaning of the results[16].

In the coupled system a propagating phonon q acquires a self-energy $P_q(\omega)$ and the dispersion relation becomes

$$\omega^2 - \omega_q^2 - 2\omega_q P_q(\omega) = 0$$

The second-order energy shift arising from admixture of the phonon with unperturbed spin states gives the coupled-mode dispersion curve shown in fig 2

$$(\omega^2 - \omega_q^2)(\omega^2 - \omega_o^2) - \frac{1}{3} c\varepsilon^2\omega_o^2\omega_q^2 \tanh\tfrac{1}{2}\beta\omega_o = 0$$

The excitations are undamped since the frequencies are real but the gap in the spectrum and velocity dispersion will still affect the thermal conductivity[12]. In higher order we must include fluctuat-ions in the admixed spin states as they are perturbed by interaction with other phonons q'. This gives rise to damping in fourth order and the physical mechanism is a scattering from q to q' as in pert-urbation theory. A general expression for the damping of the coupled excitations may be obtained but we are particularly concerned with the phonon-like parts of the dispersion curve which dominate the thermal conductivity. On these parts $\omega \simeq \omega_q + P_q(\omega_q) = \omega_q + \Delta_q - i\Gamma_q$, where Δ_q is the dispersive shift and $\Gamma_q = 1/2\tau_q$ is the phonon width. The relaxation time obtained by the coupled-mode analysis is found to be identical to that given by the perturbation result (2) for elastic scattering except that τ_q^{-1} contains an additional factor $(1 - c \tanh^2\tfrac{1}{2}\beta\omega_o)$. This arises from interference between the scatt-ered waves from different spins[8] and is important at high concentr-ations. In particular at T = 0 when all the spins are in the ground state $\tau_q^{-1} \sim c(1 - c)$, which vanishes for c = 1 since there is then only coherent scattering. The damping of the coupled modes has also been derived using Green-function equation of motion methods with rather different results for the frequency[12] or temperature[14] depen-dences though the results of Stevens and van Eekelen[13] agree subst-antially with those described above[16]. There have also been Green-function treatments of phonon scattering by a single two-level ion [7,9,17] which disagree with perturbation theory for frequencies off resonance.

Spin-spin interactions may be incorporated phenomenologically into the coupled-mode theory. Neglecting dispersion the resulting phonon damping close to resonance is then identical to that given in

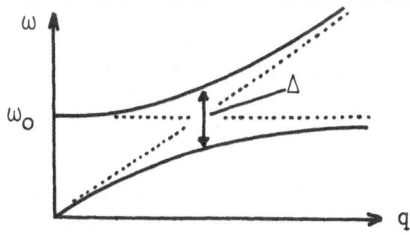

Fig 2 Dispersion curve for coupled spin-phonon modes.

eq (3) for the one-phonon direct process[11,16]. In the wings of the resonance line we have the form[16]

$$\frac{1}{\tau_q^{direct}} = \frac{\frac{2}{3} c\varepsilon^2 \omega_o^2 \omega_q^2}{(\omega_q^2 - \omega_o^2)^2} \Gamma_{ss} \tanh \tfrac{1}{2}\beta\omega_o$$

but this frequency dependence cannot be assigned more than qualitative validity. Eq (2) for the elastic scattering has a similar form with the spin-lattice width Γ_{sL} replacing Γ_{ss}[16]. Hence the criterion for coherent elastic scattering to dominate the direct absorption-emission processes is just $\Gamma_{sL} \gg \Gamma_{ss}$. This will tend to be favoured by low concentrations, when spin-spin interactions are weak, and large level splittings ω_o since spin-lattice relaxation is faster when there is a large available density of phonon states.

THERMAL CONDUCTION IN A COUPLED SPIN-PHONON SYSTEM[16]

Experiments on thermal conduction have been analyzed using the phonon relaxation times obtained from perturbation theory and also using the coupled-mode approach which takes account of dispersion. Since as ε decreases the spin-phonon damping $\Gamma_q \sim c\varepsilon^4$ decreases more rapidly than the gap $\Delta \sim c^{\frac{1}{2}}\varepsilon$ in the coupled-mode spectrum one would expect dispersion to play a more important role for weakly-coupled ions. But for strong spin-phonon scattering the band of phonons removed from the heat current can be sufficiently wide that those phonons effective in the conduction process are largely unaffected by the dispersive structure. This is the case for several iron-group ions (Cr^{2+}, Mn^{3+}, V^{4+}, Ti^{3+}, Ni^{3+} in MgO or Al_2O_3) for which the coupling is so strong ($\varepsilon \sim 10$) that very weak concentrations $c \sim 10^{-6}$ - 10^{-4} exert a dramatic effect on the conductivity[28,29]. Here the simple elastic-scattering resonance-fluorescence formula meets with considerable success in the interpretation of data. This is perhaps surprising since there is also the possibility of resonant inelastic Raman scattering[24] for multilevel ions. In contrast this formula cannot account for the temperature dependence of the thermal conductivity[28,30] of MgO:Fe^{2+} although the energy levels are thought to be well understood. It may be that there is interference between phonon-phonon and resonant scattering since the splitting $\omega_o \sim 105$ cm^{-1} is rather large. For small splittings the inhomogeneous distribution of width ~ 0.1 - 1.0 cm^{-1} due to elastic microstrains becomes important. Consequently the APR lineshapes are dominated by strain broadening and in thermal conductivity a low-frequency resonance $\omega_o \sim 0.4$ cm^{-1} in Al_2O_3:Ni^{3+} required analysis in terms of a strain broadened lineshape.[31]

In principle coupled-mode effects including dispersion are more important in thermal conduction for high concentrations of weakly-coupled ions. Theoretical treatments by Roundy and Mills[14] (large ε,

small c) and Elliott and Parkinson[12] (small $\varepsilon \sim 10^{-1}$, c = 1) have
confirmed this. But in concentrated systems spin-spin interactions
become important and experiments on rare-earth ethylsulphates[32] and
lantharum cobalt nitrate[33] have been interpreted phenomenologically
in terms of the direct-process relaxation time with a Gaussian lineshape. The spin-phonon coupling also results in a temperature dependence of the low-frequency sound velocity. This could influence
the conductivity in the boundary-scattering region but the effect
has not been unambiguously identified[9,34].

Direct experimental observation of coupled spin—phonon modes
would therefore clearly be of great interest. For this it is necessary that the coupled modes be well defined excitations. The required criterion is that the separation Δ of the two branches shown in
fig 2 be greater than the sum of the level widths[16] $\Delta > \Gamma_{sL} + \Gamma_{ss} +$
Γ_B. For sufficiently low concentrations Γ_{ss} is small and at low
temperatures the phonon damping Γ_B by processes other than spin-
phonon interaction is reduced. The criterion $\Delta > \Gamma_{sL}$ is fulfilled
by $c > 10^{-8}$ for a strongly coupled ion. The remaining requirement
is that the gap Δ be greater than the experimental resolution. Using
the expression

$$\Delta = (c\varepsilon^2 \omega_o^2 \tanh\tfrac{1}{2}\beta\omega_o)^{\frac{1}{2}}$$

we see that $\Delta \sim 3$ cm^{-1} for $c \sim 10^{-3}$, $\varepsilon \sim 10$ and $\omega_o \sim 10$ cm^{-1}. Thus
for 1000 ppm of Cr^{2+} in MgO the gap is sufficiently wide that obser-
vation by neutron scattering appears to be quite feasible[35]. Recently such a temperature dependent splitting of the phonon dispersion
curve has been observed in KCl due to resonant interaction with the
tunnelling states of CN^- molecular impurities[36]. Such experiments
with paramagnetic centres would be a useful new means of further
probing the spin-phonon interaction.

1. Kramers H A and Heisenberg W, 1925, Zeit fur Phys 31, 681.
2. Louisell W H, 1973, Quantum Statistical Properties of Radiation
   (Wiley: New York), Section 5.4.
3. Heitler W, 1954, Quantum Theory of Radiation (Clarendon Press:
   Oxford), Section 20.
4. Griffin A and Carruthers P, 1963, Phys Rev 131, 1976.
5. Seiden J, 1963, Comptes Rendus Acad Sciences 256, 393.
6. Orbach R, 1962, Phys Rev Lett 8, 393.
7. Klein M V, 1969, Phys Rev 136, 839.
8. Sheard F W and Toombs G A, 1973, Solid St Commun 12, 713.
9. Walton D, 1973, Phys Rev B7, 3925.
10. Iolin E M, 1965, Proc Phys Soc 85, 759.
11. Mills D L, 1965, Phys Rev 139, A1640.
12. Elliott R J and Parkinson J B, 1967, Proc Phys Soc 92, 1024.
13. Stevens K W H and Van Eekelen H A M, 1967, Proc Phys Soc 92, 630.
14. Roundy V and Mills D L, 1970, Phys Rev B1, 3703.
15. Fidler F B and Tucker J W, 1970, Solid St Commun 8, 2055.

16. Toombs G A and Sheard F W, 1973, J Phys C 6, 1467.
17. Joshi S K, 1975, Solid St Commun 16, 83.
18. Gehring G A and Gehring K A, 1975, Rep Prog Phys 38, 1.
19. Fletcher J R, Bluman J and Sheard F W, 1972, Int Conf Phonon
    Scattering (CEN Saclay: Paris), 228.
20. Halperin B and Englman R, preprint and Proceedings of this
    Conference.
21. Jacobsen E H and Stevens K W H, 1963, Phys Rev 131, 1976.
22. Huber D L and Van Vleck J H, 1966, Rev Mod Phys 38, 187.
23. Tucker J W, 1972, J Phys C 5, 2064.
24. Dreyfus B, 1972, Int Conf Phonon Scattering (CEN Saclay: Paris)
    207.
25. Klein M V, 1973, Phys Rev B8, 919.
26. Shen Y R, 1974, Phys Rev B9, 622.
27. Sheard F W and Toombs G A, 1971, J Phys C 4, 313.
28. Challis L J, de Goer A M, Guckelsberger K and Slack G A, 1972,
    Proc Roy Soc A330, 29.
29. de Goer A M and Devismes N, 1972, J Phys Chem Solids 33, 1785.
30. Morton I P and Lewis M F, 1971, Phys Rev B3, 552.
31. Locatelli M and de Goer A M, 1974, Solid St Commun 14, 111.
32. McClintock P V E and Rosenberg H M, 1968, Proc Roy Soc A302, 419.
33. Brock J C F and Huntley D J, 1968, Canad J Phys 46, 2231.
34. Walton D, 1970, Phys Rev B1, 1234.
35. Swain S, 1967, Ph.D. thesis, University of Nottingham.
36. Walton D, Mook H A and Nicklow R M, 1974, Phys Rev Lett 33, 412.

PHONON SCATTERING INDUCED BY JAHN-TELLER IMPURITIES

B. Halperin and R. Englman

Soreq Nuclear Research Centre

Yavne, Israel

Jahn-Teller (JT) impurity ions strongly coupled to the vibrational modes of the host crystal give rise to low-energy excitations (inversely proportional to the JT stabilization energy, E_{JT}) which are associated with the rotational motion of the lattice distortion around the ion. These excitations serve as resonant scatterers for phonons of appropriate energies as is indeed indicated by thermal conductivity or acoustic attenuation experiments. In this paper we present a theoretical study of the scattering induced by doubly degenerate JT ions at low concentration subject to strong linear coupling with a continuum of lattice modes. The Hamiltonian is /1/

$$H = \sum_\gamma \sum_i \hbar\omega_i (\tfrac{1}{2} p_{i\gamma}^2 + \tfrac{1}{2} q_{i\gamma}^2 + k_i q_{i\gamma}\sigma_\gamma), \quad (\gamma = \theta,\varepsilon). \tag{1}$$

Here k_i are (dimensionless) coupling strengths and $\sum_i k_i^2 \gg 1$. The inverse lifetime of an incident symmetry phonon is derived in the single-impurity approximation using the T-matrix method due to Klein /2/. To lowest order in the fractional concentration x of the JT ions in a crystal of N unit cells, it is

$$\tau_{i\gamma}^{-1} = -\omega_i^{-1} \, xN \lim_{\varepsilon \to 0+} \mathrm{Im}\, T(\omega_i+i\varepsilon)_{i\gamma,i\gamma}. \tag{2}$$

$T(\omega)_{i\gamma,i\gamma}$ is found from a relation between the Zubarev-Green's functions of the phonon operators for the JT coupled and uncoupled lattice /2/, which gives

$$\mathrm{Im}\, T(\omega+i\varepsilon)_{i\gamma,i\gamma} = 2\pi\omega_i^3 k_i^2 \, \mathrm{Im} \ll \sigma_\gamma ; \sigma_\gamma \gg (\omega). \tag{3}$$

The problem is thus reduced to that of calculating the Zubarev-Green's function of the electronic operators. For weak coupling ($\sum_i k_i^2 \ll 1$) these can be evaluated by the equation-of-motion method applying an RPA decoupling to terms second order in k_i.

In order to evaluate $<<\sigma_\gamma;\sigma_\gamma>>$ in the strong coupling limit, we shall first transform the Hamiltonian in such a way that only one set of modes is now strongly interacting with the electronic doublet (the quasi-molecular (Q.M.) JT system), whereas the rest of the transformed modes are left weakly coupled (the residual interaction). The chain of equations of motion is then decoupled by performing the RPA on terms of second order in the residual interaction. Thus, applying an orthogonal transformation /3/ $\tilde{q}_{j\gamma} = \Sigma_i A_{ji} q_{i\gamma}$, the Hamiltonian (1) can be separated into the following three parts

$$H = \tilde{H}_{JT} + \tilde{H}_{qu.ph.} + H' \tag{4}$$

where $\tilde{H}_{JT}$ describes a Q.M. dynamic JT coupling involving only one pair of degenerate modes $\tilde{q}_{1\theta}$, $\tilde{q}_{1\epsilon}$ say, with an effective frequency $\Omega = \Sigma_i \omega_i A_{1i}^2$ and coupling strength $K = \Omega^{-1} \Sigma_i \omega_i k_i A_{1i}$. $\tilde{H}_{qu.ph.}$ describes a continuum of transformed quasi-phonons $\tilde{q}_{j\gamma}(j\neq1)$ with frequencies $\Omega_j = \Sigma_i \omega_i A_{ji}^2$. The matrix elements A_{1i} and hence Ω and K are so chosen that the effect of the residual interaction H' on the solutions of $H-H'$ shall be as small as possible /1/. Here we will use the simple choice /3/ $A_{1i} = k_i/K$, which maximizes the JT stabilization energy $E_{JT} = \frac{1}{2} \hbar\Omega K^2$. This leads to

$$H' = \Sigma'_{j} \Sigma_{\gamma} \hbar c_j [\tilde{q}_{j\gamma}(K\sigma_\gamma + \tilde{q}_{1\gamma}) + \tilde{P}_{j\gamma} \tilde{P}_{1\gamma}], \quad (\Sigma'_{j} \text{ means } j\neq1) \tag{5}$$

with c_j obeying the sum rule $\Sigma'_j c_j^2 = <\omega^2> - <\omega>^2 \equiv <\omega>^2 \bar{\delta}$, where $<\omega^n> \equiv \Sigma_i \omega_i^n A_{1i}^2$. In what follows we shall assume that $\bar{\delta} << 1$ and take this to be the small parameter of our theory. Taking the electron-phonon coupling to be proportional to the local strain so that $k_i \propto \omega_i^{-\frac{1}{2}}$, we find for an isotropic Debye model $\bar{\delta} = 1/8$. Under the same conditions and for $\Omega_j << \omega_D$ we find /4/ $c_j^2 \simeq \omega_D^3/(6N\Omega_j)$. We also note that the density of modes $\rho(\omega)$ is practically unchanged under the transformation.

Having expressed the quasi-molecular operators in the basis of the eigenstates $|n>$ of $\tilde{H}_{JT}$, e.g. $\sigma = \Sigma_{nn'} \sigma_{n'n} b_{n'n}$ with $b_{n'n} \equiv |n'><n|$, we write down the chain of equations of motion for $<<b_{\ell'\ell};b_{n'n}>>$ and note that contributions of higher order Green's functions as $<<b_{\ell'\ell};\tilde{q}_{j\gamma}\tilde{q}_{j'\gamma'}b_{m'm}>>$ are of order $\bar{\delta}$. These are then decoupled into $<<b_{\ell'\ell};b_{m'm}>> <\tilde{q}_{j\gamma}^2>_0 \delta_{jj'} \delta_{\gamma\gamma'}$, leading to an infinite system of inhomogeneous, linear equations for the Green's functions of the Q.M. operators alone,

$$\Sigma_{m'm} [(\omega-\omega_{n'n})\delta_{m'm,n'n} - R_{m'm,n'n}(\omega)]<<b_{\ell'\ell};b_{m'm}>> = B_{\ell'\ell,n'n} \tag{6}$$

where $B_{\ell'\ell,n'n} \sim <[b_{\ell'\ell},b_{n'n}]>_0/(2\pi)+ O(\bar{\delta})$, and R is of order $\bar{\delta}$. Now, $<<b_{\ell'\ell};b_{n'n}>>$ has poles at ω near $\omega_{n'n}$ as well as near all other $\omega_{m'm}$ but with residues which are a factor of $O(\bar{\delta})$ smaller than the former. Being interested in $<<b_{\ell'\ell};b_{n'n}>>$ near $\omega_{n'n}$ we truncate the set of equations by means of a diagonal approximation, retaining in (6) $R_{n'n,n'n}$ alone, which now plays the role of a level shift operator. The expressions thus obtained for $<<\sigma_\gamma;\sigma_\gamma>>$ and hence for

τ^{-1}, Eq.(2), are valid at the vicinity of each $\omega_{m'm}$, and the level shifts (self energies of the Q.M. excitations) are correct to $O(\bar{\delta})$.

$$\tau_{i\gamma}^{-1}(\omega_i) = -(\omega_i k_i)^2 \sum_{nn'} (f_n - f_{n'}) |(\sigma_\gamma)_{nn'}|^2$$

$$\times \lim_{\varepsilon \to 0+} \text{Im } R_{n'n}(\omega_i + i\varepsilon)/[(\omega_i - \omega_{n'n} - \text{Re}R_{n'n})^2 + (\text{Im}R_{n'n})^2] \qquad (7)$$

where f_n is the population of the nth Q.M. JT level.
$\tau^{-1}(\omega)$ exhibits a sequence of resonances of Lorentzian shape centred about the shifted Q.M. JT excitations. Being interested in the attenuation rate of low-frequency phonons we shall consider the resonances near the rotational JT excitations $\omega_{M'M} = \frac{1}{2} \Omega(M'^2 - M^2)/K^2$ ($M = \pm 1/2, \pm 3/2 \ldots$). The selection rules for σ_γ restrict these to $M'=M\pm 1$, giving rise to elastic scattering only. Their widths, $\text{Im}R_{M'M}(\omega)$, give the broadening of the excitation $M \leftrightarrow M+1$ due to direct transitions from either of these levels into other Q.M. JT states, n'', with a simultaneous emission or absorption of a quasi phonon of an appropriate frequency. n'' may be a rotational state on the same (ground) radial level or on the first excited one. The former mechanism dominates at temperatures up to $kT \sim \hbar\Omega/K^2$, whereas the latter one dominates at $kT \sim \hbar\Omega$, where the higher density of modes may be utilized by M,M' to relax into the first excited radial level. Our truncation method excludes higher than one phonon processes in the broadening mechanism. At zero temperature, for a Debye model and $k_i \propto \omega_i^{-\frac{1}{2}}$, we find

$$\text{Im } R_{3/2,1/2}(\omega) = \frac{1}{4}K^{-2}\pi c^2(\omega)\rho(\omega) = \omega\pi/(8K^2) \qquad (8)$$

so that the Lorentzian shape, Eq.(7), contains an asymmetry factor proportional to the square of the phonon frequency. The narrowness of the resonance width, which is $\frac{1}{2}\pi\Omega^3(2E_{JT})^{-2}$, is due to the low density of modes on speaking terms with the low energy rotational excitation.

References
/1/ R. Englman and B. Halperin, J. Phys. C, 6, L219 (1973).
/2/ M.V. Klein, Phys. Rev. 186, 839 (1969).
/3/ M.C.M. O'Brien, J. Phys. C, 5, 2045 (1972).
/4/ B. Halperin and R. Englman, Phys. Rev. B, to be published.

RESONANCE SCATTERING OF PHONONS AT TRIGONAL

JAHN-TELLER CENTRES

M.Rueff, E.Sigmund and M.Wagner

Institute of Theoretical Physics, University of Stuttgart

7ooo Stuttgart 8o, Pfaffenwaldring 57, Germany

Since the experimental techniques for creating phonon pulses and also nearly monochromatic phonon waves are developed so far, the scattering of phonons due to crystal defects gained much interest. In the theoretical formulation of these problems many difficulties arise when the defect is of complicated nature, e.g. when the coupled electronic states are degenerate. Such a situation occurs when we study the phonon scattering by a Jahn-Teller defect.

In the present investigation we consider a trigonal Jahn-Teller centre of E-e nature. The Hamiltonian is given by ($\hbar = 1$)

$$H = \Omega \left(a_1^+ a_1 + a_2^+ a_2 \right) + \sum_k \omega_k \left(b_{1k}^+ b_{1k} + b_{2k}^+ b_{2k} \right)$$

$$+ \sum_k K(k) \left\{ \left(a_1^+ a_1 - a_2^+ a_2 \right) \left(b_{1k}^+ + b_{1k} \right) + \left(a_1^+ a_2 + a_2^+ a_1 \right) \left(b_{2k}^+ + b_{2k} \right) \right\}.$$

The b_{ik}^+, b_{ik} (i= 1,2, index of degeneracy) are the creation and annihilation operators of the degenerate phonon modes, whereas the a_i^+, a_i are the corresponding electronic operators. The dynamic of the local electronic system is governed by a quasi-spin algebra. The spin operators are defined by:

$$\sigma_z = \sigma_1 = \frac{1}{2} \left(a_1^+ a_1 - a_2^+ a_2 \right)$$

$$\sigma_x = \sigma_2 = \frac{1}{2} \left(a_1^+ a_2 + a_2^+ a_1 \right)$$

$$\sigma_y = \sigma_3 = \frac{1}{2i} \left(a_1^+ a_2 - a_2^+ a_1 \right).$$

To derive an expression for the inverse life-time of phonons of the sort (i,k), we use a Green's function formalism first developed by Klein for a two-level system.[1] The electronic part of Klein's system is described by the same quasi-spin algebra as in our model. In this situation the phonon-phonon Green's function can be traced back to Green's functions connecting the introduced quasi-spin operators. The relaxation time get the form

$$\tau_{ik}^{-1}(\omega + i\varepsilon) = \omega_k^{-2} \lim_{\varepsilon \to 0^+} \mathcal{I}m\ g_i(\omega + i\varepsilon),$$

where the Green's function $g_i(\omega)$ reads

$$g_i(\omega) = \ \ll \sigma_i\ ;\ \sigma_i \gg (\omega) \qquad (i = 1,2).$$

For the calculation of the "spin-spin" Green's functions we employ an exponential transformation of the type

$$\tilde{H} = e^{-S} H e^{S} = H + [H,S] + \frac{1}{2!}[[H,S],S] + \dots ,$$

which diagonalizes the Hamiltonian mainly. In the E-e model S is found to be[2,3]

$$S = \sum_k \frac{K(k)}{\omega_k} \left\{ \sigma_z (b_{1k} - b_{1k}^+) + \sigma_x (b_{2k} - b_{2k}^+) \right\}.$$

It could be shown that this transformation yields to exact results in the very extremal coupling regions. But also in the intermediate coupling cases the transformation leads to good results as proven by variational calculations of the eigenvalues using a great number of basis functions.

The " spin-spin" Green's function can be more easily calculated in the new coordinations. Therefore we must transform all operators, e.g.

$$\sigma_i \longrightarrow \tilde{\sigma}_i = e^{-S} \sigma_i\ e^{S}.$$

By a fourth order RPA-decoupling the interesting Green's functions are given by[4]

$$\ll \tilde{\sigma}_i\ \tilde{\sigma}_i \gg (\omega) = \left\{ -2\omega^2 \sum_k \frac{K(k)}{\omega^2 - \omega_k^2} \langle \tilde{\sigma}_i\ (b_{1k} + b_{1k}^+) \rangle_0 \right\} *$$

$$* \left\{ \omega^2 - 4 \sum_k \frac{K(k)^2 \omega_k}{\omega^2 - \omega_k^2} \langle \tilde{\sigma}_i \sum_k K(k)(b_{1k} + b_{1k}^+) \rangle_0 + \right.$$

$$\left. 4\omega \sum_k \frac{K(k)^2 \omega_k}{\omega^2 - \omega_k^2} - 4\omega^2 \sum_k \frac{K(k)}{\omega^2 - \omega_k^2} \langle (b_{1k} + b_{1k}^+) \sum_{k'} K(k')(b_{1k'} + b_{1k'}^+) \rangle \right\}^{-1}$$

$\langle \dots \rangle_0$ denotes the thermal average.

After some lengthy calculations we end up with an expression for the relaxation time

$$\tau_{ik}^{-1}(\omega) = -\frac{1}{\omega^2}\lim_{\varepsilon\to 0^+}\frac{P(\omega+i\varepsilon)}{\{[(\omega+i\varepsilon)^2 - Q^2(\omega+i\varepsilon)]^2 + (\Gamma(\omega+i\varepsilon))^2\}}$$

where

$$P = -\frac{2}{\pi}\omega^4 D_1\cdot Im\,g_1 + \frac{8}{\pi}(\Sigma\frac{K(k)^2}{\omega_k})D_2^2\cdot Im\,g_1\cdot Re\,g_3 - \frac{8}{\pi}\omega^3 D_1\cdot Im\,g_1\cdot Re\,g_3$$

$$+ \frac{8}{\pi}\omega^4 D_1\cdot Im\,g_1\cdot Re\,g_2 - \frac{8}{\pi}\omega^2 D_2^2\cdot Re\,g_1\cdot Im\,g_3 + \frac{8}{\pi}\omega^3 D_2\cdot Re\,g_1\cdot Im\,g_3 - \frac{8}{\pi}\omega^2 D_2\cdot Im\,g_2\cdot Re\,g_1$$

$$Q^2 = 4(\Sigma\frac{K(k)^2}{\omega_k})\cdot D_2\cdot Re\,g_3 - 4\omega\cdot Re\,g_3 + 4\omega^2 Re\,g_2 - 16\omega^2(\Sigma\frac{K(k)^2}{\omega_k})\cdot D_2\cdot Re\,g_4$$

$$\Gamma = -4(\Sigma\frac{K(k)^2}{\omega_k})D_2\cdot Im\,g_3 + 4\omega\,Im\,g_3 - 4\omega^2 Im\,g_2 + 16\omega^2(\Sigma\frac{K(\omega)}{\omega_k})^2 D_2\cdot Im\,g_1.$$

Γ is a measure for the width (not in Lorentzian approximation). The resonance frequency is determined by $\omega^2 - Q(\omega)^2 = 0$.
The following abbreviations are used, which determine the temperature dependence of the resonance frequency and the width of the resonance:

$$D_1 = <H_{el}^{(1)}>_0 \{-2 - 4\cdot\lim_{\varepsilon\to 0}\Gamma(\frac{\varepsilon}{2})[1 - {}_1\bar{F}_1(\frac{\varepsilon}{2};\frac{3}{2};-\frac{z}{2})]\}$$

$$D_2 = <H_{el}^{(1)}>_0 \{1 - \frac{4}{2}\int_0^{\sqrt{z}} x\cdot\lim_{\varepsilon\to 0}\Gamma(\frac{\varepsilon}{2})[1 - {}_1\bar{F}_1(\frac{\varepsilon}{2};\frac{3}{2};-\frac{x^2}{2})]dx\}$$

with $z = \Sigma\lambda_k^2\coth\omega_k/k_B T$.

$$g_1 = \Sigma\frac{\lambda_k^2\cdot\omega_k}{\omega^2 - \omega_k^2} \qquad g_2 = \Sigma\frac{\lambda_k^2\omega_k^2\coth\omega_k/k_B T}{\omega^2 - \omega_k^2} \qquad g_3 = \Sigma\frac{\lambda_k^2\omega_k^2}{\omega^2 - \omega_k^2}.$$

The results will be more simplified, if we assume the phonon dispersion relations to be of Debye form, $\omega = c|k|$, which is valid for long wave lengths. In the limits $T\to 0°K$ and $\lambda_1^2\Sigma\lambda^2\ll 1$ the half width is

$$\Gamma \sim \lambda^4\omega^2 .$$

The study of other coupling regions (especially the strong coupling limit) and the study of the temperature dependence are still going on.

1) M.V. Klein, Phys. Rev. 186, 839 (1969)
2) M.Wagner, Z. Physik 256, 291 (1972)
3) E.Sigmund and M.Wagner, Z.Physik 268, 245 (1974)
4) D.N. Zubarev, Soviet Phys.Usp. 3, 320 (1960)

MULTIMODE JAHN-TELLER EFFECTS OF IONS IN Al_2O_3

M Abou-Ghantous[∅*], C A Bates, J R Fletcher and P C Jaussaud[∅]

Department of Physics, University of Nottingham
University Park, Nottingham NG7 2RD, U K

Acoustic waves, whether they are coherent (as in APR), or of a black-body type (as in magnetothermal conductivity experiments) often suffer resonant scattering by magnetic impurity ions present in the crystal under investigation. The magnitude of the scattering is measured by the ion-lattice or spin-phonon coupling constant. It is inevitable that, in such cases, the energy levels of the magnetic ion causing the scattering are often difficult to obtain as dynamic Jahn-Teller effects (J-T-E) must be included. In addition, random internal fields and strains present in the crystal can have a considerable influence on the shape, width and peak of the observed resonance.

Most treatments of the J-T-E use a cluster model in which the magnetic ion and its immediate neighbours are considered as an isolated cluster. Even then the problem is far from simple and the extensions necessary to take account of the whole lattice are even more difficult. The resonant scattering of acoustic waves is, however, limited to transitions within the ground states of the magnetic impurity and so the method of analysis conveniently divides into two. Ions having T_1- or T_2- type exhibit ground states in an octahedral environment are treated in a different way from ions having E - type ground states.

[∅]Service BT, C E N G, BP 85, Centre de Tri. 38041 Grenoble, France.

[*]Now at Physics Department, Faculté des Sciences, Université Libanaise, Al-Hadeth, Beirut, Lebanon.

T_1- OR T_2- IONS

The usual cluster model approximation is to assume that the linear coupling between the ion and its neighbours is dominated by the E - type vibrations. The magnetic ion is then described by an effective Hamiltonian $\tilde{\mathcal{H}}$ which is obtained from the Hamiltonian $\mathcal{H}$ of the ion without J-T-E's by incorporating a reduction factor γ multiplying all T_1- and T_2- type orbital operators in $\mathcal{H}$ (1). Also present are second order contributions to $\tilde{\mathcal{H}}$ from those perturbation terms in $\mathcal{H}$ which are off-diagonal in the oscillator states. Such terms in $\tilde{\mathcal{H}}$ involve the factors f_a and f_b. It is found that:

$$\gamma = \exp(-\tfrac{1}{2}x), \quad f_a = G(\tfrac{1}{2}x) \exp(-x) \quad \text{and} \quad f_b = G(x) \exp(-x)$$

where $x = 3E_{JT}/\hbar\omega$ and $G(x) = \int_o^x [(e^u - 1)/u] du$ with E_{JT} and ω constants. Thus γ, f_a and f_b are each known if x is known. Contributions to $\tilde{\mathcal{H}}$ involving f_a and f_b can often counterbalance reductions in a given operator (such as the trigonal field) caused by γ.

In the multimode lattice model, $\tilde{\mathcal{H}}$ is most conveniently obtained from $\mathcal{H}$ by transformation methods (2). The details of the results obtained depend upon the phonon model but γ, f_a and f_b are no longer constrained by a single variable (3). In fact, the lattice sums are so complex that for calculation purposes γ, f_a and f_b are treated as independent variables.

For distorted environments further complications arise. The necessary modifications for trigonal environments have been calculated (4) and the result applied to several ions in Al_2O_3. In all cases the coupling appears to be strong and the need for a cluster/multimode model is indicated by ✓ below:

Ion Configuration	Examples	Cluster Model	Multimode Model	Reference
$3d^1 . {}^2T_2$	$T_1{}^{3+}$, V^{4+}, Cr^{5+}	✓	✓	(4), (5)
$3d^2 . {}^3T_1$	V^{3+}, Cr^{4+}	X	✓	(6)
$3d^6 . {}^5T_2$	Fe^{2+}	X	✓	(7)
$3d^7 . {}^4T_1$	−	−	−	−

For $3d^1$ ions, the data available is insufficient to distinguish between the two models. Also the coupling to T_1- or T_2- type lattice modes seems to be unimportant.

E - IONS

While the J-T-E for T_1- or T_2- type ions coupled to E - type modes can be solved exactly for a harmonic lattice, no exact solutions are possible for E - type ions. However, much progress has been made using a cluster model by again introducing reduction factors p (multiplying A_2 - type orbital operators), q (multiplying E - type orbital operators) and r. It is found that with a linear J-T coupling, $2q - p = 1$ and $r = \sqrt{2}q$ when the coupling is moderately strong. These constraints are broken if the multimode lattice model is used (8).

Ni^{3+} : Al_2O_3 is an ideal example on which the theory can be tested as analytic solutions are possible. Ni^{3+} is a strong field ion having a ^2E ground state. As it is in a trigonal environment, the reduction factor p is involved in $\mathcal{H}$ as well as q. The APR and EPR spectra have been thoroughly analysed and the proposed $\mathcal{H}$ adequately accounts for all the available data. From perturbation calculations of the electronic parameters appearing in $\mathcal{H}$, p and q have been obtained (9). It is found that for all reasonable values of the trigonal field parameters, $(2q - p)$ is much less than unity confirming the need for a multimode model of the J-T-E.

Other examples studied so far are the $3d^4$ ions Cr^{2+} and Mn^{3+} in Al_2O_3. The values of p, q and r have been deduced numerically as the best fit to APR and infra-red data. Although not complete, our analysis gives $2q - p \neq 1$ and $r \neq \sqrt{2}q$ again confirming the inappropriateness of the cluster model.

REFERENCES

(1) Ham F S, *Electron Paramagnetic Resonance*, ed S Geshwind,
        (New York: Plenum) (1972).
(2) Stevens K W H, *J Phys C: Solid St Phys*, <u>2</u>, 1934-46, (1969).
(3) Abou-Ghantous M, Bates C A, Chandler P E and Stevens K W H,
        J Phys C: Solid St Phys, <u>7</u>, 309-24, (1974).
(4) Abou-Ghantous M, Bates C A and Stevens K W H,   *J Phys C:
        Solid St Phys*, <u>7</u>, 325-38, (1974).
(5) Abou-Ghantous M, (1975) in preparation.
(6) Abou-Ghantous M, Bates C A and Goodfellow L C, (1975) in
        preparation.
(7) Bates C A and Steggles P, *J Phys C: Solid St Phys*, <u>8</u>, 2283-99,
        (1975).
(8) Halperin B and Englman R, *Phys Rev Lett*, <u>31</u>, 1052-5.
(9) Abou-Ghantous M, Jaussaud P C, Bates C A, Fletcher
        J R and Moore W S, *Phys Rev Lett*, <u>33</u>, 530-3, (1974).

DEFECT SCATTERING OF NEARLY MONOCHROMATIC ACOUSTIC PHONONS

W.E. Bron
Department of Physics, Indiana University
Bloomington, Indiana 47401 U.S.A.
F. Keilmann
Max-Plank-Institut für Festkörperforschung
Stuttgart, Germany

The frequency dependence of the scattering probability of nearly monochromatic phonon distributions off monoatomic defects has been determined experimentally in the 200 to 700 GHz range and compared with theory.

Phonon pulses of 30 to 50 nanosecond duration are generated at one end of a crystal. Narrow-band frequency distributions are achieved using the superconducting fluorescer generators[1] indicated in the insert of Fig. 1. By using successively Sn, Pb(0.5)Tℓ(0.5), and Pb superconducting films, phonon pulses are generated with frequency maxima at 287, 407 and 670 GHz. These phonons propagate through a single crystal of SrF_2 containing 0.01 mole % Eu^{2+}. The rare-earth ions act primarily as mass defects (with very small force constant changes) as can be deduced by a comparison of the crystal and dynamical parameters of the SrF_2 and EuF_2 lattices.[2,3] Time of flight measurements of the arrival of phonons at the opposite end of the crystal are made using superconducting Aℓ bolometers. All experiments are carried out with samples held near $1^{\circ}K$.

Experimental results at the three frequencies are shown in Fig. 1. Contributions to the bolometer response from unscattered phonons appear in the two ballistic peaks. All those phonons which undergo one or more scattering events contribute to the subsequent diffusive tail. Corrections to the bolometer response from scattering other than from Eu^{2+} ions have been made by subtracting the response from samples of "pure" SrF_2. Separation of the diffusive component from the ballistic component was accomplished analytically using the ramp geometry suggested by von Gutfeld.[4] The fraction of the total bolometer response in the diffusive

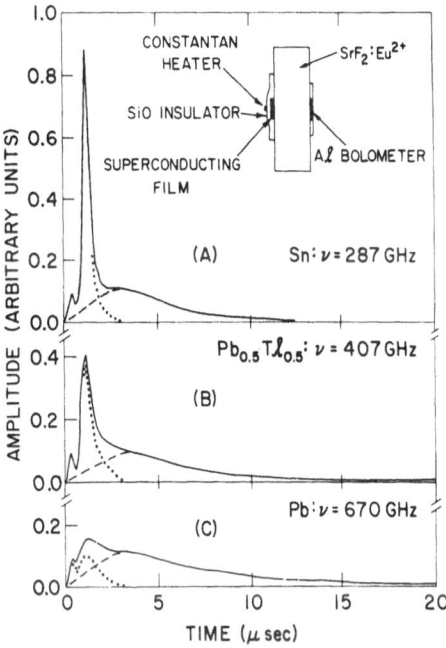

Fig. 1. Time-resolved bolometer response obtained from phonon pulses of three peak frequencies. The analytical separation indicates the ballistic and diffusive component.

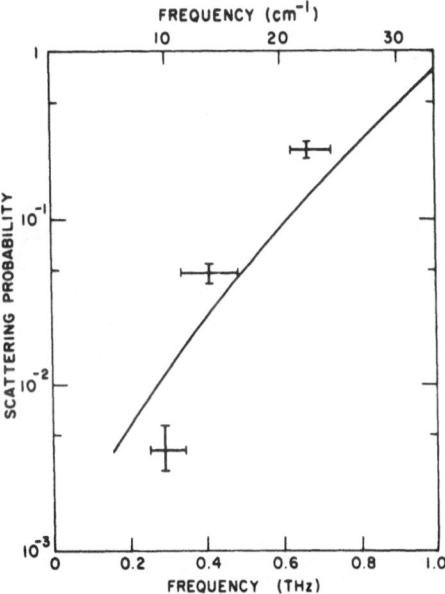

Fig. 2. Comparison of the experimental (data points) and theoretical (solid line) frequency dependence of the scattering probability.

174   W.E. BRON AND F. KEILMANN

component is taken as a measure of the scattering probability. A
comparison of Fig. 1A through 1C demonstrates that the frequency
dependence of this quantity is readily observable.

The frequency dependence of the scattering probability can
also be determined theoretically. A formalism pertinent to the
present case has been presented by Klein.[5] The scattering probability averaged over all phonons having a frequency ω is

$$ - <\tau^{-1}>_\omega = \frac{2PN}{\pi\rho^o(\omega)} \text{ trace } [\text{Im } t(\omega^2-i\epsilon) \text{ Im } g(\omega^2-i\epsilon)]. $$

In this expression τ is an effective mean free phonon lifetime,
P is the concentration of scattering sites, N is the number of
unit cells in the lattice, $\rho^o(\omega)$ is the density of states, t is
the T-matrix and g is the Green's function matrix defined in the
space of the scattering center. The above expression has been
evaluated using the eigenvalues and eigenvectors of the lattice
dynamics of SrF_2 obtained previously from a deformation shell
model,[3] and from previously obtained parameters of the disturbance.
The result of the calculation, in the frequency region of the experiment, is shown as the solid line of Fig. 2.

The scattering probability at the available phonon frequencies
is indicated by the data points in Fig. 2. The rather large uncertainties stem from the analytical separation of the diffusive
component from the ballistic component, from the subtraction of
scattering from other than the Eu^{2+} ions, and from uncertainties
in the actual band width of the phonon distributions.[1]

It is clear from Fig. 2 that the experimental methods are too
crude at this stage to afford a critical comparison between experiment and the scattering theory based on the T-matrix formalism.
Overall agreement on the relative frequency dependence of the
scattering process is, however, indicated. These experiments are
being extended to higher frequencies by generating phonons through
optical pumping, with far IR lasers, directly into the defect
induced absorption of the $SrF_2:Eu^{2+}$ crystal.

1.   R.C. Dynes and V. Narayanamurti, Phys. Rev. B 6, 143 (1972).
2.   D.H. Kühner, H.V. Lauer and W.E. Bron, Phys. Rev. B 5 4112
     (1972).
3.   D.H. Kühner and M. Wagner, Z. Physik 256, 22 (1972).
4.   R.J. von Gutfeld in Physical Acoustics, Vol. V, Edited by
     W.P. Mason, Academic Press, New York 1968.
5.   M.V. Klein in Physics of Color Centers, Ed. by W. Beall
     Fowler, Academic Press, New York 1968.

TEMPERATURE DEPENDENCE OF THE ACOUSTIC PARAMAGNETIC RESONANCE

SPECTRA FOR Cr^{2+} IN MgO AND $KMgF_3$*

J Lange and S Guha

Department of Physics, Oklahoma State University

Stillwater, Oklahoma 74074, U S A

The ground state energy minimum of some ions occurs when the electronic states couple to the lattice distortions to form vibronic states (electronic + vibrational). Under certain circumstances the vibrational coupling can contribute energy to the system which is greater than the spin-orbit splitting. In this investigation the vibronic states are characterized by their effect on the attenuation of microwave frequency sound waves. The magnetic field dependence of the attenuation resulting from the Zeeman splitting of paramagnetic vibronic states (acoustic paramagnetic resonance: APR) is used to monitor the population of particular energy levels.

Only isolated Jahn-Teller ions are considered, eliminating the complications due to cross relaxation effects. The isolation of the ions results from introducing them as a dilute substitutional impurity (100 ppm) in a cubic host lattice. Chromium impurity ions are used as substitutional impurities in either a host crystal of MgO or $KmgF_3$. The chromous (Cr^{2+}) valence state has a doublet electronic ground state in a cubic environment and exhibits a dynamic Jahn-Teller effect. This particular ground state displays two properties which are significant for the interpretation of the acoustic attentuation measurements. The spin-orbit coupling is quenched to first order which accentuates vibronic effects. Also, the electronic ground state couples only with particular vibrational strain modes which provides a selection rule for coupling of the acoustic wave to the vibronic states.

The vibronic energy levels are derived in terms of two parameters characterizing the dynamic Jahn-Teller effect which are the tunnelling splitting (δ) and the spin-orbit splitting (D). Fitting the energy level diagram to experimentally determined parameters to

evaluate the tunnelling splitting and spin-orbit coupling is the
major task of this investigation. The present approach differs
from other determinations since specific energy levels are identified by using the APR spectrum and the temperature dependence of the
APR peak heights. The specific theoretical energy levels are fitted
to the observed temperature dependence of the peak heights to assign
values to the tunnelling splitting and spin-orbit splitting. The
identification of the energy levels associated with a particular
effective "g" factor facilitates the determination of the local deviations from cubic symmetry which leads to anisotropic behaviour.

An evaluation of the Jahn-Teller tunnelling splitting and spin-
orbit coupling resulted from fitting the theoretical energy levels
for the chromous ion to the observed temperature dependence of the
APR peak height. The values for δ and D (B > 0; δ = 19 cm^{-1}, D =
2.2 cm^{-1} and B < 0; δ = 18 cm^{-1}, D = 2.3 cm^{-1}) indicate the Jahn-
Teller effect is further in the dynamic range $\delta/D \sim 8$ than previously suggested. The energy level diagram constructed from δ and D
offers transitions which are consistent with observations of acoustic relaxation, infrared absorption, and thermal resistance for the
chromous ion in MgO.

The chromous ion in KmgF$_3$ is in an environment of similar symmetry and size as in MgO but the members of the octahedral molecular
cluster are monovalent flourine ions rather than divalent oxygen.
Fitting the temperature dependence of the APR peak height leads to
quite different values of the tunnelling splitting and spin-orbit
splitting. The tunnelling splitting is estimated to be approximately 10 cm^{-1} while the spin-orbit splitting is 1.1 cm^{-1}. These
values are approximately one-half those found in MgO.

In a strictly cubic environment the "g" factor for the chromous
ion would be isotropic for the three lowest paramagnetic triplet
levels. The observed behaviour of the effective "g" factor for both
MgO and KMgF$_3$ is far from isotropic and even exhibits anisotropy
which is dependent on the type of acoustic mode used to detect the
APR signal. The anisotropy of the "g" factor is attributed to tetragonal and orthorhombic distortions of the cubic site which result
from static internal strains in the host lattice.

For the acoustic wave to induce transitions between the Zeeman
split doublet of the paramagnetic level, the acoustic strain must
perturb the doublet by introducing a small orthorhombic strain at
the tetragonal site. Only ions for which the doublet itself is not
split due to internal strains by an energy greater than $h\nu$ can be
involved in induced Δm = 2 acoustic transitions.

The anisotropy of the "g" factor is interpreted in terms of
local linear distortions of the cubic site. Application of the
theoretical development of assuming tetragonal which are larger than

the orthorhombic distortions fit the observed anisotropy for only one of the acoustic modes (longitudinal). Following the development of Ham the strain splitting of these particular chromous sites is determined to be ~ 0.2 cm^{-1} for MgO and ~ 0.03 cm^{-1} for KMgF$_3$ which is much less than previously assumed (~ 1 cm^{-1}). This suggests that a selection rule involving the interaction of the longitudinal acoustic wave with $\Delta m = 2$ transitions is acting to provide a larger coupling than to the $\Delta m = 1$ transitions.

The magnitude of the observed "g" factor varies depending on the particular triplet as well as the anion environment. For MgO the observed "g" factor is 0.8 for the lower triplet and 1.3 for the middle triplet. In KMgF$_3$ the lower triplet has a "g" factor of ~ 0.5 while the middle triplet is 0.7. These values are substantially different from the theoretical cubic site value of one. It appears the magnitude of the "g" factor for these strained sites depends on the overlap of the chromous ion charge densities with the different anion species (oxygen or fluorine) since an estimate of the strain in both crystals leads to the same value of 10^{-5}.

*Work supported by the National Science Foundation.

STRESS INDUCED CHANGES IN THE ACOUSTIC PARAMAGNETIC RESONANCE OF

CHROMOUS IONS IN MAGNESIUM OXIDE

V W Rampton and I J Shellard

Department of Physics, University of Nottingham

University Park, Nottingham NG7 2RD, U K

INTRODUCTION

Magnesium oxide is a cubic host crystal on which many investig-ations have been made. The problem of chromous ions in chromium doped MgO has been particularly difficult because of the strong coup-ling between the ion and the lattice which means that a Jahn-Teller effect occurs and also that the electronic energy levels of the ion are very sensitive to random internal strains in the lattice. The chromous ion is responsible for strong ultrasonic attenuation (King et al 1975), large thermal resistivity (Challis et al 1972) and int-ense acoustic paramagnetic resonance (a.p.r.) absorption (Marshall and Rampton 1968). Theories of the Cr^{2+} ion in MgO have been given by Fletcher and Stevens (1969) and Ham (1971) using the dynamic Jahn-Teller effect to attempt to account for the a.p.r. spectrum.

THEORY

The lowest term of the free chromous ion $3d^4$, is ^5D and the eff-ect of the cubic crystal field in MgO, the Jahn-Teller effect and spin-orbit coupling is to leave ten low lying states on five energy levels. The lowest level is a Γ_1 singlet with a triplet Γ_4 a few cm^{-1} above it. The effect of strain in the lattice is to partially remove the degeneracy of the triplet. A.p.r. absorptions are bel-ieved to occur between the states of the Zeeman split doublet which results. Only a small fraction of the total number of chromous ions contribute to the resonance; those for which the site symmetry is approximately tetragonal and for which the magnitude of the tetrag-onal distortion is such as to give, with the Zeeman splitting, the energy level separation for resonance. The effect of an applied uni-

axial stress is to move the overall internal strain distribution and
change the numbers of ions contributing to the absorption. There
is a smaller indirect effect of changing the effective g-value for
the Zeeman splitting.

EXPERIMENTS AND RESULTS

     MgO samples containing between 800 and 900 ppm of chromium were
X-ray aligned and then cut into cuboid shape. End faces were polished parallel and flat. A.p.r. experiments were carried out using
an ultrasonic pulse-echo technique with pulse width $\sim$ 0.5μs, repeat
frequency 1000 sec^{-1}, employing a quartz-rod transducer (Rampton and
Tucker 1972). Longitudinal ultrasonic waves at about 9.4 GHz propagating in the <100> direction of the sample were used. The echoes
were sampled by a boxcar detector and the output voltage (proportional to echo intensity) was used to drive the Y-input of an X-Y
recorder, the X-input being driven by a Hall-effect gaussmeter between the poles of the iron-cored electromagnet. By this means plots
of echo intensity as magnetic fields up to 2T could be obtained.
Frequency was measured using a cavity wavemeter. In the case of as-
received samples no echoes were visible at temperatures above about
3K (King et al 1975), so experiments were carried out at 2K by means
of pumping on liquid helium. It was found that by heating these
samples at 1350°C for 24 hours in Air and then 3 hours in Hydrogen
(2%), Argon (98%) echoes with the resonance characteristics of the
as-received samples were visible at 4.2K. Experiments were carried
out on a sample treated in this way. For each sample plots were
obtained of the type described above for angles θ (between the field
direction in the (010)   plane and the direction of ultrasound
propagation $\underline{k}$ <100>) ranging from 0 to 90 degrees in 10 degree steps
for three different conditions of applied external stress on the
<001> axis, zero stress, 3.4 x 10^6 kg/m^2 and 7.0 x 10^6 kg/m^2. Two
resonances were observed which were attributed to the Cr^{2+} ion, labelled the C and U lines after Marshall (1967).

DISCUSSION

     The a.p.r. spectrum is broadly consistent with the earlier exp-
erimental results of Marshall and Rampton (1968) and thus the results are in agreement with the theories of Fletcher and Stevens
(1969) and Ham (1971). There is close agreement between the values
calculated from these results for g (where g = $g_{11}/2$) = 0.98 calculated from the angular dependence of the c-line, $\bar{H}$ = $h\nu/g_{11}\beta\sin\theta$ and
Ham's (1971) theoretical value of g $\sim$ 1. The effect of uniaxial
applied stress as shown in figure 1 is to decrease the intensity of
the U-line as stress is increased. This is not in agreement with
the conclusions of Marshall and Rampton (1968). Figure 2 shows that,
within experimental error, the effect of applied stress is independ-

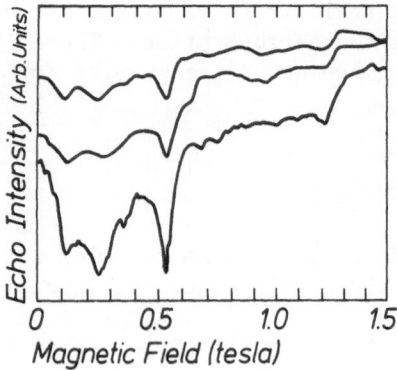

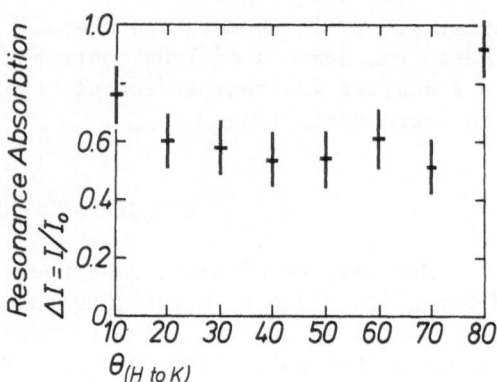

Fig 1: Ultrasonic echo intensity vs Magnetic field with $\theta = 60°$, Bottom curve - zero stress, Middle curve $3.4 \times 10^6 \text{kg/m}^2$, Top curve $7.0 \times 10^6 \text{kg/m}^2$.

Fig 2: Change in intensity of resonance absorption as a function of angle between the magnetic field and the [100] axis of the crystal.

ent of the orientation of the magnetic field. This result is in agreement with theory. We can conclude from the fact that the U-line resonance is visible at the largest applied stress that the range of random internal strains present in the unstressed crystal is at least 2.3×10^{-4}.

Future experiments are planned with a larger range of applied stresses and also to measure the change in Zeeman splitting with applied stress.

REFERENCES

L J Challis, A M de Goer, K Guckelsberger and G A Slack, Proc Roy Soc A330, 29, 1972.
J R Fletcher and K W H Stevens, J Phys C 2, 444, 1969.
F S Ham, Phys Rev B 4, 11, 3854, 1971.
P J King, S G Oates, V W Rampton and I J Shellard, Proc 2nd Int Conf on Phonon Scattering in Solids, 1975.
F G Marshall and V W Rampton, J Phys C 1, 594 1968.

ULTRASONIC RELAXATION STUDIES OF Cr^{2+} IN MgO

P.J. King, S.G. Oates, V.W. Rampton and I.J. Shellard

Physics Department, University of Nottingham

University Park, Nottingham NG7 2RD

Abstract: The relaxation peaks which are found in the microwave acoustic attenuation as a function of temperature in MgO have been studied in samples doped with V, Mn, Fe, Co and Cr before and after γ-irradiation and annealing treatments.

Introduction: The Cr^{2+} ion in MgO is extremely strongly coupled to the lattice and exhibits the Jahn-Teller effect (see Ham 1971 and the references therein). It has been much studied using A.P.R. (Marshall & Rampton 1968) and affects due to Cr^{2+} have been reported in thermal conductivity studies (Challis et al 1972). Several workers have found relaxation peaks in the microwave acoustic atten-uation at about 8K and these peaks have been studied in detail by Lange (1973) who attributes them to Cr^{2+}. This paper reports further studies of this peak, in particular in cases where the sample con-cerned has been treated either with γ rays (10 megarads over 24 hrs) or by annealing for 24 hours in air or in a 3% H_2, 97% argon mixture at 1300°C.

Experimental: The measurements reported here were made between 1.5K and 50K and at 580 MHz and 1GHz using a standard pulse echo system. The MgO samples were obtained from Spicers Ltd and were as follows: 1140 ppm (by wt) V; 840 ppm Mn; 2% Co; 910 ppm Fe; 800 ppm Cr. Each contained in addition to the main dopant a few ppm of chromium and a few tens of ppm of iron. Rods were cut and polished suitable for propagating longitudinal waves in the <100> direction, cadmium sulphide films overlaid on aluminium ground planes being used as transducers.

No peaks were found in the attenuation of transverse waves propagating in the <100> direction in any of the samples. In all 'as received' samples except that doped with iron, peaks were observed at ∿8K in the longitudinal wave attenuation. The main features of

181

the peaks were as follows. (i) The peak heights were the same order of magnitude in the chromium doped samples and in the samples which contained chromium as a trace impurity and of order $1dB\ cm^{-1}$. (ii) In a particular sample the effects of irradiation and heat treatment were reproduceable but the detailed effects depended on the main dopant present. (iii) The shape of the peak was independent of sample and sample treatment. (iv) The behaviour of the peak under annealing in our 800 ppm Cr sample was qualitatively similar to the behaviour of Cr^{2+} deduced by Anderson et al from a thermal conductivity study; in particular H_2 treatment enhanced the peak and O_2 treatment removed it. (v) In that sample the behaviour of the peak under oxygen and hydrogen annealing was the same as that of the main features of Cr^{2+} lines found in A.P.R. studies in this laboratory. It is quite clear that the attenuation peaks are closely related to the phenomena studied in A.P.R. and in thermal conductivity. The main problem in relating them to the ion Cr^{2+} is feature (ii) the lack of dependence on chromium concentration. It is known from E.P.R. studies however that in samples containing a few ppm of Cr, the Cr^{3+} concentration can be appreciably changed by treatment whereas this is not the case for more concentrated samples. It is clear from this that the Cr^{2+} concentration saturates at a few ppm and thus the peak might be expected to be of the same size in a heavily and lightly doped sample. We will thus assume that our peaks are due to Cr^{2+} in agreement with Lange, unless further evidence to the contrary arises.

The behaviour of the peaks under the different treatments is of great interest in those cases where a strong non-chromium dopant is present since it casts light upon the Cr^{2+}-Cr^{3+} valence change mechanism. In the case of the manganese doped sample the peak height was unchanged by any of the treatments to greater than 20% Mn^{2+} is known to be dominant and extremely stable in MgO. In the case of the Vanadium doped sample air (O_2) annealing enhanced the peak and hydrogen treatment produced a reduction. In this case V^{3+} is the dominant ion although smaller amounts of V^{2+} can be produced by treatment. Our measurements favour a mechanism involving a second ion X as suggested by Challis et al. (1972)

$$Cr^{2+} + X^{3+} \rightleftharpoons Cr^{3+} + X^{2+} \qquad\qquad 1$$

If the ion X is relatively stable it thus becomes difficult to produce a change in the chromium valences, otherwise changes in X produced by treatment produce corresponding changes in the chromium valences.

Analysis of the Relaxation Peaks: It is possible to deduce information from the shape of relaxation peaks, assuming a model where the attenuation is due to attempts to re-establish equilibrium between magnetic levels split by the strain of the acoustic wave. The appropriate expression for the attenuation $\alpha(T)$ is

$$\alpha(T) = 4.34\ \frac{G^2 N_0 K(T)}{\rho v^3\ kT}\ \frac{\omega^2\ \tau}{1+\omega^2\tau^2}\ dBm^{-1} \qquad\qquad 2$$

Here ρ is the medium density, v is the propagation velocity and G
is the spin-phonon coupling parameter. $N_oK(T)$ is the number of ions
in the levels split by strain and relaxing with characteristic time
τ We have put the unknown Boltzman factor $K(T) = 1$ and have analysed
$\alpha(T)$ in terms of the $\omega^2\tau/1+\omega^2\tau^2$ form to give $\tau(T)$. In general such
relaxation times will be given by

$$\frac{1}{\tau} = AT + \sum_i \frac{B_i}{e^{\Delta_i/kT}-1} \qquad\qquad 3$$

where A is a direct process and B_i, Δ_i characterise Orbach
processes involving higher levels Δ_i. It is possible to fit our $\tau(T)$
using a single Orbach term and $A = 4\times10^7$ HzT^{-1}, $B = 2\times10^{10}$ Hz $\Delta_i =$
9.5 cm^{-1}. This later value is smaller than either of those quoted
by Lange (1973) and we find no sign of a double relaxation peak.
The existing theoretical models of the Cr^{2+} ion has been examined
by Challis et al. (1972) who suggest level separations deduced
from I.R and thermal conductivity measurements. On the B<0
assumption the most likely transition to explain the attenuation
peaks is $T_1^1 \rightleftharpoons T_2^1$ an allowed transition for E symmetry phonons. The
T_2^1 levels are unlikely as relaxing levels as they are only split
to high order in strain and are some 12cm^{-1} about the ground state.
However it is not clear why the T_2^1 levels should give the dominant
Orbach processes relaxing the T_1^1 and the T_1^2 levels might be
expected to contribute. Similar problems involving the relative
sizes of matrix elements have been encountered by Challis et al (1972)
and it is clear that the Cr^{2+} ion is in detail by no means understood and much further study is needed.

References

Anderson B.R, Challis L.J, Champion D.J, Clark I.A, Jay P.R, and
   Woodhead S.M, J. Phys. C: Solid St. Phys. 8, 1472-1473, 1975.
Challis L.J, De Goer A.M, Guckelsberger K, and Slack G.A. Proc.
   R. Soc. Lond. A 330, 29-58, 1972.
Ham F.S. Phys Rev. B4 3854-69, 1971.
Lange J.N. Phys. Rev. B8, 5999-6009, 1973.
Marshall F.G, and Rampton V.W., J. Phys. C: Solid St. Phys. 1, 594-8
   1968.

CONTRIBUTION TO THE STUDY OF $3d^4$ IONS IN Al_2O_3 AND MgO BY LOW TEM-

PERATURE THERMAL CONDUCTIVITY MEASUREMENTS UNDER UNIAXIAL STRESS

By J. RIVALLIN and B. SALCE
Centre d'Etudes Nucléaires de Grenoble
Service des Basses Températures
BP 85 Centre de Tri - 38041 GRENOBLE-CEDEX (France)

It is well known that $3d^4$ ions in cubic or trigonal symetry show a dynamic Jahn-Teller effect in the fundamental orbital doublet ^5E. From previous measurements by A.P.R.[1][2], and low temperature thermal conductivity [3][4], theoretical models have been proposed for Cr^{2+} in MgO and Al_2O_3, in the limit of a small inversion splitting δ [5][6]. The coupling constant b V_E can be determined from the variation of the vibronic energy levels due to an applied stress, and this can be done by measuring the changes of the thermal conductivity K.

The systems Cr^{2+} and Mn^{3+} in Al_2O_3 and Cr^{2+} in MgO have been studied in detail [7]. The temperature range was 1.1 to 30 K ; traction or compression up to 1400 kgf/cm^2 can be applied to the sample parallel to the heat flow. Related thermal problems have been resolved and the additionnal error on K is less than 1 % [7]. Measurements of the relative change of K as a function of stress $f(\sigma)=(K-K_0)/K_0$ are done at constant temperature ; an example of results for Cr^{2+} in Al_2O_3 is given on figure 1. The set of curves at constant stress σ

TABLE 1

System	State	C^x (ppm atom)	$\hbar\omega_1$ (K)	$\hbar\omega_2$ (K)	θ^{xx} (°)	ϕ^{xx} (°)	b V_E (cm^{-1})
MgO:Cr^{2+}	virgin	1300	7.3	29	90	0	$(1.7 \pm 0.1)10^4$
Al_2O_3:Cr^{2+}	γ irradiat.	500	7.4	107	55,5	8,5	$(1.0 \pm .5)10^4$
Al_2O_3:Mn^{3+}	virgin	40	12	54	60	10,5	$(1.0 \pm .5)10^4$

x C total dopant concentration
xx θ and ϕ have been determined by M.TITEUX(Center of Research U.K)
  by X-ray diffraction.

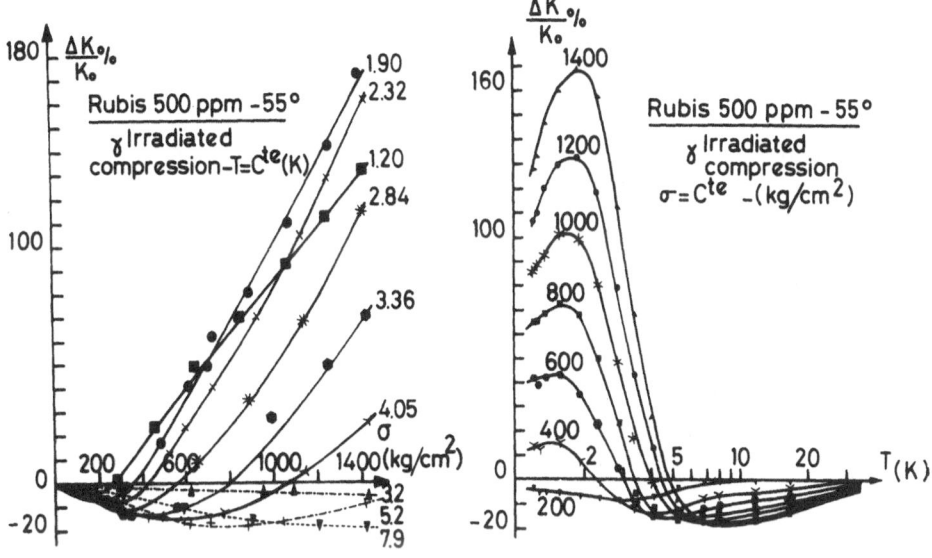

FIGURE 1   FIGURE 2

f'(T) (figure2) is deduced from the latter. The general shape of
these curves are similar for the three systems. Quantitative fits of
the set of curves f'(T) have been obtained using the Callaway model,
in the line of the previous analysis of zero-stress results where
the additionnal resonant phonon relaxation time was :

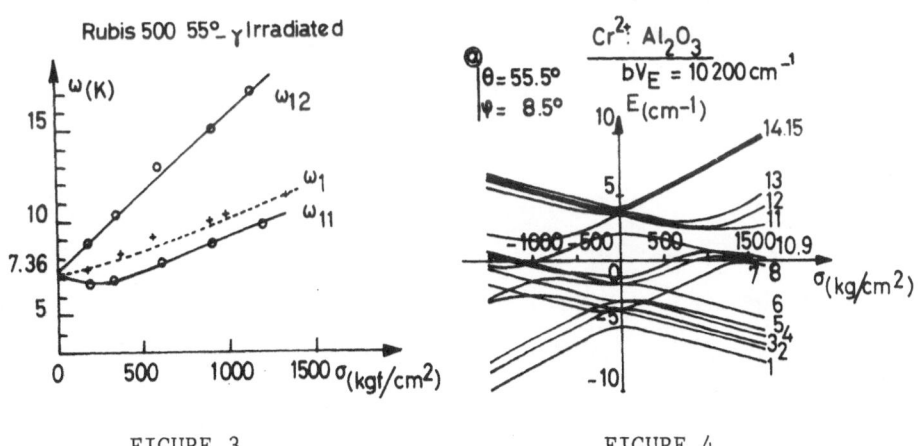

FIGURE 3   FIGURE 4

$\tau^{-1} = D_1 \omega^4/(\omega^2 - \omega_1^2)^2 + D_2 \omega^4/(\omega^2 - \omega_2^2)^2$ (ω_1 and ω_2 are given in table 1). Fits of the results with stress have been done in two ways [7], with two adjustable parameters, which are :(i) ω_1 and D_1 in the first case, (ii) ω_{11} and ω_{12}, corresponding to a splitting of the zero-stress resonance ω_1, with $D_{11} = D_{12} = D_1/2$. An example of the variations of ω_1, ω_{11} and ω_{12} as a function of σ are given in figure 3. There is a tendency to a linear variation for large stresses, and similar results are obtained for the other systems.

The level scheme as a function of stress has been calculated for each sample, by numerical diagonalization of the 15x15 matrix [5] [6], in which the terms due to the induced E-type strains have been introduced (with an arbitrary value of b V_E). The strains were obtained from the orientation of σ relative to the crystalline axis (angles θ and ϕ) and the known elastic constants of Al_2O_3 [8] and MgO [9]. An example of level scheme is given in figure 4 in the case Cr^{2+}/Al_2O_3. Comparison between experiment and theory is done in the limit of large stresses and the values of b V_E deduced from the analysis (ii) are given in table 1. The main uncertainties come from the complexity of the level scheme and the possible lack of uniformity of the applied stress [7]. We note the fair agreement with a previous estimation of b V_E in the case of Cr^{2+} : MgO [10]. Our values of b V_E are systematically smaller than those calculated in a point charge model.

[1] MARSCHALL G. and RAMPTON V.W.(1968) J.Phys.C., p 594
[2] ANDERSON R.S. and al.(1972) J.Phys.C.5, p 3397
[3] CHALLIS L.J. and al.(1969) Proc.Roy.Soc.A310, p 493
[4] de GOER A.M.(1969) J.de Physique(France) 30, p 389
[5] FLETCHER J.R. and STEVENS K.W.H.(1969) J.Phys.C.2, p 444
[6] BATES C.A. and al.(1973) J.Phys.C.6, p 898
[7] RIVALLIN J.(1974) Thèse d'état, Grenoble (France)
[8] TEFFT W.E.(1966) J.of Research of N.B.S. A 70A N° 4 p 277
[9] OLIVER and BOYD (1966) in Phonons in Perfect Lattices and in
     lattice with some defects, edited by STEVENSON R.W.H.
[10] LANGE J.N. (1973) Phys.Rev.B 8, N°12, p 5999

HEAT PULSE INVESTIGATIONS OF MgO:Cr^{2+}

J.L. Patel and J.K. Wigmore

Department of Physics, University of Lancaster
Lancaster, LA1 4YB, England

Despite the considerable quantity of data[1,2,3] that has been produced on MgO:Cr^{2+}, uncertainty still exists as to the detailed model of the dynamic Jahn-Teller effect in this system.[4,5] Specifically, is the picture one of rapid rotation (tunnelling splitting 3Γ much larger than D, the reduced spin-orbit coupling) implying only slight warping of the Mexican-hat potential, or is the system more closely approximated by the slow rotation or tunnelling model (3Γ ≃ D)?

In a further attempt to solve this problem, we have carried out phonon 'crossing spectroscopy' experiments using heat pulses; details of the experimental arrangements and preliminary observations have been reported elsewhere[6]. The magnitude of diffusive heat pulses transmitted through specimens of MgO:Cr^{2+} were monitored as a function of magnetic field up to 7T along various crystal directions, and of applied stress up to 7 kg mm^{-2} at ambient temperatures between 1.5 and 4.2K. Figure 1 shows data for three different magnetic field directions in the (011) plane, with zero applied stress. The simplest data occurred for $\underline{B}$||<100>, θ = 0°, when three fairly sharp and fairly symmetric features at 1.0, 1.6 and 2.1T were observed, in addition to a very steep gradient at 0T. (The values given in reference 6 were incorrect due to an inaccurate field calibration). Since the phonon spectrum emitted by the heat pulse generator was a smooth function of frequency, these features could be due only to crossing phenomena[7]. Furthermore, because these 'lines' corresponded to increases in phonon scattering, they had to be the result of level anticrossings, that is, of the change in scattering caused as two energy levels approached each other with varying magnetic field and were mutually repelled again without crossing. At angles other than

$\theta = 0^o$, the spectra and lineshapes became more complicated.   In
addition to one or more 'lines', other features appeared which
are best described as 'edges'.   It was not clear whether the
latter corresponded to increases or to decreases in phonon
scattering.   The $\theta = 27^o$ data were particularly interesting
because, over a range of about 3° centred on this angle, the
'line' at 0.82T changed rapidly into an 'edge'.   The angle
$\theta = 55^o$ corresponded to $\underline{B}||<111>$.

The effect of applied stress on some of the principal
features of the spectrum is shown in figure 2.   In general, the
stress produced a large decrease in the magnitude of a particular
feature without affecting either its shape or the magnetic field
at which it occurred.   We noted two exceptions.   With $\underline{B}||<100>$,
application of stress along <010> had no effect on the line at
1.0T, even on its magnitude.   In contrast, at $\theta = 27^o$, the
magnitude of the 0.82T feature actually increased and both its
shape and magnetic field position were altered by the applied
stress.

The results of the stress experiments suggest that both lines
and edges represent turning points in the magnetic field positions
of the level anticrossings as a function of strain, ε;   thus
$\partial B_{AC}/\partial \varepsilon = 0$.   It is well known that the Cr^{2+} sites are subject
to a random distribution of strains due to crystal imperfections.
The Cr^{2+} can interact in first order only with strains of
symmetry $E\theta$ or $E\varepsilon$, and for a totally asymmetric direction of $\underline{B}$,
turning points might lie anywhere in the $E\theta-E\varepsilon$ plane.   However,
for $\underline{B}||<100>$, the turning points must have reflection symmetry
about the $E\theta$ axis, and for $\underline{B}||<111>$ they must make 3-fold
symmetry in the $E\theta-E\varepsilon$ plane.

We attempted to locate the turning points by diagonalising
the energy matrix for the lowest 15 eigenstates of $MgO:Cr^{2+}$ in
the form given by Fletcher and Stevens[4] as a function of ε and $\underline{B}$,
for various values of 3Γ and D.   We find the best agreement with
the $\underline{B}||<100>$ data if we take $3\Gamma = 32cm^{-1}$ and $D = 2.5cm^{-1}$, with the
anharmonicity parameter B>0.   For these parameters, the
calculated turning points of anticrossings between the lowest four
excited states lie along the $E\theta$ axis at fields of 0, 1.2, 1.6,1.9
- 2.2T. (figure 3).   Neither $3\Gamma = 7.6cm^{-1}$, $D = 2.02cm^{-1}$,[4] nor
$3\Gamma = 16cm^{-1}$, $D = 4cm^{-1}$,[2] fit the data.   So far, however, we are
completely unable to understand the angular variation of the
spectrum.   We have not considered quantitatively the actual
distribution of internal strains, assuming only that it is broad
compared with the anticrossing from any individual Cr^{2+}.   The
effect of applied stress is then to change the number of Cr^{2+} ions
subject to a particular stress, and hence to change the magnitude
of the phonon scattering but not the field at which it occurs.
We surmise that the anomalous behaviour around $\theta = 27^o$ is due to
the breakdown of this assumption.

We are grateful to Dr. J.R. Fletcher for many valuable

discussions.   The work is supported by the Science Research
Council.

1   F.G. Marshall and V.W. Rampton J. Phys. C1 594 (1968)
2   L.J. Challis, A.M. de Goër, K. Guckelsberger, and G.A. Slack,
     Proc. Roy. Soc. A330 29 (1972)
3   J.N. Lange, Phys. Rev. B8 5999 (1973)
4   J.R. Fletcher and K.W.H. Stevens, J. Phys. C2 444 (1969)
5   F.S. Ham, Phys. Rev. B4 3854 (1971)
6   J.K. Wigmore, and J.L. Patel, Proceedings of the Lancaster
     Symposium on Microwave Acoustics (IOP) 152 (1974)
7   B.R. Anderson and L.J. Challis, J. Phys. C8 1495 (1975)

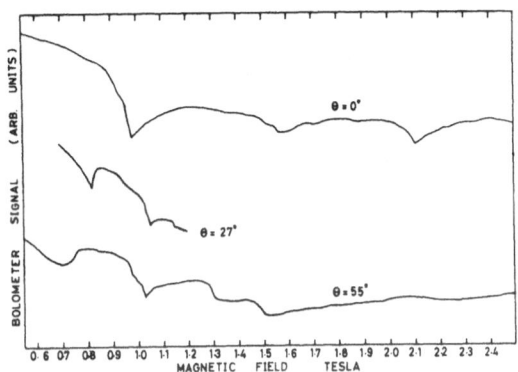

Fig. 1. Typical spectra with zero applied stress;
θ is the angle between B and<100>in (011) plane

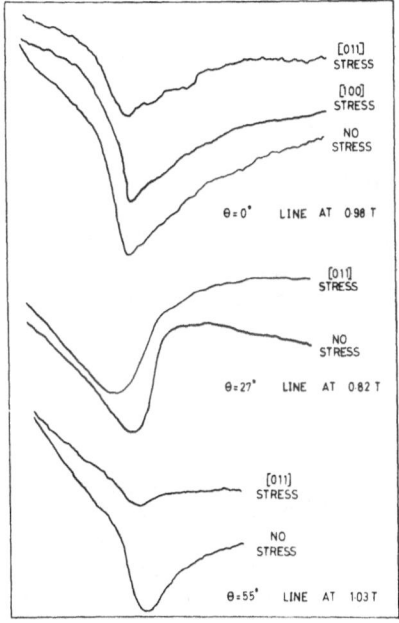

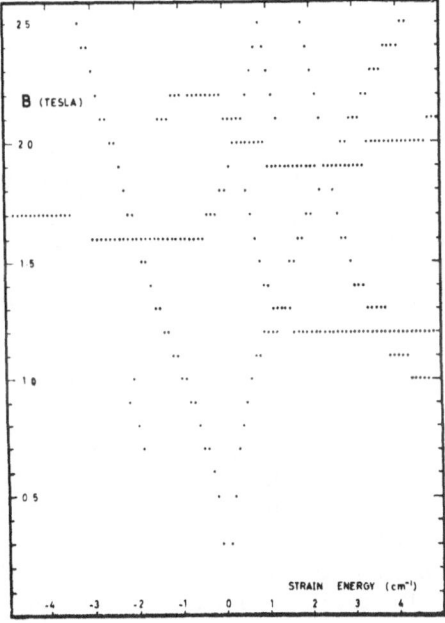

Fig. 2 Effect of stress on
some principal features

Fig. 3 Computed anticrossings
for 3Γ =32, D=2.5cm^{-1} B||<100>

THERMAL CONDUCTIVITY OF IRRADIATED, ADDITIVELY-COLORED, AND

DEFORMED MgO*

Judith B. Hartmann[+], Harold Weinstock[+], and Yok Chen[++]

[+]Physics Department, Illinois Institute of Technology
Chicago, IL. 60616
[++]Solid State Division, Oak Ridge Nat'l. Laboratory**
Oak Ridge, TN. 37830

Low temperature thermal conductivity measurements have been made on neutron-irradiated (1), electron-irradiated, additively-colored(2), and deformed MgO. Resonant scattering of phonons is observed in all these crystals at T $\sim$ 15-20 K. Correlations with optical measurements (3) indicate that this resonance is associated with anion vacancies. In addition, a resonance at T $\sim$ 1 K in neutron-irradiated crystals is attributed to defect aggregates, which are virtually absent in electron-irradiated (4) and additively-colored samples.

Figure 1 b shows the thermal conductivity of MgO irradiated by 1.5 MeV electrons to a dose of 3 x 10^{18} e/cm^2. Although not shown in this figure, measurements were also taken at lower temperatures. There is no phonon scattering at T $\sim$ 1 K. The resonance at T $\sim$ 15-20 K is much weaker than that observed in the neutron-irradiated samples, with an anion vacancy concentration of 2 x 10^{16}/cm^3 as compared to 4 x 10^{17}/cm^3 for the lowest neutron dose. Following an ultraviolet [uv] bleach of an electron-irradiated sample at room temperature, a process which has been shown (5) to partially convert F^+ to F centers [anion vacancies with one and two electrons respectively), the thermal conductivity is lower, indicating that the F center is a stronger scatterer than the F^+ center.

Whereas anion vacancies in neutron-irradiated crystals are primarily in the F^+ state (6), in additively-colored samples they are normally in the neutral F state (5). Figure 1c shows the thermal conductivity curve of an additively-colored crystal (2) with an anion vacancy concentration of 2 x 10^{17}/cm^3 entirely in the F state. A strong resonance at 20 K is observed. Following a

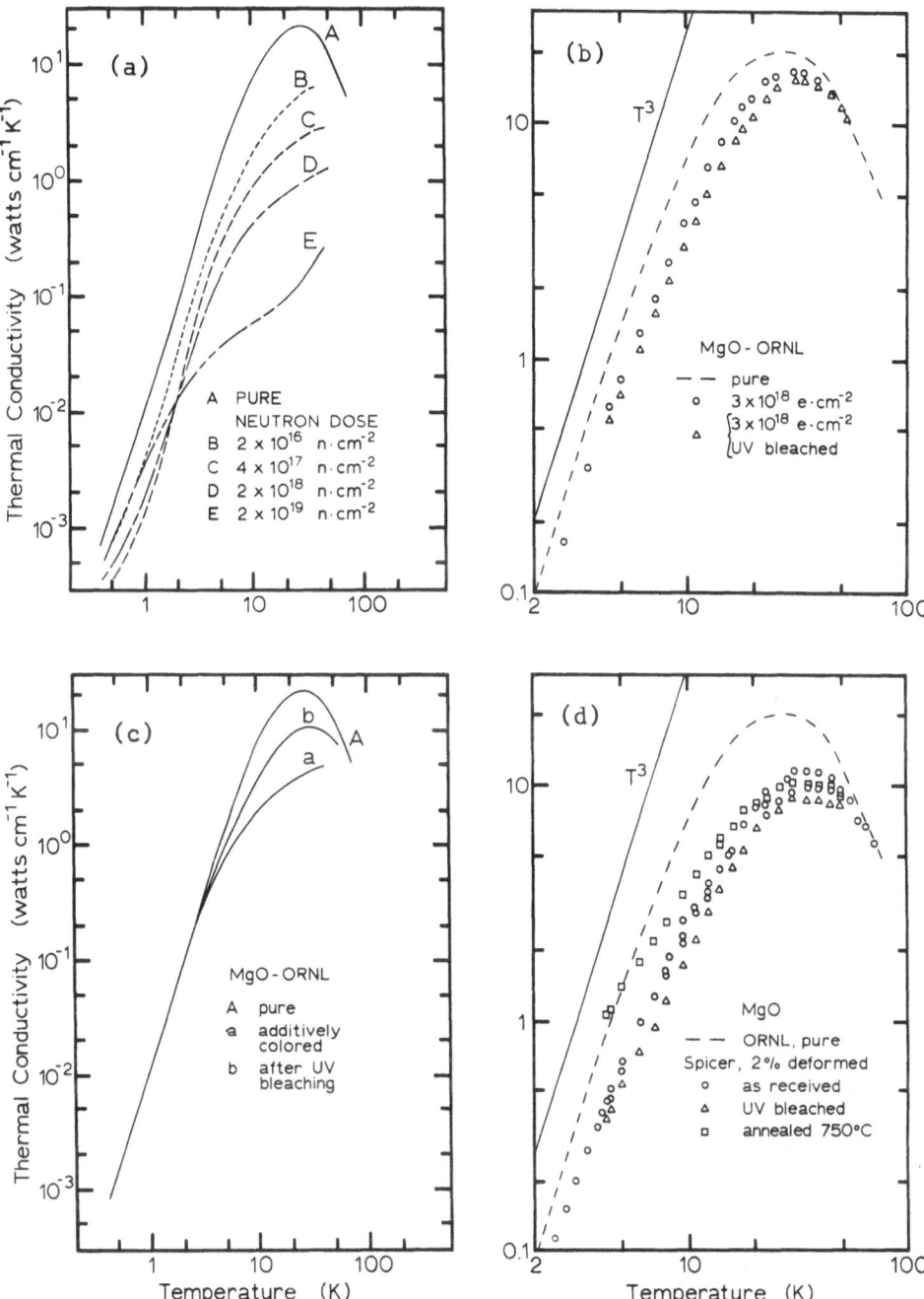

Fig. 1.   Thermal conductivity vs. temperature for MgO which has been (a) neutron-irradiated (E > 1 MeV); (b) electron-irradiated (E = 1.5 MeV); (c) additively-colored; and (d) deformed ~2% along a [100] direction.

uv bleach, which induces partial F → F+ conversion (5), the resonance becomes weaker. A reconversion of the F^+ → F centers by
annealing the crystal at 700 K restores the thermal conductivity
to its original values. Thus, thermal conductivity measurements
on irradiated and additively colored MgO indicate that anion vacancies scatter phonons at T ∿ 15-20 K and that the neutral F
center is a stronger scatterer than the F^+ center.

Figure 1d shows the results of preliminary thermal conductivity measurements on an "as-grown" sample deformed to ∿2% along
a [100] direction. The thermal conductivity decreases over a
broad temperature range, with maximum decrease occurring at T ∿
15-20 K. Optical measurements indicate the presence of a broad
deformation-induced absorption band (7) at 5.7 eV, which is probably caused by higher order point defects. Also, an MgO crystal
deformed 2% can be expected to contain ∿2 x 10^{16} F^+ centers/cm^3 (7).
In addition, dislocation dipoles and other defects are also produced during deformation. Hence, at this time it is not certain
whether it is the anion vacancies, other defects, or dislocation
dipoles that are responsible for the decrease in thermal conductivity. However, following annealing at ∿1000 K, the thermal
conductivity over part of the temperature range approaches values
obtained for the undeformed sample. At this temperature, most of
the point defects, including anion vacancies, are annealed out,
and some of the dislocation dipoles are converted into dislocation
loops (8).

*Research sponsored by the U.S. Energy Research and Development
 Administration
**Operated by Union Carbide Corporation for the U.S. Energy Re-
 search and Development Administration

(1) D.S. Kupperman, G. Kurz, and H. Weinstock, J. Low Temp. Phys.
     10, 193 (1973).
(2) D.S. Kupperman, H. Weinstock, and Y. Chen, J. Low Temp. Phys.
     14, 277 (1974).
(3) Y. Chen, R.T. Williams, and W. A. Sibley, Phys. Rev. 182,
     960 (1969).
(4) Y. Chen et al., J. Phys. C 3, 2501 (1970).
(5) Y. Chen, J.L. Kolopus, and W.A. Sibley, Phys. Rev. 186,
     865 (1969).
(6) B. Henderson and R.D. King, Phil. Mag. 13, 1149 (1966).
(7) W.A. Sibley, J.L. Kolopus, and W.C. Mallard, Phys. Stat.
     Sol. 31, 223 (1969).
(8) J. Narayan and J. Washburn, Phil. Mag. 26, 1179 (1972).

PHONON SCATTERING BY Ni^{3+} IONS IN AL_2O_3

By M. LOCATELLI and A.M. de GOËR
Centre d'Etudes Nucléaires de Grenoble
Service des Basses Températures
BP 85 Centre de Tri - 38041 GRENOBLE-CEDEX (France)

Previous work on the low temperature thermal conductivity K of Ni doped Al_2O_3 single crystals has shown that Ni^{3+} ions give rise to resonant phonon scattering at two frequencies (T_1 =.5K and T_2 = 77K). The lineshape of the phonon relaxation time for the lowest resonance was shown to be quite asymetric and supposed to be due to random strains [1]. Recent work by A.P.R. and thermally detected E.P.R. has given a more precise knowledge of the energy levels ; the zero-field splitting of the fundamental vibronic doublet is determined to be .68K and the tunnelling splitting δ is supposed to be large [2]. The present work concerns a study of the lineshapes of the resonant phonon relaxation times. In a first step, quantitative analysis of previous results have been done again using T_1 =.68K and needs the introduction of an "intrinsic" linewidth (which was previously neglected) about 10^3 smaller than the strain broadened width. Two half-lorentzians, or two half-gaussians, give good fits for two samples of similar Ni^{3+} content. The high temperature resonance is well described by an elastic scattering process. To test the validity of these lineshapes, a series of measurements in function of Ni^{3+} concentration has been done.

The thermal conductivity of several samples, listed in table I, has been measured between 50 mK and 100K ; the influence of different heat treatments has been studied on sample J_{2A}. Examples of results are given in figure 1. It was not possible to obtain good fits for all curves with two half-lorentzians or two half-gaussians for the lowest temperature resonance. Therefore we use a more complicated resonant relaxation time, based on the following physical assumptions : (i) the intrinsic process is a direct process, with a lorentzian lineshape (ii) the distribution of strain ε is gaussian (iii) the variation of the level splitting with strain is quadratic (small strains) (iv) the width of the corresponding distribution of the

TABLE 1

SAMPLE	C (ppm W)	HEAT TREATMENT (°C) T	(h) t	Gaz	SURFACE	CURVE	η_o (μ)	Cc (ppm W)	ε_o ×10⁵	G (s⁻¹) ×10⁻¹⁰	D₁ (s⁻¹) ×10⁻⁴	D₂ (s⁻¹) ×10⁻⁹
60%/c	10					A	·18	·47	32	37	1	
// b	10					B	·14	·2	29	30	1	
// c	10					C	·24	·2	32	37	2	
B	<5					D	·16	·015	27	26	·05	
J2A	10	1600	8	O₂		H	·13	10	5	·9	90	35
"	"	1100	17	H₂		F	·13	5–3	5	·9	18	11
"	"	1600	8	O₂		G	·11	10	5	·9	90	37
"	"	"	"	"	roughened	H	·16	10	5	·9	90	37
"	"	1100	17	H₂	roughened	E	·16	0			0	

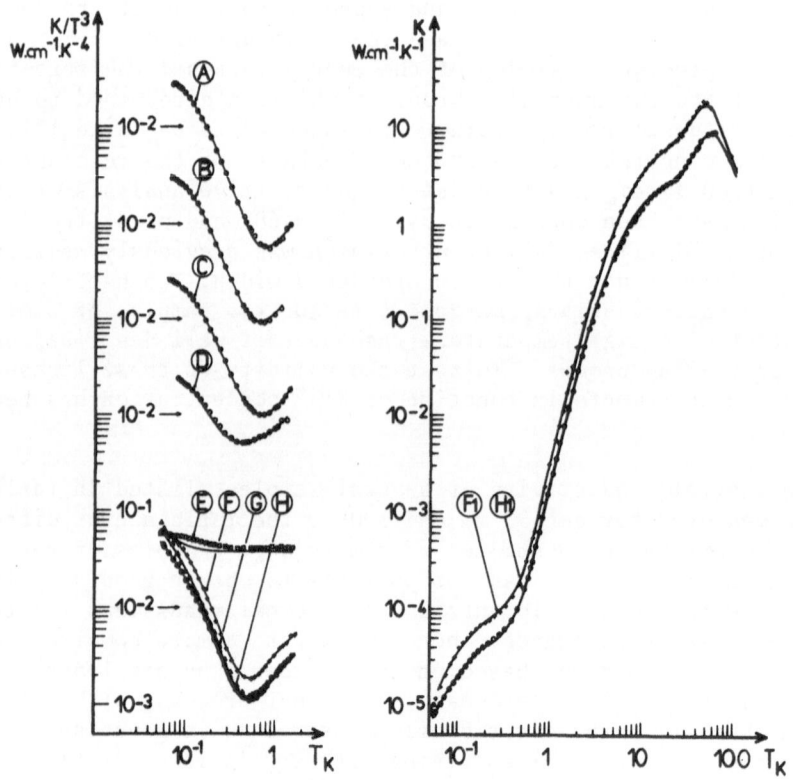

FIGURE 1   K and K/T³ versus T, Experimental points and calculated curves.

resonance frequencies is very much larger than that of the intrinsic
lorentzian. The following expression are then obtained :

$$\tau_1^{-1} = E_1 \int_{\omega_0}^{\infty} \frac{\omega}{(\omega_1-\omega)^2} \frac{\exp\left[-(\omega_1-\omega_0)/a\varepsilon_0^2\right]}{(\omega_1-\omega_0)^{1/2}} d\omega_1 \qquad \text{for } \omega > \omega_0$$

$\tau_1^{-1} = D_1 \ \omega/(\omega_0-\omega)$ for $\omega < \omega_0$; $\omega_1 = \omega_0 + a\varepsilon^2$ ($\hbar\omega_0 = .47$ cm^{-1}). On the
other hand $\tau_2^{-1} = D_2 \ \omega^4/(\omega^2-\omega_2^2)^2$, with $\hbar\omega_2 = 55$ cm^{-1}. Furthermore,
we need to consider some specular reflexion of phonons as K/T^3 is not
constant at the lowest temperature (fig.1) and is larger than any
calculated value from elastic constants [3] ; then we use for the
Casimir term, the frequency dependant expression given in [4], which
is function of the effective roughness of the surface. The fits shown
in fig.1 have been obtained by adjusting the parameters D_1, E_1,
$(a\varepsilon_0^2)$, D_2 and η_0. Values for ε_0 can be deduced as $a = 3.6 \times 10^{18}$ s^{-1}
is known from experiments under stress [5]. The values of these para-
meters and of the concentration C_c of Ni^{3+} ions deduced from E_1 are
given in table I. We note that (i) the order of magnitude of the random strains is large, especially for the as-received samples (ii) C_c
is always smaller than the total Ni content C (iii) η_0 is smaller
than the roughness observed, which is about 2μ, and is changed by
heat treatments. (iv) D_1 and D_2 increase with C_c but the theoretical
linear variation is not exactly verifyed ; the values of D_1 are somewhat uncertain as there is an interference with the Casimir term ;
further work is needed on this specular reflexion as the fit is not
quite satisfactory in the case of very low Ni^{3+} concentration
(curve E).

The most important conclusions are : (i) the value of the inversion splitting $\delta = 55$ cm^{-1} is confirmed (ii) the analysis of the low
temperature phonon resonance gives information on the magnitude of
the random strains and on the actual concentration of Ni^{3+} ions in
the crystals.

REFERENCES

[1] LOCATELLI M. and de GOËR A.M. (1974) Sol. Stat. Comm. 14, 111
[2] ABOU-GHANTOUS M. et al. (1974) J. Phys. C. 7, 2707
[3] DOULAT J. et al. This conference
[4] CAMPISI C.J. and FRANKL D.R. (1974) Phys. Rev. B, 10, 6, 2644
[5] SALCE B. This conference.

DETERMINATION OF THE LINEAR JAHN-TELLER COUPLING CONSTANT
OF Ni^{3+} in Al_2O_3 BY LOW TEMPERATURE THERMAL CONDUCTIVITY
MEASUREMENTS UNDER UNIAXIAL STRESS

B. Salce

Centre d'Etudes Nucléaires de Grenoble
Service des Basses Températures
BP 85 Centre de Tri, 38041 Grenoble-Cedex (France)

In the investigation of Jahn-Teller systems, the study of Ni^{3+}
ion in Al_2O_3 is of great interest as it is a simple one, with an ^2E
ground state [1]. From previous experiments by A.P.R., thermally detected E.P.R. [2] and low temperature thermal conductivity [3], a
theoretical model has been developed including dynamic Jahn-Teller
effect [2]. The inversion splitting δ was supposed to be large so
it was possible to consider the vibronic ground state ^2E alone, which
was described by an effective orbital momentum $T = 1/2$. In zero magnetic field, the Hamiltonian is given by :

$$\mathcal{H} = a\, p\, T_3\, S_z + (1/2)q\, V_E\, (\overline{Q}_+\, T_- + \overline{Q}_-\, T_+) \qquad (1)$$

where p and q are the Ham's reduction factors [1] The zero-field
splitting a p/2 = 0.47 cm^{-1} was deduced from experiments [2]. The
second term is a perturbation due to electric fields or strains, and
$\overline{Q}_+$ the induced E type distortions. V_E is the linear Jahn-Teller coupling constant and it is possible to measure it by studying the effect of an uniaxial static stress $\vec{\sigma}$ on the energy levels. In that
case, the splitting varies linearly with $|\vec{\sigma}|$ in the limit of large
stress and is given by :

$$\Delta W_\sigma \simeq (2/\sqrt{3})q\, V_E\, b\, (\alpha_\theta^2 + \alpha_\varepsilon^2)^{1/2} \cdot |\vec{\sigma}| \qquad (2)$$

where α_θ and α_ε are the angular parts of the induced strains and b
the average ion-ligand separation. The constant bV_E can be determi-
ned using the value of q = 0.467 [4].

We have studied in detail three samples, made by the Verneuil method ; their characteristics are given in table I. We have measured
from 1.1K to 30K, with an apparatus described elsewhere [5], the re-

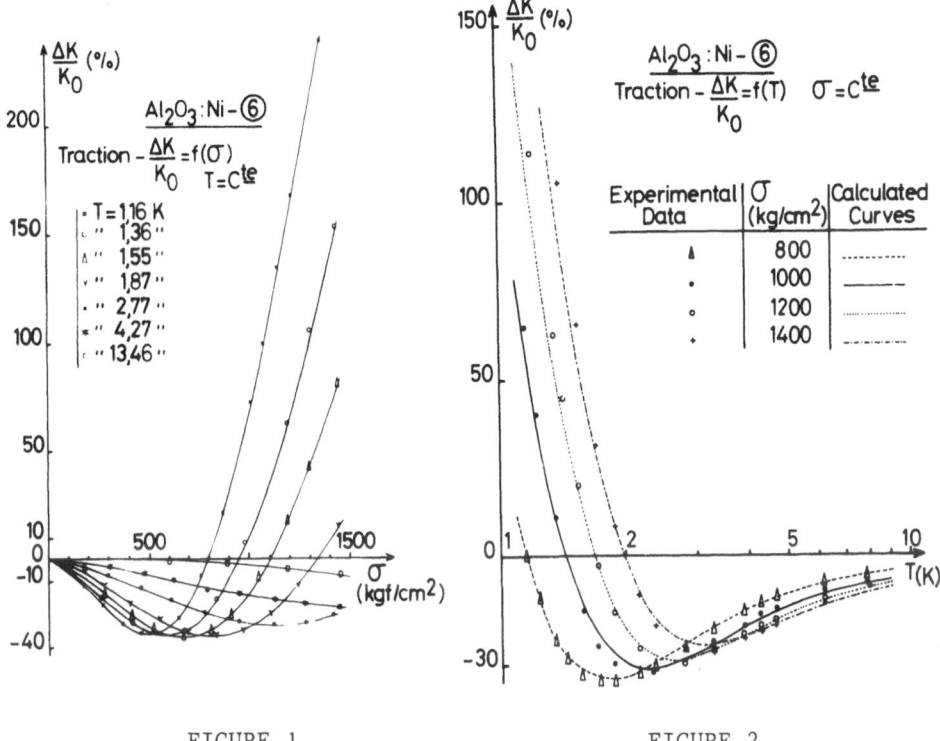

FIGURE 1   FIGURE 2

lative variation $f(\sigma)$ of the thermal conductivity at constant temperature $K(T = C^{te}, \sigma)$, when uniaxial stress $\vec{\sigma}$ is applied on the sample
parallel to the heat flow : $f(\sigma)=(K-K_0)/K_0$. Such experimental curves
are given on figure 1 as an example. From these results we deduce
another set of curves : $f'(T) = \Delta K/K_0$ with $\sigma = C^{te}$ (figure 2).

TABLE 1

Samples	C (ppm at)	θ^x (deg.)	ϕ^x (deg.)	$\alpha_\theta \cdot 10^9$	$\alpha_\varepsilon \cdot 10^9$	$\omega_1 (cm^{-1})$ $(\sigma=1000)$	$bV_E (cm^{-1})$
Ni (6)	< 10	89	8°30'	1.62	−9.53	4.46	51000 ± 10000
Ni (A)	< 5	76	− 23	2.28	4.68	6.39	50000 ± 20000
Ni (b)	10	37	indét.	∿2.4	∿5.5	4.76	35000 ± 20000

$^x\theta$ andϕ have been determined par M.TITEUX (Ugine-Kulhmann Research
Center Jarrie-France) using X Ray diffraction.

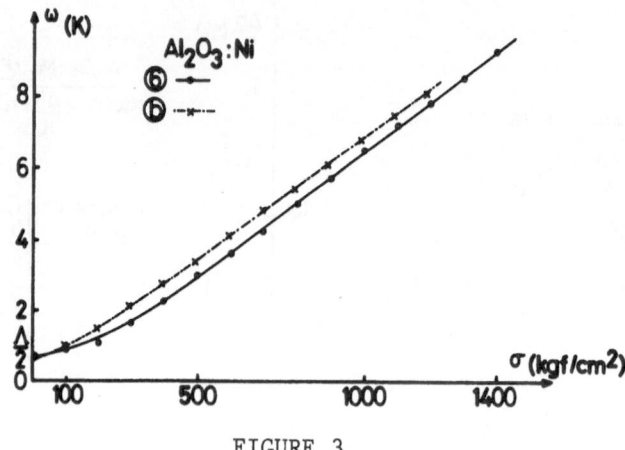

FIGURE 3

Quantitative analysis of the zero stress thermal conductivity K_0 has been achieved previously within the Callaway model by using two resonant phonon scattering terms (T_1=0.68K, T_2= 77K); the lowest one corresponds to the zero-field splitting and the inverse relaxation time τ_1^{-1} must be non symmetric [3].Possible shapes are described elswhere [6] and we use two half-lorenztian functions : τ_1^{-1}= Aω/(ω - ω_l)^2 below $\hbar\omega$ = 0.47 cm^{-1}and the same form above 0.47 cm^{-1} but the coefficient B beeing roughly 10^3 larger than A. The set of curves K(T, σ=Cte) has been fitted by adjusting ω_l, A and B(examples of fits are given in figure 2) and the variation of ω_l with σ is plotted on figure 3.

In agreement with theoretical prediction, ω_l varies linearly with stress for σ > 600 kgf/cm^2 and equation (2) can be used to deduce a value of bV$_E$; α_θ and α_ε were calculated using the known elastic constants of Al$_2$O$_3$[7] and the values of the angles θ and ϕ giving the orientation of the stress relative to the crystalline axis. The values of θ,ϕ,α_θ,α_ε, bV$_E$ are given in table I. We obtain a mean value bV$_E$ = 45000 cm^{-1} $\pm$ 15000 cm^{-1}, quite compatible with a previous estimation [8]. We hope to improve the precision of this determination by using good single crystals with convenient orientations.

REFERENCES

[1] HAM F.S.(1968) Phys.Rev.166,2,307
[2] ABOU-GHANTOUS M. and al.(1974) J.Phys.C.7, 2707
[3] LOCATELLI M. and DE GOER A.M.(1974) Sol.Stat.Comm.14,111
[4] ABOU-GHANTOUS M. and al.(1974) Phys.Rev.Letters 33,N°9,530
[5] RIVALLIN J.(1974) Thèse d'Etat, Grenoble
[6] LOCATELLI M. and DE GOER A.M. This Conference
[7] TEFFT W.E.(1966) J.Research N.B.S.A.70A, N°4, 277
[8] HAM F.S.(1972)in Electron Paramagnetic Resonance edited by
    S.GESCHWIND(N.Y.Plenum) p 1.

PHONON SPECTROSCOPY IN AL_2O_3 DOPED WITH
TRANSITION METAL IMPURITIES

H. Kinder[+] and W. Dietsche[+]

Institut für Festkörperforschung der KFA Jülich

We have studied Al_2O_3 with a variety of transition metal impurities by using monochromatic phonons in a frequency range of 120 to 870 GHz. This frequency range was achieved with superconducting tunneling junctions made of PbBi alloy. Transition metal impurities in Al_2O_3 are interesting because of their low lying electronic states which have been extensively investigated in the past by thermal conductivity and infrared techniques. We can obtain additional information because of the high resolution as compared to the conventional heat conductivity and because of the different selection rules as compared to infrared measurements.

The first example of this class of experiments was V^{3+} which has been previously described in detail.[1] With the higher frequency range which is now available we have also investigated the V^{4+} state. This state can be obtained by γ-irradiation.[2] Fig.1 shows the result of a transmission experiment. The intensity of transverse phonons propagating in the c-direction is plotted as a function of frequency. The resonant scattering dip at 860 GHz is attributed to V^{4+}. This has a single d electron whose lowest orbital doublet state in the trigonal crystal field is split by spin orbit coupling. The transition has been also observed by infrared spectroscopy.[3] A quantitative determination of the "spin"-phonon coupling constants was not yet possible because the concentration of V^{4+} ions was very low.

The γ-irradiation also leads to the appearance of

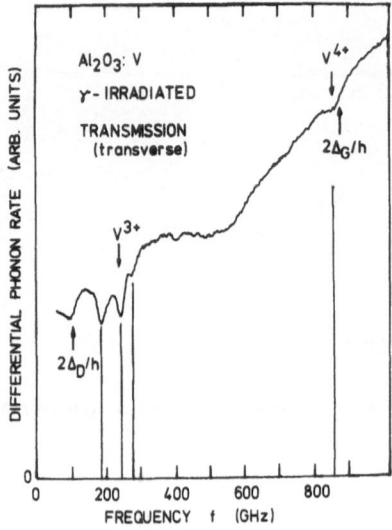

Fig.1: Transmission spec-
trum of sapphire with V-
ions after γ-irradiation.
The lines at 860 GHz and
250 GHz are due to V^{4+}
and V^{3+}, respectively.
The two additional low
frequency lines and the
band at 500 to 600 GHz
are not identified yet.

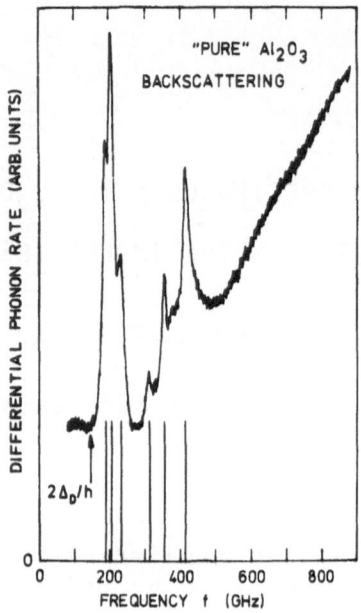

Fig.2: Resonance fluorescence
spectrum of unknown impurities in nominally pure sapphire. The sharp peaks demonstrate a very good frequency
resolution.

two absorption lines in addition to the V^{3+} line at lower
frequencies (194 and 288 GHz). These lines could be possibly explained by a locally changed environment of some
of the V^{3+} ions. However, they could also be due to the
"unknown impurities"[4] whose valency may have changed by
γ-irradition. The corresponding absorption lines show always up in our purest Al_2O_3 samples. They disappear by
doping with V,Cr,Fe, or Ti, and are diminished by annealing at 1200K in O_2, probably by the reverse valency change.
Fig.2 shows the resonance fluorescence spectrum of these
unknown impurities as obtained by a PbBi junction generator. The resonance fluorescence is observed by a detector which is close to the generator junction on the same
crystal surface.[1] In this arrangement, the resonances
lead to peaks in the backscattered (reemitted) intensity,
if the phonon mean free path is always larger than the
distance between generator and detector.

      In the case of Al_2O_3:Mn^{3+} a dynamic Jahn-Teller ef-
fect with a complicated level scheme was expected theo-

retically.[5] However, only one weak line was observed by
IR spectroscopy.[6] Fig.3 shows our results[2] with a sample
of 15 mm length containing about 10ppm Mn^{3+}.

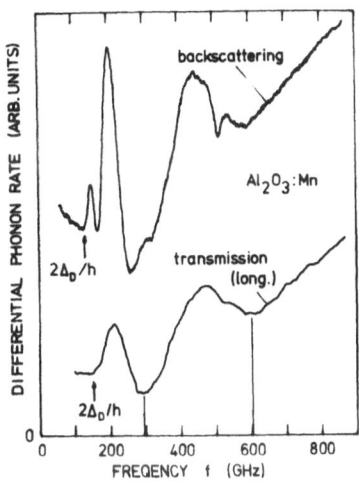

Fig.3: Transmission of longitudinal phonons (lower trace) and
backscattering of phonons (upper
trace) for Mn^{3+} in sapphire. The
complex structure and the strong
scattering are consistent with a
Jahn-Teller effect.

In transmission, only a weak
longitudinal and no transverse
pulse was observed. The transmission spectrum for longitudinal
phonons is shown as the lower
trace. Because of the low intensity, a poor frequency resolution
was used. The main absorption
line at 300 GHz is consistent

with the IR 9.3 cm^{-1} line. The result for the backscattering geometry is shown as the upper trace of Fig.3. A
good frequency resolution revealed a far more detailed
structure. Due to an extremely strong scattering at the
resonances the phonon mean free path was shorter than the
distance between the junctions. Therefore, the centers of
the resonance lines fold back to yield minima instead of
peaks expected for weaker scattering.[7] This makes an unambiguous identification of the transitions difficult as
yet. However, the strength of the scattering and the complexity of the spectrum qualitatively confirm the Jahn-
Teller effect. Measurements with crystals containing less
Mn^{3+}ions and using smaller junctions are in preparation.

+ present address: Physik Department der Technischen
Universität München, 8046 Garching, West-Germany

1. H.Kinder, Z.Physik 262, 295, (1973)
2. in cooperation with A.M.de Goër, CEN Grenoble
3. J.Y.Wong, M.J.Berggren,and A.L.Schawlow, J.Chem.Phys.
   49, 835 (1968)
4. H.Kinder, Int.Conf.on Phonon Scattering in Solids,
   ed.by H.J.Albany (CEN Saclay 1972), p.284
5. C.A.Bates, P.C.Jaussaud, and W.Smith,J.Phys.C6,898
   (1973)
6. J.H.M.Stoelinga, P.Wyder, L.J.Challis, and A.M.de Goër,
   J.Phys. C6, L486 (1973)
7. K.W.H.Stevens, L.J.Challis, private communication

AN INVESTIGATION OF FREQUENCY CROSSING EFFECTS IN PHONON

SCATTERING BY V^{3+} IONS IN Al_2O_3

L J Challis and D L Williams

Department of Physics, University of Nottingham

University Park, Nottingham NG7 2RD, U K

V^{3+} ions in Al_2O_3 have been studied by a variety of techniques which have shown that in zero magnetic field the lowest states form a doublet ~ 8 cm^{-1} above a ground state singlet. They can be described by the spin Hamiltonian

$$\mathcal{H} = D(J_z'^2 - \tfrac{2}{3}) + g_{11}\beta H_z J_z' + g_\perp\beta(H_x J_x' + H_y J_y') \qquad (1)$$

with J' = 1. In general resonant phonon scattering occurs at three frequencies ω_{10}, ω_{-10} and ω_{-11} where $\hbar\omega_{ij} = |E_i - E_j|$, the difference in energy of two levels. In a magnetic field these frequencies change, a 'frequency crossing' occurs when two of these frequencies become equal. From (1) it can be shown that frequency crossings occur at
H = 0; $H = H_0(\tfrac{1}{2}(2+\alpha^2)\cos^2\theta - \tfrac{1}{2}\alpha^2)^{-\tfrac{1}{2}} = H_0 f(\theta)$; $H = 3H_0$, for $\theta = 0$;
where θ is the angle between the field and the c-axis, $g\beta H_0 = (1/3)D$ and $\alpha = g_\perp/g_{11}$. The positions of the three crossings are shown in the detail of fig (1) for $\theta = 0$. The crossing at $H = H_0 f(\theta)$ is a pure frequency crossing (no level crossing occurs), so it should give rise to a minimum in the thermal resistivity[1].

Measurements have been made on a rod containing 650 at.ppm of vanadium[2], the axis of the rod was along the a-direction. The magnetoresistance was plotted continuously against field for directions in the b-c plane and at temperatures between 1 and 4K. Examples of the resistivity in the region of the frequency crossing are shown in fig (2). When θ is small, the crossing signal consists of a central minimum with two or more minima on each side but as θ increases the minima broaden and the structure changes.

When $\theta = 0$, the central peak occurs at $H = H_0 = 30.71 \pm .3$kOe

where the main uncertainty is in the magnet calibration. Now
$D = 3g_{11}\beta H_o$ and if we assume $g_{11} = 1.915 \pm 0.002$[3] we obtain
$D = 8.13 \pm 0.08$ cm^{-1} which is in good agreement with other values
such as the value obtained from infra-red spectroscopy of $8.25 \pm$
0.02 cm^{-1} [4]. A plot of H^{-2} against $\cos^2\theta$ is linear, fig (1), in
agreement with the expression derived from the spin Hamiltonian.
From the slope of the line, $(2 + (g_\perp/g_{11})^2)/2H_o^2$, and using the
measured value of H_o and $g_{11} = 1.915$ we obtain $g_\perp = 1.46 \pm 0.12$.
This value is significantly smaller than previous values which lie
close to the infra-red value of 1.74 ± 0.02[4].

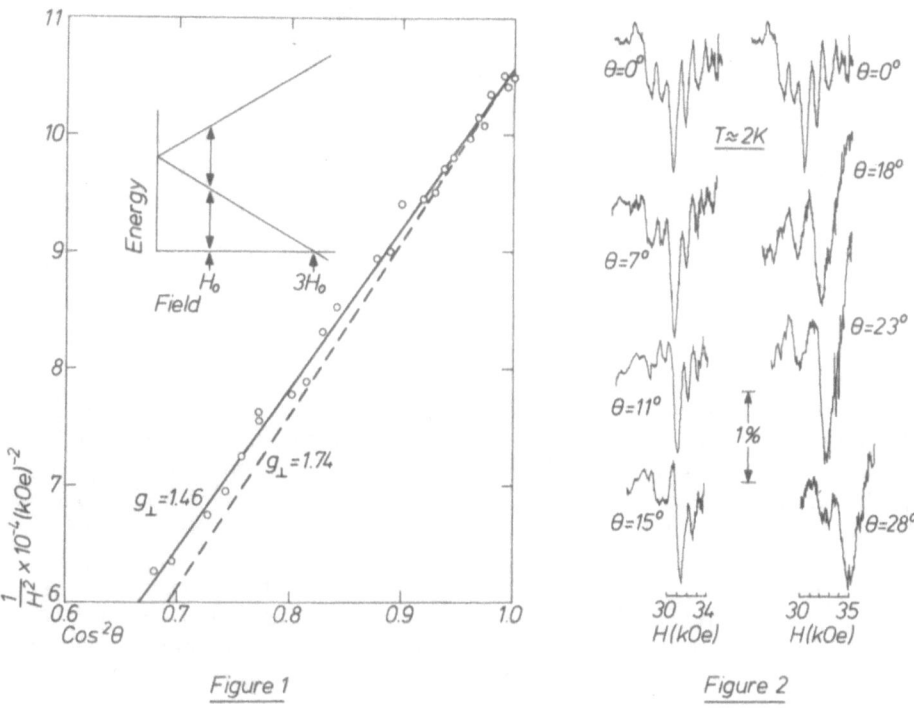

Figure 1   Figure 2

We have not yet identified the cause of the structure of the
frequency crossing signal although some information on this was
obtained from making measurements on the same sample after γ-irradiation. The effect of this is to increase the area of the satellite minima at a particular temperature by ∿ 20% relative to the
area of the central minimum. The area of the whole signal is decreased by ∿ 10% which we associate with the conversion of V^{3+} to V^{4+}
produced by γ-irradiation of similar samples[5]. This could suggest

that the satellite minima are due to V^{3+} ions in sites differing
from those of the central minimum by the presence of nearby charges. The separation of the minima suggests crystal field splittings of order 0.5 cm^{-1} indicating that the effects observed are due
to charges many lattice spacings away. This could perhaps account
for the apparently large proportion of 'distorted' sites. Measurements on other specimens are planned and work is in progress to
identify the nature of the distortion. We note that the frequency
crossing technique can be rather sensitive to ions of low abundance.
As an illustration of this in the present case let us suppose that
the sites differ only in the value of D. The minima occur when the
upper transition frequency ω_{-11}, which is the same for all sites,
equals ω_{-10} which depends linearly on D, i.e. at H_o, $H_o(D'/D)$ etc.
Now when kT << D, the ω_{-11} transition is very weak because of dep-
opulation. We suppose it is much weaker than the ω_{-10} transition
corresponding to the D and D' sites. In this limit, the depth of
the minimum is the same for the D and D' ions and the ratio of
their widths varies only as the square root of their abundance
ratio. This enhancement feature is possibly the reason why the
structure has been seen in frequency crossing but not in work
using infra-red or monochromatic phonons[6].

An attempt has also been made to look for the crossing effects
expected at high fields. For $\theta = 0$, the $|-1>$ level should cross
the $|o>$ level at H = $3H_o \approx 92kOe$. In practice strains or any small
misalignment of the field will result in level anticrossing, i.e.
the two levels approach closely but are repelled by each other.
Recent work[7] has shown that the effect of state mixing in the region of the anticrossing is to produce a Lorentzian maximum in the
thermal resistivity. In the V^{3+} system, the transition frequency
at the anticrossing is 2D, i.e. 16.3 cm^{-1} so that we should expect
the anticrossing maximum to be largest at T ∿ 5K. High field
measurements have been made in the region of the anticrossing and
at temperatures from 4 to 8K but we have so far been unable to
detect the Lorentzian signal. This could suggest that the phonon
matrix elements between the $|1>$ state and the two other states are
similar so that state mixing has only a small effect.

1. e.g. Berman R, Brock J C F and Huntley D J, 1963, Phys Lett 3,
        310; Anderson B R and Challis L J, 1975, J Phys C 8, 1475.
2. We are very grateful to Mme de Goer for the loan of this
        specimen.
3. Zverev G M and Prokhorov A M, 1961, Soviet Physics JETP 13, 714.
4. Joyce R R and Richards P L, 1969, Phys Rev 179, 375.
5. de Goer A M and Devismes N, 1972, J Phys Chem Solids 33, 1785.
6. Kinder H, 1972, Phys Rev Letts 28, 1564.
7. Anderson B R and Challis L J, 1974, J Phys C 7, L440; and 1975,
        8, 1495.

AN INVESTIGATION OF THE POINT DEFECT SCATTERING OF
PHONONS BY MINUTE CONCENTRATIONS OF Cr^{4+} IN
CORUNDUM BY THERMAL TECHNIQUES

M.A. Brown†, I.A. Clark* and W.S. Moore*
†Department of Physics
Loughborough University of Technology
*Department of Physics
Nottingham University, England

We report the results of a number of different
experiments on a series of magnesium-doped Al_2O_3
specimens I-IV which were manufactured by Ugine
Kuhlmann (38-Jarrie, France) and for which the manufac-
turer's analysis of impurities is shown in Table 1. The
original thermal conductivity experiments of Brown (1973)
revealed the presence of strong resonant phonon
scattering at ∿19K in specimen I and further thermal
conductivity and thermally-detected EPR experiments
(Brown 1975, Clark and Moore 1975) have identified the
low-lying levels of the 3d' ion Cr^{5+} as responsible for
the resonant scattering.

During the course of the EPR experiments, Cr^{4+}
was always detected in very small amounts (∿1 ppm) and
so an attempt was made to find the distribution of the
total chromium between the possible valency states 3+
4+ and 5+. It can be seen from the relative concentrations $[Cr^{4+}]$ of Cr^{4+} measured by EPR and normalized
to specimen I in Table I that this implies that only a
very small fraction of any chromium in excess of a few
parts per million is converted to Cr^{4+} and stabilized
by the presence of magnesium (presumably as Mg^{2+}).

An even smaller fraction of the chromium (∿ 0.2ppm)
was found to be in the 5+ state in specimen I, and in
specimen III Cr^{5+} was detected and its concentration
estimated to be less than 0.02 ppm. The Cr^{3+} concentration was examined by conventional EPR, when it was
found that signals could be obtained from only specimens
II and IV and that the ratio of the Cr^{3+} intensity seen

205

Table I

conc-atomic ppm

Specimen No.	Mg	Cr	Fe	$[Cr^{4+}]$	A
I	∿30	<1	<3	1.0	1.0
II	∿80	∿80	<3	4.6	4.5
III	∿30	<1	30	0.4	0.5
IV	∿9	7	<3	1.6	2.0

for these was almost that which would be obtained assuming that all the chromium was in the 3+ state.

It seems therefore that the presence of magnesium at a concentration exceeding that of all other impurities tends to stabilize the higher valency states of chromium.

Phonon scattering in 'pink' ruby and so called 'orange ruby' has been the subject of a study by Brown et al (1972). For the 'pink' rubies, apart from a rather large amount of point defect (Rayleigh) scattering in the most dilute specimen, the additional point defect scattering found as the chromium concentration was raised could be accounted for satisfactorily by assuming that the majority species, Cr^{3+}, in these specimens behaved as a simple mass defect scatterer, obeying quantitatively the standard Klemens (1951) formula. When a similar analysis of the specimens I-IV was attempted, and the measured thermal conductivity data was fitted to the full Callaway (1959) model, a most striking correlation appeared. The relative coefficient of the point defect scattering term, A, normalized to specimen I in Table I is seen to follow, well within experimental error, the concentration of Cr^{4+} as measured by thermally detected EPR This is unexpected, especially since the two sets of measurements were made independently, but even more unexpected is the fact that when the Klemens formula is used with the known concentration of Cr^{4+} ions to estimate the mass defect of the scattering centre, a value corresponding to the total mass of five unit cells of Al_2O_3 per unit cell is obtained. Even if all the impurities Mg + Cr + Fe are assumed to act as simple independent mass defect scatterers, a value is obtained for the mass defect coefficient A which is ten times smaller than its value for specimen I. Clearly the scattering centre cannot be simple, but must be the same centre that was responsible for the anomlously large observed point defect term observed in the weakest of the orange rubies of the series measured by Brown et al (loc cit). The iron present cannot per se be responsible for the large point defect scattering, since specimen III with the largest iron concentration shows the least scattering, and in addition it is known that there is less iron in specimen IV than in I or II.

There remains the problem of the detailed structure of the scattering centre, and any model has to explain not only the correlation with measured Cr^{4+} concentration and the very high apparent mass defect, but also the large value of the point defect scattering found in so-called 'pure' Al_2O_3 (Brown 1972, 1975). Until further experiments are done with other Al_2O_3 specimens, it is difficult to separate the effects of magnesium and other possible lattice defects such as cation vacancies on the stabilization of Cr^{4+}. The effects seen by us may simply be a reflection of the tendency of single chromium ions to stabilize in their 4+ state, naturally occurring or magnesium-induced micro-regions of imperfection in the Al_2O_3 lattice. The large mass defect could then be accounted for by the resultant complex centre formed in terms of the coherent scattering of several mass defects and the associated lattice strain.

References

BROWN, M.A. de GOER A.M., DEVISMES N, and VILLEDIEU M., (1972) Proc. Int. Conf. Phonon Scattering in Solids, Paris, 272-6.
BROWN M.A., (1973), J. Phys. C, 6, 642-649.
BROWN M.A., (1975) in preparation for J. Phys. C.
CALLAWAY, J., (1959) Phys. Rev., 113, 1046-51.
CLARK I.A. and MOORE W.S., (1975) in preparation for J. Phys. C.
KLEMENS P.G., (1951) Proc. Roy. Soc. A208, 108.

A STUDY OF PHONON-SPIN INTERACTION IN THE SYSTEM $3d^1$ in Al_2O_3

By N. DEVISMES and A.M. de GOËR
Centre d'Etudes Nucléaires de Grenoble
Service des Basses Températures
BP 85 Centre de Tri - 38041 GRENOBLE-CEDEX (France)

The low lying energy levels of Ti^{3+} and V^{4+} ions in Al_2O_3 are well known and are described by a model including dynamic Jahn-Teller effect in the fundamental orbital triplet $^2T_{2g}$, the coupling to Eg modes being only considered [1]. These systems are suitable to study phonon-spin interactions, as transitions in zero magnetic field between the Kramers doublets can be observed by low temperature thermal conductivity measurements [2]. Previous results have been analysed within the Callaway model with an additional phonon relaxation time :

$$\tau^{-1} = D_1 \ [\omega^4/(\omega^2-\omega_1^2)^2 + (D_2/D_1) \ \omega^4/(\omega^2-\omega_2^2)^2] \ ;$$

the frequencies ω_1 and ω_2 corresponding to the zero field splittings [3] are given in table I. This is an approximate expression for elastic scattering processes [4], as derived in the case of phonon scattering by acceptors in semiconductors [4]. A similar calculation for d^1 ions in Al_2O_3 gives a theoretical expression of D_1 and D_2, then the validity of the elastic scattering process can be tested and orbit lattice coupling constants deduced from the experimental values of D_1.

TABLE I

	$\hbar\omega_1$ cm^{-1}	$\hbar\omega_2$ cm^{-1}	$(\frac{D_2}{D_1})_{exp}$	$(\frac{D_2}{D_1})_{th}$	$(aV_E)_{exp}$ cm^{-1}	$(aV_E)_{th}$ cm^{-1}
V^{4+}	28.1	53	50	3	9 000 ± 1 000	16 000 cm^{-1}
Ti^{3+}	37.8	107.5	100	8	> 6 000	20 000 cm^{-1}

The thermal conductivity K of single crystals Al$_2$O$_3$: V and Al$_2$O$_3$: Ti has been measured between 1.4 and 100K. Samples with a total concentration of Vanadium varying in a larger range than previously [2] have been studied [59 to 2800 ppm weight]. γ irradiation was used to increase the V^{4+} concentration. Examples of experimental results are given in figures 1 and 2 and solid lines have been calculated by adjusting the coefficient D$_1$ alone ; the ratio D$_2$/D$_1$, given in table I, was kept constant and all the experimental curves have been such fitted [5].

Theoretical phonon relaxation time for elastic scattering process by d^1 ions has been derived in the line of [4], with the following hypothesis : (i) the two doublets involved in each resonance are considered as isolated, (ii) the coupling is supposed to be only with E$_g$ modes, then the perturbation hamiltonian is written :
$\mathcal{H}_{int}$ = V$_E$ ($\bar{Q}_2$E$_\varepsilon$+$\bar{Q}_3$E$_\theta$) [6], E$_\theta$ and E$_\varepsilon$ are orbital operators, $\bar{Q}_2$ and $\bar{Q}_3$ the E-type distorsions of the cluster induced by the phonons, and V$_E$ is the coupling constant. We use the vibronic eigenstates given in [7], and the following expression of τ^{-1} was obtained :

$$\tau^{-1} = K(aV_E)^4 \frac{N}{V} \underset{i=1,2}{\Sigma} \frac{1+\exp\ (-\hbar\omega_i/kT)}{Z} (\omega_i^2+\omega^2) \frac{\omega^4}{(\omega^2-\omega_i^2)^2} \quad (1)$$

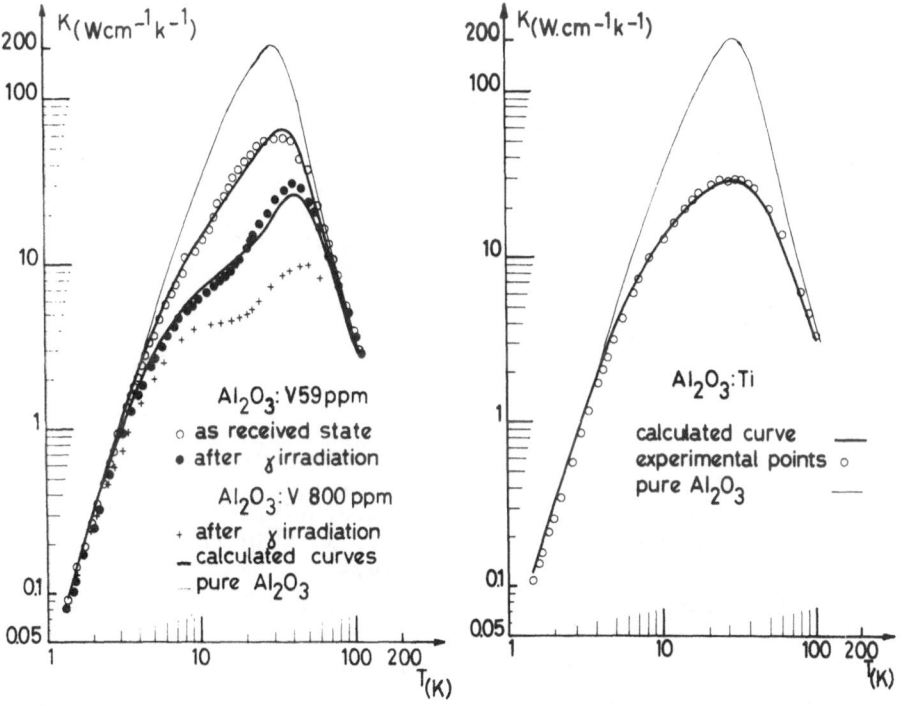

FIGURE 1   FIGURE 2

Z is the partition function, N/V the number of ions per unit volume, a the mean ligand-ion separation and K a constant depending only on the properties of the matrix. Everything is known, in principle, except V_E. We note two differences with the expression used to fit the experimental results : (i) there is an additional frequency dependance in (1), so that the comparison can be done only at resonance, (ii) the theoretical coefficients $D_{1,2}$ depend on temperature and the "experimental" ones do not. In fact, at the temperature where the effect of a resonance is maximum on the K(T) curves, nearly $T_{1,2} = \hbar\omega_{1,2}/3.8k$, the exponentials are very small ($Z \approx 1$) and this temperature dependance can then be neglected, so that

$$D_{1,2} = 2K \ \frac{N}{V} \ (aV_E)^4 \ \omega_{1,2}^2 \tag{2}$$

at resonance. The ratio D_2/D_1 is equal to ω_2^2/ω_1^2 and these theoretical values are given in table I. They are larger than one, as was found experimentally but this is a lack of quantitative agreement, which can be due to the simplified form used in the analysis. Then we have tested the influence of the additional term in ω^2 on the curve fitting and found that the experimental ratio D_2/D_1 can be decreased down to about 6 [5], so that the final agreement is not bad. On the other hand, the experimental values of D_1 are not very much changed and can be used to deduce a value of (aV_E) from expression (2), if the concentration N/V is known. In the case of V^{4+} ions, the concentrations are known from a systematic study of γ irradiation effects by three technics [5] : N/V is 1% of the total vanadium concentration and the linear variation of D_1 with N/V has been verified. In the case of Ti^{3+} ions, we can give only a lower limit of (aV_E), supposing that all the titanium ions are in the trivalent state. These "experimental" values of (aV_E) are given in table I, as well as the values calculated in a point charge model, which are larger, but of the same order of magnitude. This is satisfactory, in view of the number of simplifying assumptions done in both cases.

REFERENCES

[1] MACFARLANE R.M. and WONG J.Y, Phys. Rev. 166, 250 (1968)
[2] de GOËR A.M. and DEVISMES N, JPCS, 33, 1785 (1972)
[3] JOYCE R.R. and RICHARDS P.L, Phys. Rev. 179, 375 (1969)
[4] SUSUKI K. and MIKOSHIBA N, Phys. Rev. 3, 2550 (1971)
[5] DEVISMES N. Thèse d'Etat Grenoble 1975
[6] BATES C.A, CHANDLER P.E. and STEVENS K.W.H, J. Phys. C : Solid
     St. Phys, 4, 2017 (1971)
[7] ABOU-GHANTOUS M, BATES C.A. and STEVENS K.W.H, J. Phys. C : Solid
     St. Phys, 7, 325 (1974).

TEMPERATURE DEPENDENCE OF THE LIFETIME OF 29 CM^{-1}

PHONONS IN RUBY BETWEEN 3 AND 17 K

G. Pauli, W. Eisfeld and K. F. Renk

Universität Regensburg, Fachbereich Physik

8400 Regensburg, W.-Germany

By applying an optical phonon detection method (1) we measured the lifetime of 29 cm^{-1} phonons in ruby between 3 K and 17 K. From a blackbody spectrum of heat pulses we detect only the 29 cm^{-1} phonons: these are absorbed due to electronic transitions between the R_1- and R_2-levels of optically excited Cr^{3+} ions in ruby and give rise to an increase δR_2 of the R_2-fluorescence intensity (fig.1b). It has been shown (1,2) that for a high concentration of excited Cr^{3+} ions the 29 cm^{-1} phonons can be captured by resonant scattering in a small detector volume (fig. 1c). Therefore, the decay of the δR_2 signal after heat pulse injection is a measure of the lifetime of 29 cm^{-1} phonons.

The R_2 radiation transmitted through a monochromator was detected by the photon counting technique. Time resolved spectra were obtained with a time to pulse height converter triggered by the heat pulse generator. Under the condition of strong resonant capture at constant optical pump intensity we measured the time dependence of the R_2-fluorescence for different crystal temperatures. We find (fig. 1a) a time-independent background (strongly dependent on crystal temperature), a steep increase in intensity (δR_2) caused by the injection of the phonon pulse and a decay of δR_2 corresponding to the finite lifetime of the captured 29 cm^{-1} phonons.

Our preliminary results (fig.2) show that between 3 K and 9 K the decay time of δR_2 is temperature-independent while above 9 K it decreases with increasing temperature.

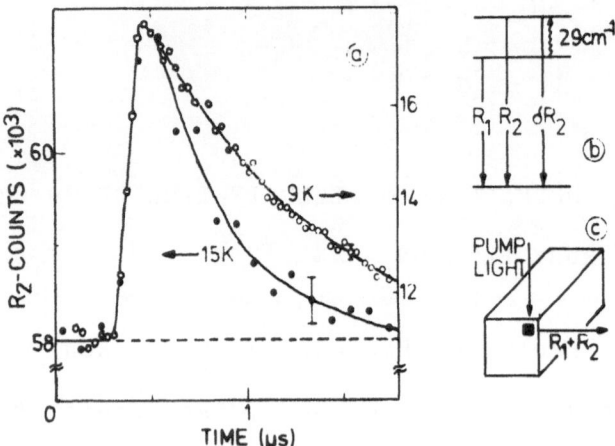

Fig. 1. (a) R_2-fluorescence after phonon injection (dashed
line: thermal background); (b) optical detection of
29 cm^{-1} phonons; (c) arrangement (10x10x17 mm^3 ruby crystal with 0.01 mol % Cr^{3+}, 1 mm^2 constantan heater H).

These results for the phonon lifetimes may be interpreted by a simple phonon decay model. The low temperature decay time is probably the lifetime for spontaneous
decay of a "longitudinal" phonon because transverse
phonons cannot decay (3,4). Due to resonant scattering
the captured 29 cm^{-1} phonons change polarization and
direction, therefore, we probably measure the decay time
of the longitudinal phonon with the shortest decay time.
The experimental low temperature value for the phonon
lifetime agrees within an order of magnitude with
theoretical estimations (3,4).

The observed temperature dependence of the phonon
lifetime can be explained by thermally stimulated phonon
decay: assuming that a 29 cm^{-1} phonon decays into two
phonons of half the frequency, one obtains for the phonon
lifetime

$$\tau(T) = \tau(0) \cdot [1 + 2\bar{n}(14.5 \text{ cm}^{-1}, T)]^{-1}$$

where $\bar{n}$ is the thermal occupation number for 14.5 cm^{-1}
phonons at temperature T and $\tau(0)$ is the lifetime at low
temperature. The calculated curve (dashed line in fig. 2)
explains within the experimental errors the observed temperature dependence.

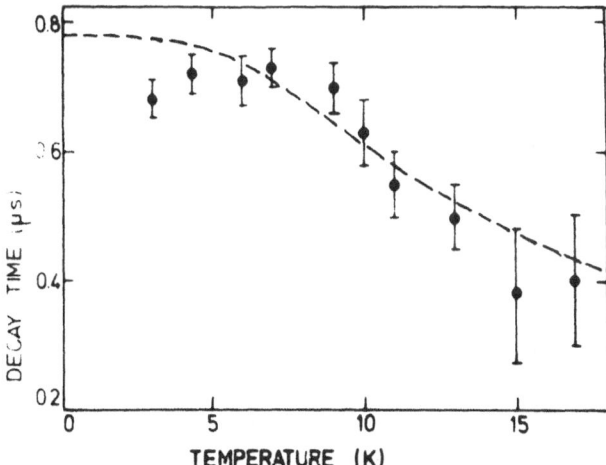

<u>Fig. 2.</u> Decay time of δR_2: experimental temperature de-
pendence (points) and phonon decay model (dashed).

    Our low temperature value for the phonon decay time
is lower by a factor of three compared with published
values (1,2). The discrepancy is probably due to wall
effects in the formerly used smaller crystals.

    Our results indicate that for the temperature range
up to 17 K the lifetime of 29 cm⁻¹ phonons in Al_2O_3 is
governed by phonon decay.

REFERENCES

(1) K.F. Renk and J. Deisenhofer, Phys. Rev. Letters
     <u>26</u>, 764 (1971).
(2) K. F. Renk and J. Peckenzell, J.Physique <u>C4</u>,103 (1972).
(3) R. Orbach and L.A. Vredevoe, Physics <u>1</u>, 91 (1964).
(4) P.G. Klemens, J. Appl. Phys. <u>38</u>, 4573 (1967).

A STUDY OF SPIN-PHONON INTERACTIONS BY SPIN-LATTICE RELAXATION PROCESSES

C A Bates, A Gavaix[†], P Steggles, H Szymczak*
A Vasson[†] and A-M Vasson[†]
Physics Department, University of Nottingham

University Park, Nottingham NG7 2RD, England

When a particular transition in a paramagnetic impurity has been saturated, the return to thermal equilibrium with the lattice does not necessarily follow a simple exponential. Even when it does the relaxation time τ_1 so defined cannot usually be predicted by theory, one problem being to take adequate account of the lattice dynamics. The situation is often even more complicated when the ground state of the monitored ion is not a simple doublet and when other impurities are present in the sample. The return to thermal equilibrium is then seriously affected by cross relaxation both within the monitored spin system and also to other fast relaxing ions. In such a situation it is not easy to relate the τ_1 measured experimentally to particular spin-lattice transition probabilities.

An extensive experimental programme of τ_1 measurements on many ruby samples containing various extra magnetic impurities and a careful theoretical analysis has removed many of these problems (1, and references therein). In fact these measurements themselves provide a powerful technique for detecting and studying ions which are strongly coupled to the lattice and hence exhibit dynamic Jahn-Teller effects.

We have made measurements of τ_1 by pulse saturation at 9.3GHz for different transitions within the Cr^{3+} : Al_2O_3 ground states

[†]Laboratoire de Radioélectricité et Théorie du Solide, Université de Clermont-Ferrand, BP45, 63170 Aubière, France

*Polish Academy of Sciences, Instytut Fizyki PAN, Warszawa, Lotników 32/40, Poland

and looked at how τ_1 varies with temperature T (in the liquid-
helium range) and with χ (the angle between the Zeeman field H
and the crystal c axis). Figure 1(a) shows a typical τ_1 against
χ plot. It illustrates two fundamental observations: (i) the
presence of resonant dips in τ_1 and (ii) a general non-resonant
reduction in τ_1 for all χ.

Cross relaxation within the Cr^{3+} spin system (e.g. 2) can
involve two or more spins and will often lead to a minimum in τ_1.
Examples of such processes are indicated in figure 1(a). Resonant
cross relaxation to an impurity ion F (fast relaxer) is also
possible. In the case of figure 1(a) this ion is Fe^{2+}. It has
been postulated that when such an effect cannot be explained by
a dipole-dipole interaction it is due to a quadrupole-electric-
field effect or to virtual phonon exchange (1). The shape of
such a dip in τ_1 is always in agreement with the resonance lineshape of the ion F involved.

The non-resonant effect at all χ is due to a process which
has been previously called cross-spin-lattice relaxation (3, 4).
This is, in fact, a second-order process involving both the F-
lattice hamiltonian and the dipole-dipole hamiltonian for the
Cr^{3+}-F interaction. A phonon of the appropriate frequency to
conserve overall energy is either emitted or absorbed.

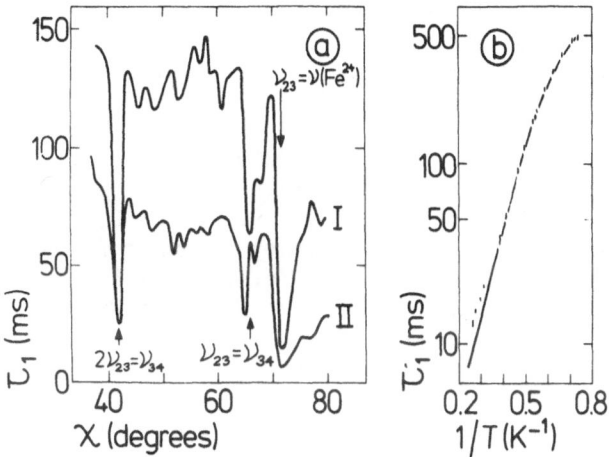

Figure 1(a) τ_1 against χ for the 2→3 high-field transition in
Cr^{3+} : Al_2O_3 at 4.2K. (The four Cr^{3+} levels have energies $E_n > E_{n+1}$).
Both samples are doped with Fe and Si. The concentration of Fe^{2+}
in II is about twice that in I. (b) τ_1 against $1/T$ for Cr^{3+} :
Al_2O_3 with zero Zeeman field for a sample doped with V. The experimental points are shown plus a full curve representing
$1/\tau_1 = 1.19T + 2250 \exp(-11.95/T)$ s^{-1}.

A resonant process involving F presents no problem in defining a time typical of cross relaxation since the Cr^{3+} rate equations are dominated by this process. From the T dependence of τ_1 under these conditions it has been possible to calculate the zero-field splitting D of the F ground state with excellent results (1, 5). In the case of non-resonant cross relaxation all Cr^{3+} transitions can couple to F and the rate equations must be solved rigorously including both cross-relaxation and Cr^{3+}-lattice probabilities. A two-phonon Orbach-type process of the coupled Cr^{3+}-F system is important and it is expected that the temperature dependence of the slowest exponential would follow $(\alpha T + \gamma e^{-\Delta/T})$ where Δ is typically the energy of the F transition involved. Figure 1(b) shows τ_1 against $1/T$ for $F \equiv V^{3+}$ in the simpler situation when $H = 0$ so that $\Delta = D = 11.95K$ in agreement with other measurements of D. There is a slight deviation from the above temperature dependence of $1/\tau_1$ for $T \lesssim 2.5K$.

An ion which is strongly coupled to the lattice experiences Jahn-Teller effects. One method of dealing with these is to use transformations which replace the ion-lattice coupling by a constant (e.g. 6). This at first presents a problem in relaxation calculations because the very coupling responsible for relaxation has thus been removed within the determined states of F. It transpires however that the main contribution to τ_1 is from the E_θ, E_ε components of the lattice momentum acting through the spin-orbit coupling and trigonal-field perturbations (7). Finally, it is usual to treat the crystal as an elastically isotropic material. Even for cubic crystals the neglect of elastic anisotropy leads to errors of an order of magnitude or more in calculations of the spin-phonon coupling constants (8). In Al_2O_3 the estimated errors are ~30% (9).

1. Bates C A, Gavaix A, Steggles P, Vasson A and Vasson A-M, 1975, *J Phys C: Solid St Phys*, 8, 3000-16.
2. Standley K J and Vaughan R A, 1969, *Electron Spin Relaxation Phenomena in Solids* (London: Hilger) ch 3.
3. Bloembergen N and Pershan P S, 1961, *Advances in Quantum Electronics*, ed J R Singer (New York: Columbia UP) pp 373-87.
4. Gill J C and Ivey P A, 1973, *J Phys C: Solid St Phys*, 6, L240-2.
5. Bates C A, Gavaix A, Steggles P, Vasson A and Vasson A-M, 1974, *Proc XVIII Congr AMPERE* (Nottingham Organizing Committee) pp 417-8.
6. Abou-Ghantous M, Bates C A and Stevens K W H, 1974, *J Phys C: Solid St Phys*, 7, 325-38.
7. Bates C A and Steggles P, 1975, *J Phys C: Solid St Phys*, 8, 2983-99.
8. Bates C A and Szymczak H, 1975, *J Phys C: Solid St Phys*, 8, 2502-8.
9. Szymczak H, 1973, *Acta Phys Polon*, A43, 649-55.

INTERMEDIATE LATTICE COUPLING THEORY OF THE <111> PARAELASTIC

AND PARAELECTRIC TUNNELING SYSTEM (XY$_8$)

P.Gosar and T.Kranjc

Department of Physics and Institute J.Stefan, University

of Ljubljana, 19 Jadranska, 61000 Ljubljana, Yugoslavia

We have investigated the effect of the "phonon dressing" on the energy splittings and the lattice relaxation rates of the <111> paraelastic and paraelectric tunneling system (XY$_8$) (1) in a host lattice of cubic symmetry. In paraelastic and paraelectric systems the dipole-phonon interaction is generally very strong. The case is considered when the small-polaron binding energy E$_b$ of the oriented dipole is comparable with the tunneling matrix elements. The symmetry group of the dipole-lattice hamiltonian $\hat{H}$ is O$_h$. It has a subgroup D$_{2h}$ which is Abelian group and has therefore only one-dimensional irreducible representations. As the symmetry group of $\hat{H}$ contains an Abelian group one can exploit this fact in a similar way as this is done in the perfect crystal physics where the group of lattice translations is also Abelian. First, new normal coordinates for the displacements of ions which transform as the irreducible representations of the D$_{2h}$ group are introduced. The phonon annihilation operators a(m,λ) for the new normal modes are the following linear combinations of the annihilation operators a($\hat{G}$λ) for the plane wave modes:

$$a(m,\lambda) = 8^{-1/2} \sum_{G} \chi_m(G)\, a(\hat{G}\lambda). \tag{1}$$

Here λ denotes the wave vector and the polarization of the lattice plane wave in the first octant of the Brillouin zone. G are symmetry operations of the D$_{2h}$ group. m = 1-8 specifies the one-dimensional representations and $\chi_m(G) = \pm 1$ are group characters.

In terms of the symmetrized phonons (1) the dipole-lattice hamiltonian can be written as

$$\hat{H} = - \sum_G \sum_i w(i,1)|\hat{G}i\rangle\langle\hat{G}1| + \sum_{m\lambda} \omega(\lambda)a^\dagger(m,\lambda)a(m,\lambda)$$

$$- \sum_G \sum_{m\lambda} \omega(\lambda)\chi_m(G)|\hat{G}1\rangle\langle\hat{G}1|\left[z(m,\lambda)a^\dagger(m,\lambda)+z^*(m,\lambda)a(m,\lambda)\right], \quad (2)$$

where the first term describes the tunneling processes between 8 localized states $|i\rangle$ of the dipole. The state $|1\rangle$ corresponds to the dipole orientation $\langle111\rangle$. The second and the third term represent the lattice and the interaction hamiltonian, respectively. The coupling parameters $z(m,\lambda)$ for the elastic dipole-lattice and the electric dipole-lattice interaction are different from zero only for the representations B_{1g}, B_{2g}, B_{3g} and B_{1u}, B_{2u}, B_{3u}, respectively. In particular, for the interaction of the elastic dipole with the elastic continuum we find

$$z(B_{1g},\lambda) = - i(4/3)v_o(\lambda_1-\lambda_2)C_{44}\left[V\rho\, \omega^3(\lambda)\right]^{-1/2}$$

$$\cdot \left[e_x(\lambda)q_y(\lambda) + e_y(\lambda)q_x(\lambda)\right], \quad (3)$$

where $v_o(\lambda_1-\lambda_2)$ is the elastic dipole strength. C_{44} is the shear constant, V the volume, and ρ the density of the medium. $\underline{e}(\lambda)$ and $\underline{q}(\lambda)$ are the unit polarization vector and the wave vector of the plane wave mode λ, respectively. In the derivation of (2) we exploited the property of Abelian groups: $\chi_m(GG_1) = \chi_m(G)\chi_m(G_1)$.

$\hat{H}$ commutes with the operators

$$\hat{P}(G) = \left(\sum_i |\hat{G}i\rangle\langle i|\right) \prod_{m\lambda} \left[\chi_m(G)\right]^{a^\dagger(m,\lambda)a(m,\lambda)} \quad (4)$$

which are a counterpart to the translation operators $\exp(i\underline{P}.\underline{R})$ in the perfect crystal case. Consider a wave function $|\psi_m\rangle$ of symmetry m which is an eigen-function both of $\hat{H}$ and all $\hat{P}(G)$. It may be easily shown that the projection of the wave function $|\psi_m\rangle$ on the localized state $|1\rangle$ satisfies the equation

$$\hat{H}_{ph} \langle1|\psi_m\rangle = E(m) \langle1|\psi_m\rangle ,$$

$$\hat{H}_{ph} = - \sum_G w(\hat{G}1,1)\chi_m(G) \prod_{s\lambda}\left[\chi_s(G)\right]^{a^\dagger(s,\lambda)a(s,\lambda)}$$

$$+ \sum_{s\lambda}\omega(\lambda)\{a^\dagger(s,\lambda)a(s,\lambda) - \left[z(s,\lambda)a^\dagger(s,\lambda) + z^*(s,\lambda)a(s,\lambda)\right]\}. \quad (5)$$

New hamiltonian (5) does not contain the dipole coordinates and depends explicitely on the symmetry m of the total wave function $|\psi_m\rangle$. The corresponding eigen-values of (4) are $\chi_m(G)$.

A linear superposition

$$<1|\psi_m> = \sum_G f(G)\; \chi_m(G)\; \{\Pi_{s\lambda}\; [\chi_s(G)]^{n(s,\lambda)}\}$$

$$\cdot \exp\{x \sum_{s\lambda} \chi_s(G)[z(s,\lambda)a^\dagger(s,\lambda) - z^*(s,\lambda)a(s,\lambda)]\}|n> \quad (6)$$

of the displaced oscillator wave functions is now used in the variational calculation of the eigen-states and energies of $\hat{H}_{ph}$. Here, $|n>$ stands for the phonon occupation number eigen-state of the phonon field with $n(s,\lambda)$ phonons in the mode s,λ. The variational parameter x is a measure of the deformation of the lattice around the dipole. In the small polaron approximation $x = 1$ and $f(G \neq E) = 0$. The expansion parameters $f(G)$ and the energy $E(m)$ are obtained from the solution of the Schrödinger equation (5) in which the terms which correspond to the non-diagonal transitiuns $n(s,\lambda) \to n(s,\lambda) \pm 1$ in the phonon field $|n>$ have been omitted. Such non-adiabatic transitions are of minor importance in the energy calculations, but must be taken into account in the calculations of the relaxation rates. The parameter x is then determined by minimizing the energy.

We present the results for the case of the elastic dipole – lattice interactions when the tunneling $w(j,i) = w$ takes place only between the nearest neighbour sites. The parameter x and the binding energy $E(m)$ are obtained from

$$x = 1 - \frac{1}{2} a^2 [x^2 - \epsilon ax + a^2 + (x - \frac{1}{2}\epsilon a)(x^2 - \epsilon ax + a^2)^{1/2}]^{-1},$$

$$E(m)/E_b = x^2 - 2x + 2(x-1)(2x - \epsilon a)(3x-1)^{-1}; \quad a = 3w/2E_b, \quad (7)$$

where ϵ is 1 and -1 for the tunneling states of symmetry g and u, respectively. The renormalized tunneling splitting of the energy levels of the (XY_8) system $\Delta = 2w\exp\{-(4/3)x^2 \sum_{m\lambda} |z(m,\lambda)|^2 [2\bar{n}(\lambda)+1]\}$, $\bar{n}(\lambda)$ being the thermal equilibrium number of phonons in the mode m,λ, and the lattice relaxation rates are very sensitive to the renormalization when E_b is not much greater than w. If $a < 1$, we have approximately $x = 1 - a^2/4$. The details of the relaxation rate calculations will be published elsewhere.

REFERENCE

(1) Bridges F., Critical Reviews in Solid State Sciences 5, 1 (1975).

PHONON SPECTROSCOPY OF OH⁻ and Li⁺ TUNNELING STATES

IN ALKALI HALIDES

R. Windheim and H. Kinder[x]

Institut für Festkörperforschung, Kernforschungsanlage Jülich, 517 Jülich, West Germany

Phonon spectroscopy with superconducting tunnel junctions is a new method for the analysis of meV energy states in solids [1] . We present here the direct measurement of OH⁻ and Li⁺ tunneling states in alkali halides by resonant scattering of monochromatic phonons. It is known that the substitutional OH⁻ or Li⁺ ions can perform a tunneling motion among regions of the barrier potential minima which arise from interactions of the defect with the lattice. The tunneling leads to a multiplet of tunneling levels [2] with certain transition probabilities for phonons.

For the energy range of 0.165 to 1.14 meV we used Sn-oxide-Sn junctions as phonon generators, and Al-oxide-Al junctions as detectors. The phonons were propagated in the [100] crystal direction for [001] stress, in the [010] direction for [011] stress and in the [1T0] direction for [111] stress.
The OH⁻ system with the largest tunneling splitting was NaCl:OH [3] . For OH⁻ concentrations of approximately 0.01 ppm, we observed strong resonant scattering at about 0.4 meV which was much stronger for longitudinal than for transverse phonons. Fig. 1 shows the differential phonon rate measured by the Al detector as a function of generated phonon energy [1] . The parameter of the different traces was the [001] stress. At zero stress only a single line was observed which is shown in the two lowest traces for two samples. With increasing stress the line split into one (a) which was shifted to higher energies and a second one (b) which was nearly constant.

In KCl:OH which is a prominent and well investigated OH⁻
system [2] a few ppm OH⁻ showed strong resonance scattering of phonons. Because of the smaller tunneling splitting, we could observe with our Al detectors only the
line (a) which appeared under sufficiently high stress
up to 300×10^6 dyn·cm⁻². For KBr:OH with a still smaller
tunneling splitting, the stress-induced resonance was
weak: 4 % intensity loss for 70 ppm OH⁻. In NaF with
100 ppm OH⁻, no resonance scattering was observed for
[001] stress from zero to 200×10^6 dyn·cm⁻².

The behavior of the resonance lines is completely consistent with an octahedral crystal field at the OH⁻ lattice site, which leads to the level scheme indicated in
Fig. 2 (inset). We identify the lines (a) and (b) with
the $A_{1g} \rightarrow E_g$ transition which is the only one allowed
for phonons and which splits under [001] stress. In Fig. 2
the measured resonance energies for NaCl:OH are plotted
versus stress. The solid line $E_a(S)$ represents a least
squares fit of the expected transition energy [3] to the
data of line (a), assuming constant tunnel matrix elements and neglecting overlap integrals. This yielded a
90 degree tunneling parameter $\Delta = 0.138$ meV and a stress-
splitting parameter $\alpha = 6.09 \times 10^{-24}$ cm³. With the same
parameters the second line, which is nearly constant,

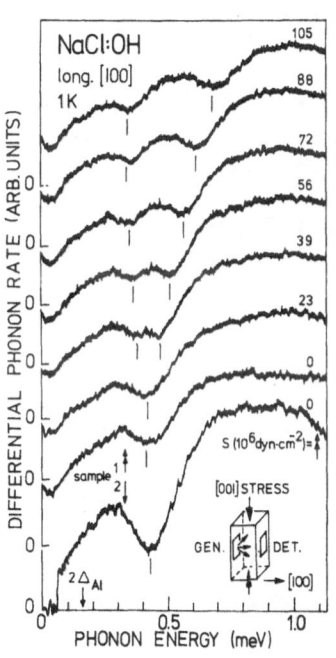

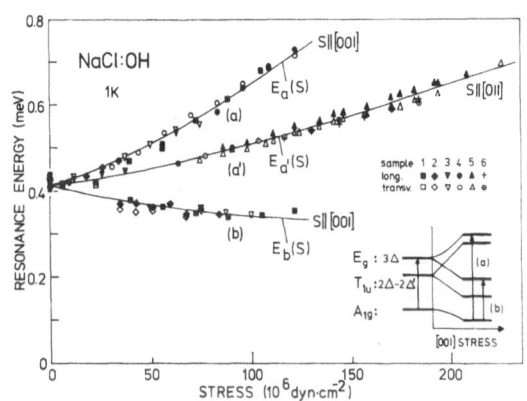

Fig.2: Resonance energy as a function of stress. The Al detector sensitivity threshold was 0.165 meV for
sample 1-3 and 0.43 meV for sample
4-6. Solid lines: theoretical. Inset: level scheme in octahedral crystal field under [001] stress.

Fig.1: Differential phonon rates as
a function of emitted phonon energy
for NaCl:OH. The resonance line
splits under the [001] stress.

and the line $E_a(S)$ for $[0\overline{1}1]$ stress are also shown, which nicely fit the data. The value of α for NaCl:OH compares with that in other host materials [4]. A comparison of Δ with the IR measurements [5] of the $A_{1g} \longrightarrow T_{1u}$ transition shows that the 180 degree tunneling is probably weak.

For KCl:OH the parameters were Δ = 0.019 meV and α = 5.7×10^{-24} cm^3. For KBr:OH the stress-splitting parameter was α = 7.8×10^{-24} cm^3. These results are in agreement with data from other methods [2], [4].

Also, Li$^+$ tunneling states in KCl showed strong resonance scattering of phonons, depending on the Li isotope (^{6}Li or ^{7}Li) and stress direction. For zero stress one weak resonance was found which did not change under [001] stress. With increasing [011] stress, this line was shifted to higher energies and became stronger. Further, at about $100 \times 10^6 \cdot$dyn cm^{-2} a second line appeared whose frequency and absorption strength was about twice as much as that of the first line. The stress splittings for both of the lines were larger for the heavier isotope ^{7}Li. For [111] stress we also observed two lines with still larger stress splittings.

The dependence of the stress directions, i.e. the largest splitting in the [111] stress direction and no splitting for [001] stress, confirm the well established model of an [111] Li off center site. On the other hand the stress splitting measured by more indirect techniques [6], 4×10^{-24} cm^3, is consistent only with the lower line we observed under stress. However, the second line we observed with a stress splitting of about 9×10^{-24} cm^3 could probably not be observed by these techniques and seems not to be understood by the existing models.

[x] Present adress: Physik-Department der TU München, 8046 Garching, West Germany

References

1 H. Kinder, Phys. Rev. Lett 28 (1972) 1564
   H. Kinder, Z. Phys. 262 (1973) 295

2 Narayanamurti and R.O. Pohl, Rev. Mod. Phys. 42 (1970) 201

3 R. Windheim and H. Kinder, Phys. Lett. 51A (1975) 475

4 H. Härtel, Dissertation, Stuttgart 1966

5 R.D. Kirby, A.E. Hughes and J.A. Sievers, Phys. Rev. B2 (1970) 481

6 N.E. Byer and H. Sack, J. Phys. Chem. Solids 29 (1968) 677

DISCUSSION ON MAGNETIC IONS AND MOLECULAR DEFECTS

Spin-Phonon Interactions   F W Sheard   Page 154

J Jackle: You have given a formula for the temperature dependence
of the elastic scattering which remains finite at T=0. At T=0
there is no difference between a two level system and a harmonic
oscillator because only one upper level is important. The
Hamiltonian is purely linear in the case of a harmonic oscillator
and you get the coupled modes involving two eigenstates.

F W Sheard: I do not accept that argument. There is a difference
even at T=0 between a two level system and a harmonic oscillator
which can be shown by comparing the perturbation theory step by
step. In the coupled mode theory we find a concentration dependence
$C(1-C)$ in the limit T=0 due to the random distribution of ions.
The damping goes to zero only when C=1 and there is translational
symmetry.

K W H Stevens: I agree that there is a big difference between a
two level system and a harmonic oscillator. In one case two
ladders of energy levels are coupled from two harmonic oscillators
and in the other case a pair of levels is coupled to a ladder of
levels.

G A Gehring: Kjems, Hayes and Smith have recently made the neutron
experiment you mentioned on the paramagnetic phase of thulium
vanadate. They observed the anticrossing between the spin and
phonon levels over a wide range of magnetic fields.

L J Challis: Mm de Goer and I have applied to do a similar experiment on Cr^{2+} in MgO at I.L.L.

H M Rosenberg: You have used a Lorentzian line shape whereas one
does not know what line shape one should use and it is critical for
fitting experiment to theory.

F W Sheard: I do not think anyone knows how to give a rigorous
theory of line shape due to spin-spin interactions.

L J Challis: Most calculations in second-order perturbation theory
agree with your result except for one by Walton which is temperature dependent. Can you comment?

F W Sheard: Walton obtained a temperature dependent phonon relaxation time for elastic scattering by using the temperature-independent T=0 limit for the spin-lattice broadening.

K Lassman: How are inelastic Raman processes to be accounted for?

F W Sheard: I do not think any detailed work has been done on this.
Experiments are often made on multilevel systems where resonant
Raman scattering can occur if more than two levels are strongly
coupled to phonons.

****

Phonon Scattering Induced by Jahn-Teller Impurities   B Halperin
and R Englman   Page 163

K W H Stevens: Do you know of any experimental evidence for the
rotational levels you have described?

B Halperin: We have assumed only linear coupling whereas real
systems have also third order or anharmonic coupling which gives
another splitting making comparison between theory and experiment
more difficult. I believe that at least one of the observed
resonances of Ni^{3+} in Al_2O_3 corresponds to the lowest one which I
have described.

****

Defect Scattering of Nearly Monochromatic Acoustic Phonons
W E Bron and F Keilmann   Page 172

H Kinder: What normalisation procedure did you use to compare the
three different sources of phonons?

W E Bron: The voltage applied to the heat pulse generator was
adjusted so that the maximum in the generator spectrum corresponded
to the energy gap of the superconducting filter. The whole
detected signal was integrated and normalised to unity and then
the ratio of ballistic phonons to total phonons was taken.

H Kinder: Narayanamurti and Dynes used a tunnel junction detector
with their fluorescer generator and hence did not observe the low
frequencies to which your bolometer detector would respond. Are
you not just observing the shift in the thermal maximum as you
increase your heater voltage?

W E Bron: The Narayanamurti and Dynes generator has a skew distribution near 2Δ of the filter and I believe this gives the frequency response I have shown.

****

Temperature Dependence of the Acoustic Paramagnetic Resonance Spectra for Cr^{2+} in MgO and $KMgF_3$ J Lange and S Guha Page 175

V W Rampton: Did the a.p.r. spectrum, in particular the temperature dependence of the intensity, depend on the ultrasonic power?

J Lange: No power dependence of the intensity is observed for the chromous ion but the usual complex behaviour is observed for the ferrous ion.

J R Fletcher: Did you make any assumption about the relative magnitude of the strain splittings and the spin-orbit splittings?

J Lange: I assumed that the strain splittings were negligible in comparison with the spin-orbit splittings.

C A Bates: In your analysis of Cr^{2+}: MgO did you allow the random internal strain to vary as a function of the angle of the magnetic field?

J Lange: No anisotropy of the strain was considered and only an order of magnitude determined from the g-factor anistropy.

****

Stress-Induced Changes in the Acoustic Paramagnetic Resonance of Chromous Ions in Magnesium Oxide   V W Rampton and I J Shellard
Page 178

R Buisson: The low field line which looks to be due to Fe^{2+} seems also to be reduced by the strain. So, the reduction of the Cr^{2+} lines intensity is, at least partially, due to another mechanism than the strain induced change of the number of Cr^{2+} ions seen by APR.

I J Shellard: It does appear that the intensity of the Fe^{2+} line is reduced when stress is applied. An applied stress would be expected to split the Fe^{2+} triplet so we should not be surprised if the Fe^{2+} line intensity is reduced.

J K Wigmore: Do your data agree in detail with that of Lange?

I J Shellard: The work was done at a higher frequency than that of Lange so a detailed comparison is not possible. In general the spectra appear to be consistent.

****

Ultrasonic Relaxation Studies of Cr^{2+} in MgO   P J King, S G Oates,
V W Rampton and I J Shellard   Page 181

J K Wigmore: In order to compare your measured transition energy
with a theoretical model, you should not use the zero-stress
eigen-values. Our calculations show that, in general, the energy
distribution for a particular transition - due to the random
stresses in the crystal - has a maximum which does not occur at
the zero stress value.

P J King: This point is extremely important both in interpreting
the relaxation data and, for example, the effects observed in the
thermal conductivity.

R Buisson: Resonant phonon absorption between strain split levels
at zero magnetic field can be responsible for part of the attenuation
observed. Did you take into account of this possibility?

P J King: Yes. We estimated that the size of this effect would be
small at the relatively low frequencies being used in the relaxation
measurements.

****

Heat Pulse Investigation of MgO:Cr^{2+} J L Patel and J K Wigmore
                                                            Page 187

R Buisson: Could you tell us what is the spectrum (frequency of
the peak for instance) of your pulses?

J L Patel: The experimental results are not very sensitive to the
heater excitation temperature, which is a few degrees Kelvin at an
ambient temperature of 1K. Thus the frequency spectrum is a thermal
distribution with a peak at a few hundred gigahertz.

****

Thermal Conductivity of Irradiated, Additively-Colored and Deformed
MgO   J B Hartmann, H Weinstock and Y Chen   Page 190

H J Albany: Have you done optical measurements in order to corre-
late the thermal conductivity behaviour as a function of neutron
dose with the introduction of colour centres?

H Weinstock: Yes. The strength of the 20K resonance scales quite
well with F center optical density.

A M de Goer: Have you done any measurements after annealing the neutron irradiated specimens?

H Weinstock: No.

****

Determination of the Linear Jahn-Teller Coupling Constant of Ni^{3+} in Al_2O_3 by Low Temperature Thermal Conductivity Measurements under Uniaxial Stress   B Salce   Page 196

C A Bates: 1. Did you allow A and B to change with the applied stress? 2. Does the externally applied strain exceed the internal strain at the half peak position?

B Salce: 1. Yes, there is a change in A and B, but in very different ways. The change for A is a rapid increase. The change for B is a slow decrease. This change is in agreement with a new calculation by Fletcher on the relaxation time variation with applied stress. 2. Yes, the internal strains are equivalent to stresses of about $\sim$ 200 kgf/cm^2 and the maximum applied stress is 1400 kgf/cm^2.

****

Phonon Spectroscopy in Al_2O_3 Doped with Transition Metal Impurities
H Kinder and W Dietsche   Page 199

J K Wigmore: What is the frequency resolution obtained using your PbBi generator?

H Kinder: The frequency resolution depends on both the modulation amplitude and the sharpness of the density of states singularity. The amplitude can be made rather small with low impedance PbBi junctions and the density of states is very sharp for alloys. Therefore, a resolution of better than 1% of $2\Delta_G$ could be observed experimentally.

K Renk: The absorption line you report for Al_2O_3:V^{4+} seems to be at a higher value than the known far-infrared value. What is the reason?

H Kinder: We don't know the reason yet, but we cannot exclude a simple experimental error in measuring the generator voltage. In principle, one can determine the frequency with the accuracy given by the frequency resolution, i.e. better than 1% of $2\Delta_G$.

M A Brown: A comment:- after Professor Kinder's remarks about large lines being observed due to unknown ions I would like to mention the recent work by Brown, Clarke and Moore (to be published) on Cr^{5+} in Al_2O_3 formed by adding Mg to a nominally 'pure' Al_2O_3 powder. Estimated concentration of Cr^{5+} is 0.02ppm and the effects observed are large. It is possible that other ions may be affected by Mg and that you are observing them?

H Kinder: From our results of doping and annealing 'pure' crystals we also infer that the valency of the unknown impurities changes. We estimate a concentration of these impurities of 1 ppm or less. We have not yet investigated the influence of Mg impurities but I suspect them to be present in our samples.

C A Bates: Comment: We are in the process of modifying the energy level diagrams for $3d^4$ ions in Al_2O_3. The Mn^{3+} work is complete but Cr^{2+} is still to be done.

****

An Investigation of Frequency Crossing Effects in Phonon Scattering by V^{3+} Ions in Al_2O_3 L J Challis and D L Williams   Page 202

C A Bates: Recently published work using high frequency EPR methods by Pontnau and Adde have given a single value for D. Have your results been obtained on one specimen only or on several?

D L Williams: So far several experiments have been made but on one sample only, both before and after γ-irradiation. Measurements on other samples will be made shortly.

****

A Study of Phonon-Spin Interaction in the System $3d^1$ in Al_2O_3
N Devismes and A M de Goer   Page 208

W Forkel: I would like to remark, that in the phonon spectroscopy experiments which Professor Eisenmenger reported on, I have also recently covered the frequency range up to 120 cm^{-1} in order to see the 107.5 cm^{-1} line of $Al_2O_3:Ti^{3+}$. The result might be interpreted as a continuous scattering background increasing with frequency and including a rather weak and broad dip near 107 cm^{-1}. This also indicates that the ratio D_2/D_1 might be much smaller than 100.

N Devismes: I agree completely as this ratio $D_2/D_1 = 100$ has been obtained with the approximate phonon relaxation time and the theoretical value is nearly 8.

****

Temperature Dependence of the Lifetime of 29 cm^{-1} Phonons in Ruby
Between 3 and 17K   G Pauli, W Eisfeld and K F Renk   Page 211

J K Wigmore: What is the effect of boundaries on your measurements?
Can you distinguish specular from diffuse scattering of the phonons
at the walls?

G Pauli: We have measured the decay time for the crystal immersed
in liquid He and in gaseous He at the same crystal temperature.
Within our experimental accuracy we found no difference for the
decay time in the two cases.

****

A Study of Spin-Phonon Interactions by Spin-Lattice Relaxation
Processes   C A Bates, A Gavaix, P Steggles, H Szymczak, A Vasson
and A-M Vasson   Page 214

D J Meredith: When a small number of fast relaxing impurities are
responsible for the relaxation of a greater number of paramagnetic
centres, the spatial diffusion problem leads to a non-exponential
recovery of the spin resonance signal in a pulse experiment. How
were your values of τ_1 measured?

P Steggles: The rate equations indicate the presence of three
exponentials in the return to equilibrium of the monitored transition.
If the experimental result does not show a single exponential we
consider the tail of the trace, that is the slowest exponential,
and try to relate the relaxation time so defined to our theoretical
model.

R Buisson: Could you give more comments on the proposed mechanism
for the cross relaxation and explain why the usual dipole-dipole
interaction is not active?

P Steggles: In some instances of resonant cross relaxation - for
Cr^{3+} to Fe^{2+} for example - the relevant transition in the fast
relaxer is magnetic-dipolar forbidden unless the states are sufficiently mixed by the Zeeman field. (The minimum in τ_1 corresponds
to both ions being at sites for which random strains are very close
to or equal to zero). Coupling would then be strongly field
dependent and this is not observed experimentally. We have been
able to show that the coupling of the Fe^{2+} cluster to the electric
field due to the electric quadrupole moment of the Cr^{3+} ion is of
sufficient strength to explain the experimental results. We have
also found that virtual phonon exchange may be important. In the
case of non resonant cross relaxation, the magnetic dipole-dipole
interaction is likely to be the most important form of coupling.

Phonon Spectroscopy of OH⁻and Li⁺ Tunnelling States in Alkali
Halides   R Windheim and H Kinder   Page 220

J G Collins: Can your technique give information about the state
of CN⁻impurities in NaCl?

R Windheim: We did some measurements on KCl:CN, 1 ppm, which show
a large zero field splitting and also the largest stress splittings
in (111) stress directions. This suggests an (111) oriented CN⁻
dipole.

J P Harrison: Were you able to confirm the 40% isotope shift of
the tunnelling levels for KCl:Li?

R Windheim: Under zero stress only a weak resonance was observed
for phonons in the (100) or (110) direction which makes a direct
measurement of the isotope shift uncertain. On the other hand we
have up to now no theory to fit our data of the two well defined
Li resonance lines under stress which show different stress
splittings. However, if we try to extrapolate our data linearly
to zero stress, we obtain in fact about 40% isotope shift of the
tunnelling levels for KCl:Li.

PHONONS IN COOPERATIVE SYSTEMS

G.A. GEHRING

Department of Theoretical
Physics, University of Oxford
12 Parks Road, Oxford OX1 3PQ

This paper will review the theoretical and experimental work which has been done on some rare earth crystals which show the cooperative Jahn Teller phase transition[12345] Particularly simple examples are TmVO$_4$ and TmAsO$_4$ in which the ground state is an orbital doublet in the high temperature (tetragonal) phase so the theory is developed for them.

I.   The Cooperative Jahn Teller Effect

The degeneracy of the Tm^{3+} doublet may be split by a magnetic field, $\Delta = g\beta H$, or a local distortion, Q_n, we use a pseudo spin notation for the doublet, $S_n^z = \pm 1$, to obtain a single ion Hamiltonian.

$$H_n = \Delta S_n^X + A_n Q_n S_n^Z \tag{1}$$

Writing the distortion in terms of phonon coordinates and a bulk strain, ε, we obtain the Hamiltonian;

$$H = \sum_{\underline{k},s} \hbar\omega_{ks}(a^+_{ks} a_{ks} + \tfrac{1}{2}) + \sum_{\underline{k},s} \xi_{\underline{k},s} (a^+_{ks} + a_{-ks}) S^Z_k$$

$$+ \Delta \sum_n S^X_n + \eta\varepsilon \sum_n S^Z_n + \tfrac{1}{2} N\Omega C_o \varepsilon^2 \tag{2}$$

Here $N\Omega$ is the volume of the sample, C_o is the relevant elastic constant and η and ξ_{ks} are coupling constants.

This Hamiltonian is simplified by a transformation to new strain and phonon coordinates defined relative to the distorted ground state.

$$\gamma^+_{k,s} = a^+_{ks} + \frac{\xi_{ks}}{\hbar\omega_{ks}} S^z_k, \quad \varepsilon' = \varepsilon + \frac{S^z_o}{\Omega c_o} \tag{3}$$

$$H = \sum_{\underline{k},s} \hbar_{\omega ks}(\gamma^+_{\underline{k}s} \gamma_{\underline{k}s} + \tfrac{1}{2}) - \tfrac{1}{2} \sum_k J(k) S^z_k S^z_{-k}$$

$$- \mu\langle S^z\rangle \sum_n S^z_n + \sum_n \Delta S^x_n + \tfrac{1}{2} N\Omega c_o \varepsilon'^2 \tag{4}$$

where

$$J(\underline{k}) = 2\sum_s \frac{|\xi_{sk}|^2}{h\omega_{ks}} - \frac{2}{N} \sum_{k,s} \frac{|\xi_{sk}|^2}{\hbar\omega ks} \quad \text{and} \quad \mu = \frac{n^2}{\Omega c_o} \tag{5}$$

The elastic constants may be calculated by including a stress energy, σ, in equation (4)[4],

$$1/c_\Lambda = \lim_{\sigma\to o} \frac{\partial\varepsilon}{\partial\sigma}\Big)_\Lambda = \lim_{\sigma\to o}\{\frac{1}{c_o} - \frac{n}{\Omega c_o}\frac{\partial S^z_o}{\partial\sigma}\Big)_\Lambda\} = \frac{1}{c_o}\{1 + \mu X_\Lambda\} \tag{6}$$

Here Λ represents the thermodynamic conditions and X_Λ is the susceptibility which diverges at the phase transition so forcing C_Λ to zero. The frequencies of the elementary excitations may be calculated from equation (4). In the random phase approximation [125] these are given by,

$$\hbar^2\omega\{\omega^2 - \omega^2_E(k)\} = 4\omega^3\Delta\langle S^x\rangle \sum_s \frac{|\xi sk|^2}{\hbar\omega ks(\omega^2 - \hat{\omega}ks^2)} \tag{7}$$

where $h^2\omega^2_E(k) = 4\{(J(o)+\mu)^2\langle S^z\rangle^2 + \Delta\{\Delta - J(k)\langle S^x\rangle\}\}$ (8) $\omega_E(k)$ is the RPA frequency of the transverse Ising model. At low wave vector such that $\omega_{ks} \ll \Delta$ one branch corresponds to the acoustic phonon,

$$\omega^2 = \bar{\omega}^2_{ks} = \frac{\omega^2_{ks}\,\omega^2_E(k)}{\omega^2_E(k) + 4(S^x)|\xi ks|^2/\hbar^2\omega ks} \tag{9}$$

At the phase transition $\lim_{k\to o}\omega_E(k)\to o$ and the soft mode behaviour is transferred to the acoustic phonon. The elastic constant calculated from equation (6) is in agreement with equation (9) when χ is calculated using RPA or mean field theory.

However equation (9) holds for all values of Δ including $\Delta = o$ for which equation (10) yields uncoupled

modes and therefore no change in the elastic constant.
This contradiction arises because the RPA calculation
excludes relaxation which is included in the thermodynamic treatment[5,6]. Sandercock et al [5] and
Pytte[6] have noted that this will give rise to a central
peak whose width goes to zero at the phase transition[7]
This has been observed by neutron scattering in TbVO$_4$
and TmVO$_4$[8] in zero magnetic field. For $\Delta=o, T>T_D$ Pytte
(9) finds the acoustic susceptibility in the RPA of the
form,

$$\chi^{-1}_{a_c} = \omega_k^2 - \omega^2 - \frac{i\omega}{\tau a} - (\omega_k^2 - \bar{\omega}_k^2)(1-i\omega\tau_s T/(T-T_o))^{-1} \quad (10)$$

where τ_a and τ_s are the lifetimes of the phonon and the
pseudospin excitations and T_o is the clamped transition
temperature. This expression yields a frequency dependent elastic constant, a central peak and shows that the
pseudospin damping gives rise to strong ultrasonic
attenuation near to T_D. Similar results hold for
$\tau_s\Delta<1$, i.e. overdamped excitation of the transverse
Ising Hamiltonian, but results of the form of equation
(7) should hold for $\tau_s\Delta>1$.

II Second Order Couplings

The second order coupling between a pseudo spin
and two phonons can be written as;

$$\sum_{kk'} \sum_{ss'} A^{ss'}_{kk'} (a^+_{ks} + a_{-ks})(a^+_{k's'} + a_{-k's'})S^z_{-k-k'} \quad (11)$$

This type of interaction gives rise to the following
effects;
(i) Raman scattering of phonons from the random distribution of the pseudo spins above the transition temperature. This is an important scattering mechanism for
$\Delta=o$, $T>T_D$. (19)
(ii) Splitting of the degenerate E phonon modes below
the transion temperature (2).

(iii) A central peak; this is an anharmonic effect due
to the coupling of the pseudo spins to the phonon
density(11) and is in addition to the central peak
discussed in I; it may be observed if Δ is made large
enough for the RPA solution to be a good approximation.

III Effects of Domain Walls

Most crystals will not form monodomain samples
when they are cooled through the phase transition, and

they certainly break into domains if they are subject
to a non uniform stress. The domain walls are efficient
ultra sonic scatterers so that it is often impossible to
obtain an echo unless the sample has been forced into a
single domain. In addition the movement of domain walls
can mean that the ultrasonics are measuring the susceptibility of the sample to domain wall motion rather than
the bulk susceptibility [12].The Brillouin scattering
experiments are possible on multidomain samples although
the lines are broader than above the phase transition [5].

The scattering of phonons from domain walls is also
believed to be important in reducing the thermal conductivity in the ordered phase of $TmVO_4$ [9, 13].

IV   Effects of Random Strains

The Jahn Teller ions are very sensitive to random
strains. These would be included in the Hamiltonian
equation [4] as $\eta \Sigma_n \epsilon_n S_n^z$. In a large magnetic field the
random strains have two effects: the single site energies
become $\epsilon_n^2 = \Delta^2 + \epsilon_n^2 \eta^2$ and $(S_n^x) = \Delta/\Sigma_n \tanh \epsilon_n/2kT$. It is
found that the leading contribution to the phonon scattering
from the random variations in ϵ_n varies as Δ^{-6} and from
(S_n^x) as Δ^{-4} in large magnetic fields. The experimental
results of Page and Rosenberg [14] on ultrasonic absorbtion in $TmVO_4$ do show a leading Δ^{-4} behaviour at high
fields but for lower fields the experimental attenuation
falls below the Δ^{-4} behaviour which is not expected.

V   Summary of the Experimental Situation

There is now good experimental evidence for the
mixed modes described in section I. The elastic constants have been measured as a function of temperature
and magnetic field [15] and as a function of frequency
for another crystal, $TbVO_4$, which has a degenerate
doublet [5]. The mixing region has been studied by neutron
scattering [7,8]. All this is in addition to the observation of the electronic properties[1] by absorbtion
spectroscopy, electronic Raman scattering and infra red
absorbtion. The agreement between theory and experiment
is very satisfactory. This is due to the fact that the
interactions via the bulk strain are very important;
these are long range and so the molecular field and RPA
theories should work well. A corollary of this is that
one should expect these systems (Ising Hamiltonian with
long range forces) to show classical critical exponents.
This has been confirmed by the measurement of β for
$TbVO_4$[16] (but not for $DyVO_4$ for which the long range
forces are believed to cancel out).

The measurements of the relaxation times and damping
of the collective excitations is more difficult. The
most unambiguous result of critical slowing down is the
neutron scattering measurement of the width of the central peak in TbVO$_4$(8). Other methods of looking at the
relaxation times come from ultrasonic attenuation(14)
and thermal conductivity(13).

The ultrasonic measurements of attenuation are
hard to do because the attenuation becomes so large for
these systems. This is because the strain is the order
parameter and the acoustic phonon is the soft mode. Or,
expressing this another way, the pseudospin-phonon
coupling is of a similar strength to the interactions
which cause the phase transition.

This is in contrast to the magnetic systems in
which the phonons are coupled not to the order parameter
but to a less singular function - the energy density or
another bilinear combination of magnetic operators.Also
the spin phonon coupling is generally much weaker than
the exchange interactions.

Low frequency ultrasonic measurements (15) found
that the echo for the 'soft'phonon in TmVO$_4$ was lost
for T<4K where T_D=2.1K and no signal could be observed
below the phase transition for this mode. The modes
were also studied as a function of magnetic field at
low temperatures: if $\Delta>\Delta_c$= (J(o)+μ) the system reverts
to the high symmetry phase: the phonon frequency falls
to zero at $\Delta=\Delta_c$. Again the signal was lost when Δ
approached Δ_c from above.

High frequency ultrasonics (10GHz)(14) are attenuated much more. The echoes can only be observed in a
magnetic field (>0.6T) and the attenuation is almost
independent of temperature. This means that the Raman
scattering cannot be responsible and the data may be
explained in terms of random strains as discussed in
section IV. Thus so far there is no evidence that
there is a maximum frequency for attenuation corresponding to $\omega\tau_s$~1. However the high temperature value of τ_s
that was measured for TbVO$_4$ was 173 GHz(5) so maybe
the frequencies are still too low. Thermal conductivity
measurements have been made on TmVO$_4$ as a function of
temperature and magnetic field (13). The important
effects are the Raman scattering process, equation (11),
the damping of the phonons due to the admixture of the
pseudo spin mode, equation (10), the reduction in the
density of states at the anticrossing (17) and, below

T_D, scattering from domain walls. In zero magnetic field above T_D the splitting is zero and the Raman process dominates. The contribution of this process to the phonon scattering varies as the susceptibility $\chi(q)$. This fits the data satisfactorily(9) without any evidence of non-molecular field behaviour. Below T_D it seems that the domain wall effects are dominating the thermal resistivity(9) and masking the, more interesting, dynamic effects.

Acknowledgements

The author is very grateful for the many interesting and useful discussions held with A.P. Donegan, K. Loftus, J. Page, M. Parsons and H.M. Rosenberg and also to the authors of references 7 and 8 for permission to quote unpublished work.

References

1) G.A. Gehring and K.A. Gehring Rept. of Progr. in Phys. 38, 1(1975)

2) R.J. Elliott, R.T. Harley, W. Hayes and S.R.P. Smith Proc. Roy. Soc. A328, 217 (1972)

3) E. Pytte and K.W.H. Stevens Phys. Rev.Lett. 27, 862 (1971)

4) G.A. Gehring, A.P. Malozemoff, W. Staude and R.N. Tyte J. Phys. Chem. Sol. 33, 1487 (1972)

5) J.R. Sandercock, S.B. Palmer, R.J. Elliott, W.Hayes, S.R.P. Smith and A.P. Young J. Phys.C5, 3126 (1972)

6) E. Pytte Phys. Rev. B8,3954 (1973)

7) M.T. Hutchings and S.R.P. Smith (to be published)

8) J.K. Kjems, W. Hayes and S.H. Smith (to be published and private communication)

9) A.P. Donegan private communication

10) R.A. Cowley and G.J. Coombs J. Phys. C6,121 and 143 (1973)

11) A.P. Young and R.J. Elliott J.Phys. C7,2721(1974)

12) M.E. Mullen et al Phys.Rev. B10, 186 (1974)

13) M.W.S. Parsons: Thesis Oxford University (1973)

14) J. Page and H.M. Rosenberg Proceedings of this Conference

15) R.L. Melcher, E. Pytte and B.A. Scott Phys. Rev. Lett. 31, 307 (1973)

16) R.T. Harley and R. Macfarlane (to be published)

17) R.J. Elliott and J.B. Parkinson Proc. Phys. Soc. 92, 1024 (1967)

ULTRASONIC ATTENUATION IN THULIUM VANADATE AT 10 GHz AT LIQUID

HELIUM TEMPERATURES

J. H. Page and H. M. Rosenberg

Clarendon Laboratory, University of Oxford

Oxford, OX1 3PU, England. U.K.

Abstract

    The ultrasonic attenuation in $TmVO_4$ has been measured at 10 GHz as a function of magnetic field up to 4.5 T at 4.2, 3.0 and 1.8 K using a pulse–echo technique for longitudinal waves. The attenuation is very high at low fields and no echoes can be observed, but at high fields it varies as $(field)^{-4}$. The results are discussed in the light of current theories.

    Thulium vanadate has a ground state doublet which splits, due to a Jahn–Teller distortion, into two singlets separated by 2.97 cm^{-1} at T_D = 2.15 K. A magnetic field also splits the doublet, and when applied below T_D, it drives the crystal back to the high symmetry phase at a critical value B_D (B_D = 0.617 T at 0 K)[1,2]

    In an earlier paper[3] which was primarily concerned with the variation of the elastic constant C_{66} as a function of field, we also drew attention to the fact that the attenuation of ultrasonic pulses decreased as the field was increased. The present paper reports the results of further experiments in which the attenuation of ultrasonics at 10 GHz has been studied in more detail at 4.2, 3.0 and 1.8 K.

    The attenuation is so strong that no echoes can be observed at low fields at temperatures both above and below T_D. The lowest field at which echoes can be detected is about 0.6 T which even below T_D is a field in which the high-symmetry phase has already been restored.

237

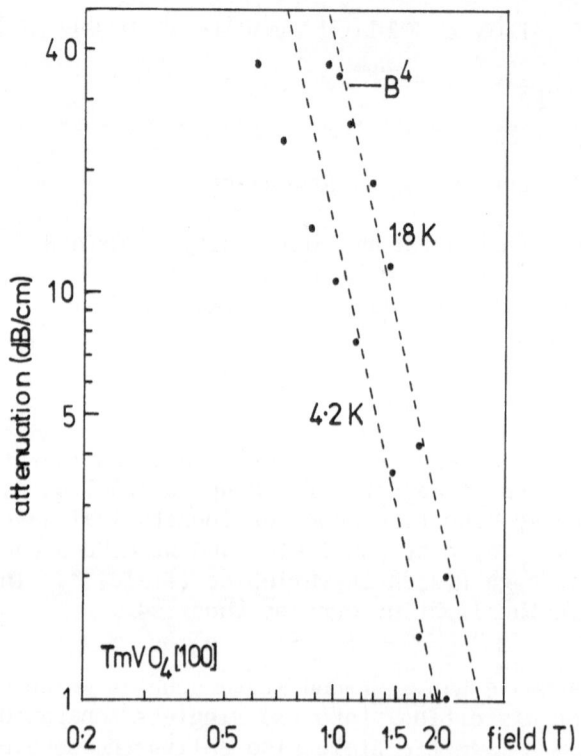

The attenuation of 10 GHz longitudinal ultrasonics in $TmVO_4$, along the [100] direction, as a function of magnetic field, at 1.8 and 4.2 K. This illustrates the (field)$^{-4}$ dependence at high fields.

The attenuation is almost temperature independent. In the high field limit it was found that it varied as $(field)^{-4}$, although it could also be described in the form $\exp(-g\beta B /\Delta)$, where Δ is about 2.5 K. A typical series of results is shown in the figure. The same field dependence was found for both the [100] and [110] propagation directions.

In addition to experiments on the concentrated material, measurements were also made on specimens containing 80% Tm and 5% Tm in $LuVO_4$. The 80% specimen exhibited the same effect as the concentrated material. Only in the 5% specimen was the attenuation mechanism sufficiently reduced that single ion transitions between the levels of the doublet ground state could be observed.

The attenuation of the ultrasonics is of course not a resonant phenomenon such as that observed in acoustic paramagnetic resonance. Such a resonance should occur at about 0.06 T but the attenuation is too strong for it to be detected. The attenuation which we observe is believed to be due to local variations in the elastic constants which are caused by random strains. These will scatter the ultrasonics in a manner analogous to the scattering of light when it passes through regions of varying refractive index. As the field is increased these local strains are removed and the attenuation is reduced. Such a mechanism will give rise to an effect which is not temperature dependent.

A fuller discussion of this phenomenon is given in the paper by G. A. Gehring[4].

References

1. A. H. Cooke, S. J. Swithenby and M. R. Wells, Solid State Commun., 10, 265 (1972).
2. P. J. Becker, M. J. M. Leask and R. N. Tyte, J. Phys. C. 5, 2027 (1972).
3. J. H. Page and H. M. Rosenberg, Symposium on Microwave Acoustics (ed. E. R. Dobbs and J. K. Wigmore) p. 141, (Inst. of Physics 1974).
4. G. A. Gehring, Proceedings of this conference.

THE THERMAL CONDUCTIVITY OF HOLMIUM PHOSPHATE, A STRONG ISING

ANTIFERROMAGNET WITH A LARGE HYPERFINE INTERACTION

M W S Parsons

Department of Physics, University of Nottingham

University Park, Nottingham NG7 2RD, U K[*]

Ho^{3+} is a non-Kramers ion, the ground state being split by the crystal field into doublets with the two lowest pairs separated by 66 cm^{-1}, and with only one non-zero g value in the ground state, g_c, of 16.7[1]. The nuclear spin is 7/2 and the total spread of the hyperfine octuplet is 1.75K[2]. $HoPO_4$ is tetragonally symmetric, space group $I4_1/amd$, and orders antiferromagnetically at 1.391K[3]. In view of the single non-zero g value, it can be regarded as an almost perfect Ising system.

The metamagnetic phase transition from the antiferromagnetic to the paramagnetic state below the Neel temperature will be observable in the magnetothermal resistivity by applying a magnetic field along the preferred axis to a single crystal[4].

Thermal conductivities were measured using a 'Searles Bar' technique, a $^3He/^4He$ dilution refrigerator and a superconducting solenoid. The heat current and the magnetic field were in all cases along the preferred (c) axis.

The thermal conductivity as a function of temperature is plotted in fig 1. The dotted line indicates T^3 behaviour, and it is noted that the conductivity is broadly T^3, with significant deviations. Between 2K and 1K the conductivity falls more rapidly, before returning to a T^3 dependence, at a conductivity of twice the extrapolated T^3 data, at 0.4K. At 0.2K the conductivity falls faster than T^3 over a narrow temperature region before continuing as T^3 down to 0.12K. Fig 2 shows the reduced thermal resistivity (relative to the value in zero field) plotted against the applied magnetic

---

[*]Work carried out at the Clarendon Laboratory, Oxford, England.

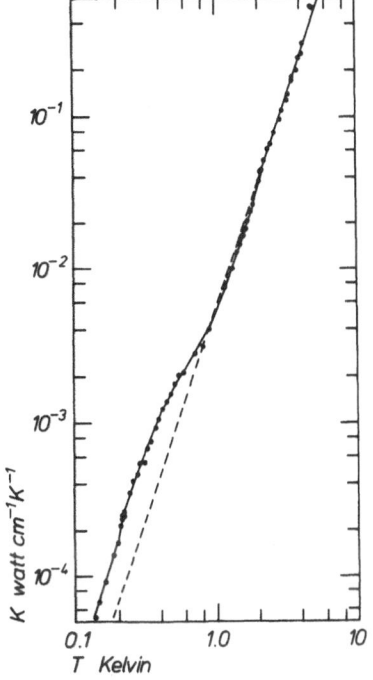

Fig 1: The thermal conductivity of HoPO$_4$ as a function of temperature.

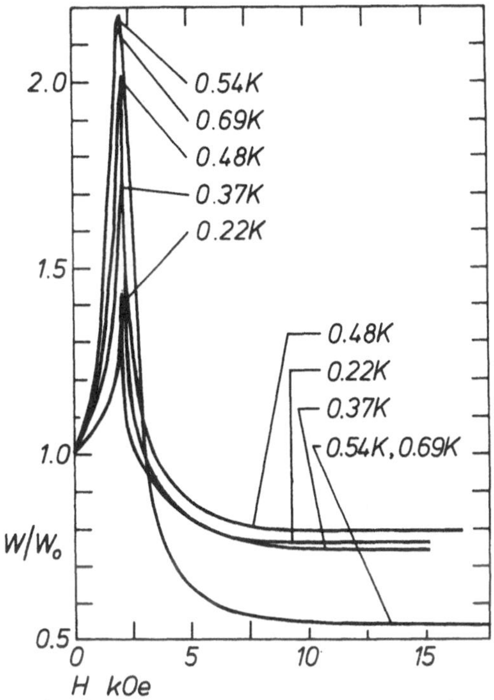

Fig 2: The reduced thermal resistivity of HoPO$_4$, as a function of magnetic field at various temperatures.

field (the samples were needle-like, and so there were no demagnetising corrections). The resistivity rises to a sharp maximum and then falls away to a constant value, lower than the zero field resistivity. The maxima are well defined at 1.85 ± 0.05kOe, and the magnitudes of the maxima appear to reach a maximum value at about 0.6 K. Measurements were made in both rising and falling fields, and no evidence of hysteresis was observed.

Above T_N the ground state is a degenerate doublet, with a hyperfine splitting of 0.25K between each level in the hyperfine octuplet[2]. As T_N is reached from above the degeneracy of the doublet is raised by the exchange interaction, to a limiting value Δ at low temperatures. Phonons may be scattered by spins inside this electronic doublet as the degeneracy is raised and inside the lower hyperfine octuplet at much lower temperatures.

Elliot and Parkinson[5] have shown that for a non-Kramers ion the dominant thermal phonons are found at energy ∿ 2.5kT, in the limit of neglecting the intrinsic line width. A somewhat approximate estimate of Δ can then be obtained by observing the temperature of the maximum deviation from T^3 of the conductivity as the crystal orders. This occurs at 1.5 ± 0.3K leading to an estimate of Δ of 3.7 ± 0.7K. A somewhat crude estimate that is likely to be on the low side, as no account has been taken of the hyperfine levels on the energies involved.

A considerably better estimate of Δ can be obtained by calculating the Lorentz and dipole fields and obtaining the exchange field, H_E, from the critical field for the metamagnetic phase transition, H_c, by equating the Gibbs free energies in the two phases. This gives

$$H_c = H_E - H_{12} - \tfrac{1}{2}H_L$$

H_{11} and H_{12} are the intrasublattice and intersublattice dipole fields and H_L the Lorentz field. Evaluating these over a 40Å Lorentz sphere gives values for H_{11}, H_{12} and H_L of +0.12, −1.88 and ±2.18kOe respectively (a field is regarded as positive when it is in the same direction as the spins on the source sublattice). H_E is found to be 2.15 ± 0.05kOe. The doublet is in an effective field $H_E + H_{11}$ − H_{12}, which leads to

$$\Delta = g_c\beta(H_E + H_{11} - H_{12})$$

Hence Δ = 4.46K or 3.10 ± 0.07cm^{-1}, and the exchange constant J is found to be given by J/k = −0.58 ± 0.05K. These values are in excellent agreement with those obtained from specific heat and magnetic moment data[3].

The drop in conductivity at 0.2K is tentatively identified as being due to a spin phonon resonance between the lowest and third lowest hyperfine levels (i.e. $\Delta M_I = 2$). This agrees with what is known of the splitting of the hyperfine levels, but it is not clear why this transition should be observed and not for $\Delta M_I > 2$.

It would seem likely that the resistivity increase at H_c is due to critical spin fluctuations at the phase change, such a mechanism being expected to have a temperature dependence similar to point defect scattering, T^4. Hence it will decrease relative to the T^3 boundary scattering as the temperature falls, as is observed.

It is a pleasure to acknowledge helpful discussions from Dr H M Rosenberg and Mr M R Wells, and experimental assistance from Mr K V Loftus.

1. P J Becker, H G Kahle and D Kuse, Phys Stat Solid 36, 695 (1969).
2. J M Baker and B Bleaney, Proc Phys Soc A68, 936 (1955).
3. A H Cooke, S J Swithenby and M R Wells, J Phys C 6, 2209, (1973).
4. M J Metcalfe and H M Rosenberg, J Phys C 5, 450, (1972).
5. R J Elliot and J B Parkinson, Proc Phys Soc 92, 1 (1967).

Thermal Conductivity of $PrCl_3$ from 0.1 to 1K in Zero and 2.2

Tesla Magnetic Field

J.P. Harrison and J.P. Hessler

Physics Department, Queen's University

Kingston, Ontario, Canada

The thermal conductivity of $PrCl_3$ has been measured to learn more about the 0.4K phase transition which has been attributed to a co-operative Jahn-Teller crystallographic distortion[1]. The studies presented here support this attribution and furthermore provide evidence for heat conduction via the soft phonon modes.

Praseodynium trichloride is an ionic solid with hexagonal crystal structure. The Pr^{3+} ion occupies a site of C_{3h} symmetry[2] The lowest Russel-Saunders multiplet is $^3H_4$ which when split by the crystal field gives a doubly degenerate non-Kramers ground state[3]. The nearest excited state is a singlet at 33 cm^{-1}. Specific heat measurements of $PrCl_3$ showed two anomalies, a sharp λ-anomaly at 0.4K, and a broad anomaly at 0.85K[4]. Both were ascribed to antiferromagnetism, three dimensional and one dimensional respectively. Subsequent nuclear quadrupole resonance on the Cl^{35} nucleus confirmed a phase transition at 0.4K which lowered the symmetry of the crystal structure[1]. However since it was found that the phase transition was non-magnetic it was speculated that there was a co-operative Jahn-Teller crystallographic distortion. Because of the orbital degeneracy of the ground state, a Jahn-Teller distortion is possible[5]. Motivated by the work of Parsons and Rosenberg on the Jahn-Teller systems $DyVO_4$ and $TmVO_4$[7], we undertook thermal conductivity measurements on $PrCl_3$.

Figure 1 shows the thermal conductivity in zero field and a field of 2.2 Tesla. The upper solid line is the limit to the thermal conductivity based upon boundary scattering of the phonons[7]. To calculate this limit the Debye temperature was assumed to be the same as $LaCl_3$, that is $150^\circ K$[8]. It is clear that over most of the

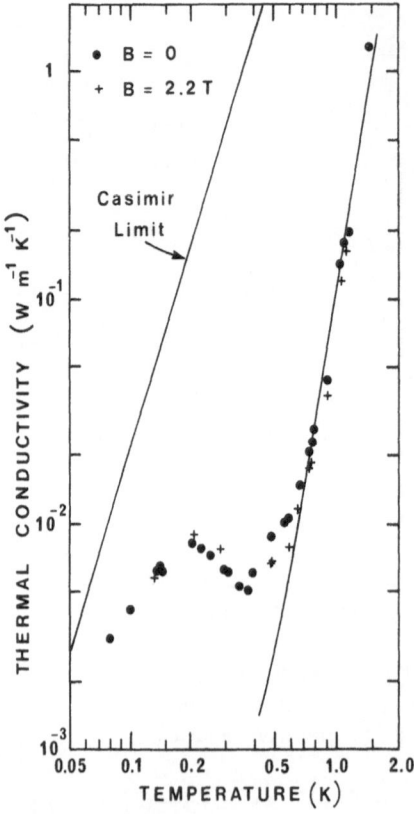

Figure 1. Thermal conductivity of a single crystal of PrCl$_3$ in zero field and 2.2 Tesla. The heat flow was 45° to the symmetry axis. The field was parallel to the heat flow. The sample cross section was 2.9 mm x 7.6 mm. The solid lines are discussed in the text.

temperature range, the conductivity is strongly depressed below this Casimir limit. Also the behaviour as a function of temperature is quite striking. Similar temperature dependencies have been measured for RbMnF$_3$(9) and for UO$_2$(10). The behaviour in the former was attributed to magnon conductivity below the dip and in the latter to strong phonon scattering near the antiferromagnetic phase transition. For the present case, the variation is reminiscent of resonant scattering of the phonons$^{(11)}$. However, in this temperature range resonant scattering cannot produce a temperature dependence with negative slope. Note that below 0.4K the field causes an increase in conductivity and above 0.4K, a decrease. Interpretation of the results was aided by measurements of the specific heat in a magnetic field of 2.2 Tesla. There is not room here to describe fully these results. Basically the 0.4K sharp anomaly moved down in temperature by 5%, and the broad anomaly moved up by about 5%.

The following model is proposed for the thermal properties of PrCl$_3$, and the results are interpreted in terms of it. The ground state doublet is split even above the 0.4K transition. This doublet is then responsible for the upper specific heat anomaly and the strong depression in the thermal conductivity above 0.5K. The lower solid line on figure 1 is the calculated conductivity for resonant scattering of phonons by a doublet with splitting k$_B$ x 1.5K (1.1 cm^{-1}). The doublet then splits further in a magnetic field which accounts for the negative change in conductivity above 0.5K. The required 6% increase in splitting is in accord with the specific heat and within 25% with the static susceptibility. A similar doublet splitting or broadening has been proposed to explain the broadening of some optical transitions in PrCl$_3$(12), and was based upon ion pair interaction parameters(13). Next, it is proposed that the Jahn-Teller distortion does take place at 0.4K, further removing the ground state degeneracy. Since the change in a magnetic field in both the thermal

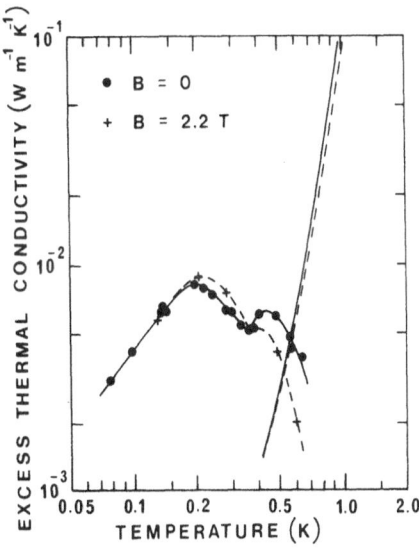

Figure 2. The thermal conductivity remaining when the conductivity due to resonantly scattered "Debye mode" phonons has been subtracted. This conductivity is attributed in first approximation to the conductivity of the soft phonon modes associated with the structural phase transition.

conductivity and specific heat are positive below 0.4K and negative above 0.4K, then it follows that the excitations responsible for the specific heat must be the heat conductors. Figure 2 is a first approximation of this contribution, obtained by subtracting the resonantly scattered phonon contribution above 0.4K. No account has been taken of phonon conduction below 0.4K, or increased phonon scattering by fluctuations above 0.4K. The behaviour of the "soft phonon mode conductivity" is understandable in terms of a mode velocity which drops to zero at Tc. We believe this to be the first firm evidence of heat conduction via soft phonon modes associated with a crystallographic distortion. However it does not pass the Hartmann test of producing a conductivity above the Casimir limit[9].

We wish to acknowledge the financial support of the National Research Council of Canada, and valuable discussions with Professor R.J. Elliott and Dr. D.R. Taylor.

1. J.P. Hessler and E.H. Carlson, J. Appl. Phys. 42, 1316 (1971).
2. W.H. Zachariasen, Acta Crystal. 1, 265 (1948); B. Morosin, J. Chem. Physics. 49, 3007 (1968).
3. G.H. Dieke, Spectra Levels of Rare Earth Ions in Crystals (H.M. Crosswhite and H. Crosswhite, ed.), Interscience, N.Y. 1968.
4. J.H. Colwell, B.W. Mangum, and D.B. Utton, Phys. Rev. 181, 842 (1969).
5. G.A. Gehring and K.A. Gehring (to be published).
6. M.W.S. Parsons and H.M. Rosenberg, Int. Conf. on Phonon Scattering in Solids, Paris, (H.J. Albany, ed.) p. 326 (1972).
7. H.B.G. Casimir, Physica, 5, 495 (1938).
8. F. Varsanyi and J.P. Maita, Bull. Am. Phys. Soc. 10, 609 (1965).
9. J.B. Hartmann, Ph.D. Thesis (Cornell University, 1974).
10. K. Aring and A.J. Sievers, J. Appl. Phys. 35, 1496 (1967).
11. For a review, see R.O. Pohl, International Conference on Localised Excitations in Solids, (R.F. Wallis, ed.), Plenum Press, N.Y. 1968, p. 434.
12. K.R. German and A. Kiel, Phys. Rev. Letters, 33, 1039 (1974).
13. J.W. Culvahouse, D.P. Schinke and L.G. Pfortmiller, Phys. Rev. 177, 454 (1969).

THERMAL CONDUCTIVITY OF GdCl$_3$, A REPRESENTATIVE MAGNETIC INSULATOR

G. S. Dixon

Oklahoma State University

Stillwater, Oklahoma 74074, U.S.A.

Gadolinium trichloride is a ferromagnetic insulator with a Curie temperature (T$_c$) of 2.2 K. In common with many other lower temperature magnetic materials, it has a multi-sublattice structure, and magnetic dipole interactions which play a role comparable to that of exchange in producing the ordered magnetic state. Despite these complexities GdCl$_3$ is an especially appealing system in which to try to understand the magnon-phonon interactions. Its equilibrium magnetic properties have been exhaustively studied[1,2], and it has been found that these can be completely accounted for by magnetic dipole forces and isotropic exchange which is antiferromagnetic between nearest neighbors and ferromagnetic between second neighbors.

The thermal conductivity of GdCl$_3$ was measured parallel to the c-axis as a function of temperature from 0.3 to 175 K and in magnetic fields up to 35 kOe. For temperatures less than 1 K the thermal conductivity was observed to be saturated at the highest fields[3] indicating that in this regime the heat conduction was entirely due to phonon, both scattering and conduction by the magnons being negligible. It was possible to account for the data in this regime and in the paramagnetic phase in zero field by a Debye model for phonon heat conduction including only boundary, point defect, and phonon-phonon scattering. The boundary length agreed with the external sample size to within 5% indicative of the excellent quality of these samples.

Because of the rather detailed information that one has on the magnetic interactions in GdCl$_3$ and the apparent absence of complications from lattice defects, an attempt to account for the experiments by model calculations for the magnon-phonon interactions was

246

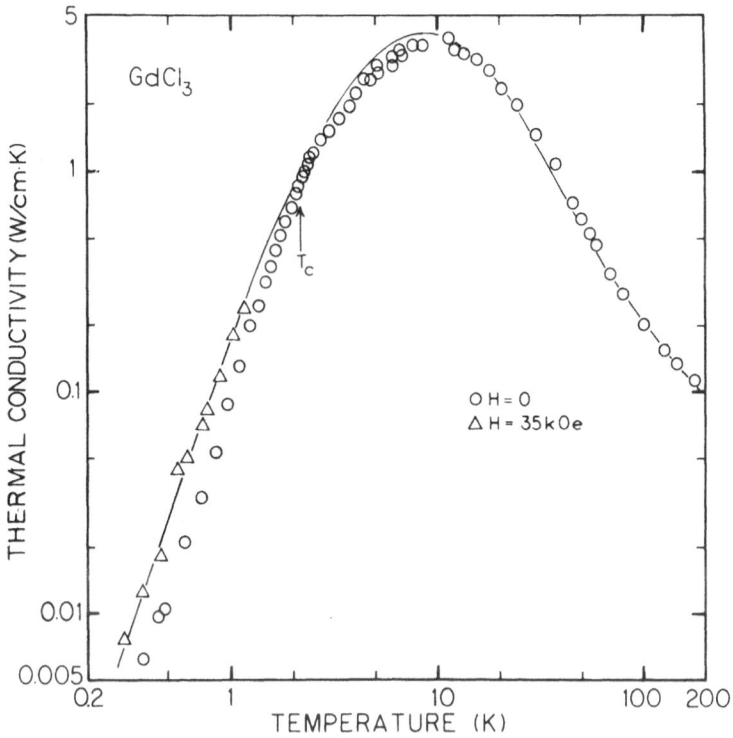

Fig. 1. Thermal conductivity of GdCl₃ in zero magnetic field and in a saturating field. The curve is a Debye model fit described in the text.

made. The spin-lattice coupling was taken to arise from moderation of the isotropic exchange and magnetic dipole forces by the phonon strain field. Both one magnon one phonon and two magnon one phonon processes were included. The coupling constants arising from the dipolar forces were calculated by performing the appropriate lattice sums; those arising from exchange were estimated from the pair spectra results of Hutchings, Birgeneau and Wolf.[2] The one magnon one phonon processes were taken into account by diagonalizing the spin lattice Hamiltonian including the terms bilinear in the magnon and phonon operators.[4] The two magnon one phonon terms were included as relaxation times calculated from perturbation theory. The solid curve in Fig. 2 is a "best fit" to the zero field data using these relaxation processes. It was found that the magnon-like branches of the excitations are of negligible importance as heat carriers throughout the range of these experiments. The dipolar contribution to the phonon relaxation rate agreed with that calculated by the lattice sums to within a factor of 2. The exchange contributions are substantially smaller than the expectation values of the pair spectra experiment but are nevertheless within their limits of un-

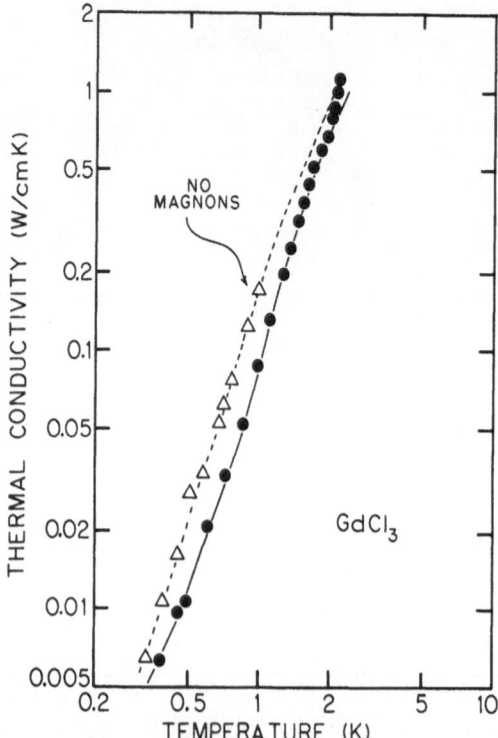

Fig. 2.   Effect of magnon-phonon scattering on the thermal conductivity.

certainty. This smaller value is also consistent with the absence
of significant critical scattering near T_c. Details of the calculations will be published elsewhere.

REFERENCES

1. R. Clover and W. P. Wolf, Solid State Commun. 6, 331 (1968).
2. M. T. Hutchings, R. J. Birgeneau, and W. P. Wolf, Phys. Rev.
    168, 1026 (1968).
3. G. S. Dixon, J. J. Martin and N. D. Love, A.I.P. Conf.,Proc.
    18, 1073 (1974).
4. M. A. Savchenko, Fiz. Tverd. Tela 6, 864 (1964).

SPIN-PHONON INTERACTIONS IN EXCHANGE INDUCED VAN VLECK PARAMAGNETS

C. M. Care and J. W. Tucker

Department of Physics, Sheffield University

Sheffield S3 7RH, England

The simplest model that has been proposed to describe an exchange-induced Van Vleck paramagnet is the singlet-singlet model whose Hamiltonian in the paramagnetic regime and in the absence of an applied field can be reduced to that of an Ising model in a transverse field. It has been suggested (Hoenerlage 1973) that the difference in temperature dependence of the spectrum of collective excitations as predicted by this model and that observed experimentally is in part due to spin-phonon interactions. In this paper we obtain via diagrammatic perturbation theory the frequencies and lifetimes of the collective spin-phonon modes for this system. The system we consider is described by the Hamiltonian

$$H = -\Gamma \sum_i S_i^z - \tfrac{1}{8} \sum_{ij} J_{ij} (S_i^+ + S_i^-)(S_j^+ + S_j^-) + \sum_q \omega_q (c_q^+ c_q + \tfrac{1}{2})$$

$$+ N^{-\frac{1}{2}} \sum_{iq} \tfrac{1}{2}(\varepsilon\Gamma\,\omega_q)^{\frac{1}{2}} (c_{-q}^+ + c_q)(S_i^+ - S_i^-) \exp(iq.R_i) \qquad (1)$$

The dispersion relation for the coupled spin-phonon excitations may be determined from the poles of the generalised spin susceptibility $\underline{G}$ defined by Stinchcombe 1973. $\underline{G}$ may be expanded diagrammatically and its poles occur when $\det(1-\underline{M}.\underline{I})=0$ where $\underline{M}$ is a polarization matrix that includes all those diagrams which cannot be separated into two parts by the breaking of a single interaction (either phonon or exchange). The interaction matrix $\underline{I}$ has only two components

$$I^{xx} = \beta\, J_q \qquad\qquad I^{yy} = -2\beta\varepsilon\Gamma\, \omega_q^2/(\omega_q^2 - \omega^2) \equiv \beta\, A \qquad (2)$$

The poles of $\underline{G}$ are then given by

$$(1-\beta AM^{yy})(1-\beta J_q M^{xx}) - \beta^2 AJ_q M^{xy}M^{yx} = 0 \tag{3}$$

This leads to well-known results in the limits of either $J(q)$ or ε tending to zero. The matrix elements of $\underline{M}$ which appear in (3) were evaluated to first order in ε using the techniques of Vaks et al. (1968). $\underline{M}$ was taken to be

$$\underline{M}^{\alpha\beta} = \qquad\qquad\qquad\qquad + \qquad\qquad\qquad \tag{4}$$

where the broken line contains a phonon propagator and the hatched-circle a vertex part. Insertion of this into (3) gives after a considerable amount of algebra

$$(\nu_q^2 - \omega^2) - A\nu_q^2(R/\Gamma) - [2\varepsilon R - 3\pi i \varepsilon(\omega/\omega_D)^3 P/R]K_q = 0 \tag{5}$$

with

$$K_q = J_q \Gamma R - \Gamma^2(J_q \Gamma R + \nu_q^2 AR/\Gamma)/(\Gamma^2 - \omega^2)$$

Here $\nu_q = [\Gamma^2 - J_q \Gamma R]^{\frac{1}{2}}$ is the frequency of the collective spin excitations, $R = \frac{1}{2}\tanh(\frac{1}{2}\beta\Gamma)$ and $P = (d/d\beta\Gamma)R$. To lowest order the term involving $K(q)$ can be neglected and this is used to eliminate ω in $K(q)$ to give an approximate expression valid when ω is not close to Γ.

$$\nu_q^2(1+2\varepsilon R) - \omega^2 - A\nu_q^2(R/\Gamma) - 3\pi i \varepsilon(\omega/\omega_D)^3 P\nu_q^2/R = 0 \tag{6}$$

The real part of (6) is identical to the result obtained by Hoenerlage from an equation of motion approach. Outside the immediate vicinity of the crossover region where $\omega(q) = \nu(q)$ this equation has two solutions for ω one near $\omega(q)$ and the other near $\nu(q)$ whose imaginary parts (which are inversely proportional to the lifetime of the associated excitations) are found to be

$$\mathrm{Im.}(\omega) = 3\pi\varepsilon^2\nu_q^4\omega_q^4 P[\omega_D^3(\omega_q^2 - \nu_q^2)^2]^{-1} \qquad \text{near to } \omega_q$$

$$\mathrm{Im.}(\omega) = 3\pi\varepsilon\nu_q^4 P/(2\omega_D^3 R) \qquad\qquad \text{near to } \nu_q \tag{7}$$

In the region where ω is close to Γ it is not possible to make much progress because an infinite number of terms contributes in this region. The leading terms can be shown to be diagrams made up of two basic elements linked in chains by the free phonon propagator. In Vaks notation these elements are

$$\text{(a)} \qquad \longrightarrow \quad = [\beta(\Gamma - \omega)]^{-1}$$

(b)

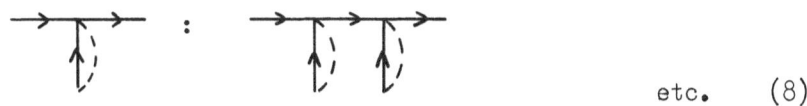

etc.   (8)

Unfortunately the pre-multiplying factors for each diagram are complicated as contributions come from several vertex parts. A result
may be obtained in the limit P → 0 (i.e. when T → 0) when the series
reduces to

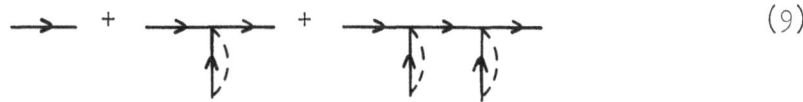

(9)

This may be summed to give a renormalised propagator with its frequency shifted from Γ to $\Gamma(1+\varepsilon/4R)$. At zero temperature this is
the same frequency shift as that obtained by Elliott and Parkinson
(1967) but unfortunately it does not seem possible to extend the
result to finite temperatures. Nevertheless the approximation will
be good provided P is very small, which requires $\tanh^2(\tfrac{1}{2}\beta\Gamma)$ to be
close to unity. Because the transition temperature in exchange-
induced paramagnets is determined not only by Γ but also by J it is
possible in certain materials for this condition to be satisfied at
temperatures of interest. For example, in Pr_3Tl where $J/\Gamma \approx 1.72$
it is found in the vicinity of the transition point $(1/\beta=0.19\Gamma\approx1/\beta_c)$
that $\tanh(\tfrac{1}{2}\beta\Gamma)\approx0.99$ so P is indeed small. However by the time $1/\beta$
has increased to 0.8Γ $\tanh^2(\tfrac{1}{2}\beta\Gamma)$ is 0.308 so the result is no
longer satisfactory. Some improvement in the theory is therefore
necessary if the exact temperature dependence of the excitations
in the vicinity of Γ is required over a large temperature range.

References

Elliott, R. J. and Parkinson, J. B. (1967) Proc. Phys. Soc. 92,
    1024.
Hoenerlage, B. (1973) Z. Physik 260, 403.
Stinchcombe, R. B. (1973) J. Phys. C. 6, 2459.
Vaks, V. G., Larkin, A. I. and Pikin, S. A. (1968) Sov. Phys. JETP
    26, 188.

MAGNETIC FIELD DEPENDENT THERMAL CONDUCTIVITY IN $FeCl_2$ AND $FeBr_2$

D. Petitgrand and G. Laurence

Laboratoire de Physique des Matériaux, D. Ph.E.P.

C.E.N. de Saclay, 91190 Gif-sur-Yvette   (France)

$FeCl_2$ and $FeBr_2$ are two antiferromagnetic insulators whose magnetic properties have been widely investigated these last years (1)-(6). Besides, other properties, such as anomalous thermal conductivity (7), pressure (4)-(5) and strain (6) dependence of magnetic properties and change of the lattice constants at the Neel temperature (8), have suggested the presence of a large magnetoelastic interaction in these two compounds. A first measurement of the thermal conductivity in $FeCl_2$ (7) has revealed a pronounced dip in the thermal conductivity, in a wide temperature range, the minimum being at 17 K, a value lower than the Neel temperature. In a previous paper (9), we showed that this anomalous behaviour could be accounted for by making the assumption of a resonant interaction between magnons and phonons. We now report measurements of the thermal conductivity in magnetic field of $FeCl_2$ and $FeBr_2$. The measurements were made using a steady state heat-flow method in a pumped helium cryostat between 1 K and 50 K. Magnetic fields up to 75 kOe were applied with a superconducting solenoid ; the sample being rotated in the cryostat so as to apply the field either parallel or perpendicular to the c axis.

The variations of the thermal conductivity with temperature when the field is applied along the c axis are shown on fig 1 for $FeCl_2$ and fig 2 for $FeBr_2$. In the case of $FeCl_2$ the effect of applying a field of 10 kOe is to reduce the thermal conductivity between 2 K and T_N (24 K) ; a typical value of κ (10 kOe)$/\kappa$ (0) being 0.7 in the range 5 K to 10 K. When fields higher than 10kOe are applied, the thermal conductivity increases and reaches the

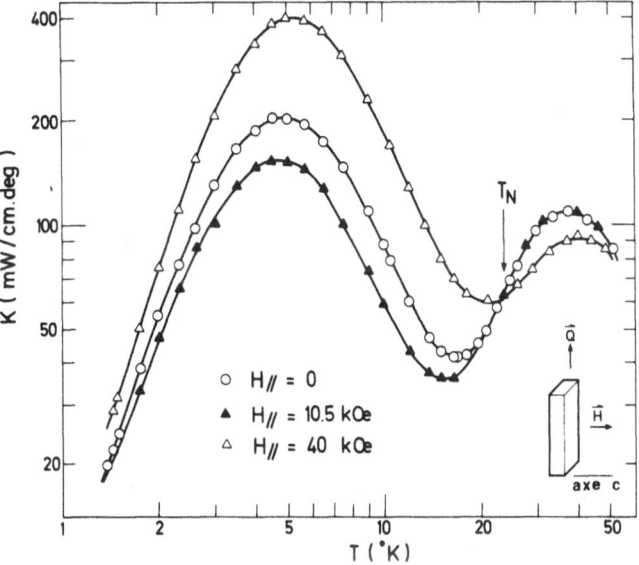

Fig 1 - Thermal conductivity of FeCl$_2$ in parallel fields

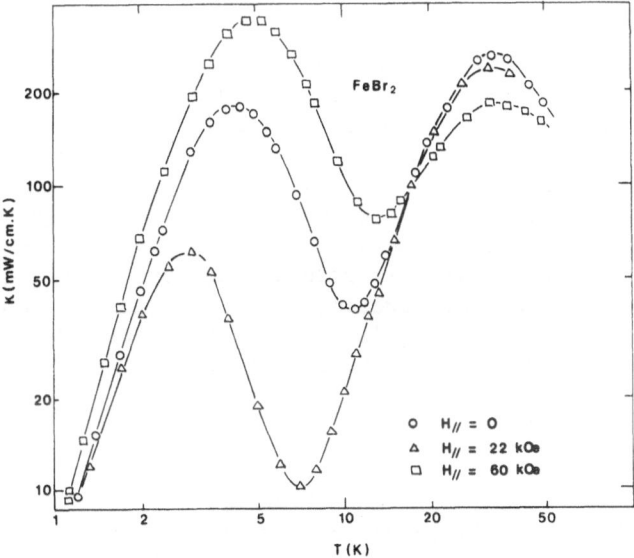

Fig 2 - Thermal conductivity of FeBr$_2$ in parallel fields

zero field value around 20 kOe. For a field of 40 kOe, the ratio $\kappa(H)/\kappa(0)$ is 1.9 at 5 K. For temperatures higher than T_N, the effect of a 10 kOe field is undetectable, while a 40 kOe field lowers the thermal conductivity between 24 K and 50 K. In the case of $FeBr_2$ the behaviour is quite similar ; however the conductivity decreases with applied field up to 26 kOe ; the ratio κ (26 kOe)/κ(0) being 0.05 at 6 K. A typical plot of $\kappa(T)$ for H = 22 kOe is shown on fig 2. For higher fields the conductivity increases : at 60 kOe the curve $\kappa(T)$ stays above the zero field one for T < 17 K and beneath it for T > 17 K. The maximum of the ratio κ(65 kOe)/κ(0) is obtained at 6 K where it reaches 2.7.

When the magnetic field is applied perpendicularly to the c axis, the behaviour of κ is quite different. In both materials the conductivity is always a decreasing function of the field at a given temperature. This lowering is large for T < T_N and weak for T > T_N. In $FeCl_2$ and $FeBr_2$ the ratio $\kappa(H_\perp)/\kappa(0)$ is 0.5 for a field of 40 kOe and a temperature of 4 K. However in $FeBr_2$ we have been able to make measurements at higher fields and we have found that the conductivity falls rapidly, the ratio $\kappa(H_\perp)/\kappa(0)$ reaching 0.09 for a field of 75 kOe.

The behaviours of κ in the two materials appear to be very similar if we keep in mind that both undergo a magnetic phase transition in a parallel magnetic field of 10.5 kOe for $FeCl_2$ and 29 kOe for $FeBr_2$, whereas they stay in a continuous phase in perpendicular fields smaller than 75 kOe. In the antiferromagnetic phase the application of a parallel or perpendicular magnetic field lowers the conductivity. This is consistent with the interpretation given earlier in terms of magnon-phonon resonant interaction (9) : because of the splitting of the two antiferromagnetic spin-wave modes induced by the magnetic field, the phonons which are concerned by the interaction lie in a wider energy band, thus the thermal conductivity is reduced. Moreover, for a given magnitude of the field, the lowering of κ is larger for a parallel field than for a perpendicular one which agrees with the fact that we expect the splitting to be larger for $H_\parallel$ than for $H_\perp$.

(1) P. CARRARA, Thesis, Orsay, (1968), unpublished.
(2) A.R. FERT et al., J. Phys. Chem. Solids, 34, 223, (1973).
(3) M.C. LANUSSE et al., J. Phys. (France), 33, 429, (1972).
(4) C. VETTIER et al., C.R.Acad. Sc. Paris, 275, 915, (1972).
(5) C. VETTIER et al., Phys. Rev. Let., 31, 1414, (1973).
(6) J.A. NASSER, J. HAMMAN, Phys. Stat. Sol., 56, 95, (1973).
(7) G. LAURENCE, Phys. Lett., A 34, 308, (1971).
(8) R. KLEINBERGER, (private communication).
(9) G. LAURENCE, D. PETITGRAND, Phys. Rev. B, 8, 2130, (1973).

ATTENUATION AND DISPERSION OF FIRST SOUND NEAR THE LAMBDA LINE OF

He^3 - He^4 MIXTURES[†]

A. Ikushima*, D. B. Roe and H. Meyer

Duke University, Department of Physics

Durham, North Carolina 27706 U.S.A.

We have measured the attenuation and velocity of sound near the superfluid transition of several liquid He^3-He^4 mixtures at saturated vapor pressure, ranging in concentration from X_3=0.9% He^3 to 36%. For these measurements, two acoustic cells were used covering the range of 1-15 MHz and above 12 MHz respectively. The pairs of quartz transducers were spaced approximately 2 mm apart. Sound frequencies ranged from 1.0 to 36 MHz for the 36% mixture, but so far the other mixtures were studied only at 12 and 36 MHz. For the sound velocity experiments, we used a pulse time-of-flight phase-comparison technique. The attenuation was measured by monitoring the amplitude of the received sound pulse.

The classical Navier-Stokes attenuation is assumed to be nearly constant over the small temperature interval around the superfluid transition. The critical attenuation is then the difference between the background pulse amplitude far from the transition and the attenuated amplitude near the transition. Fig. 1 shows this critical attenuation for several mixtures. Qualitatively the peak becomes lower and broader as X_3 increases but higher and broader as the frequency increases.

We analyze the attenuation data using a model applied first[1] to pure He^4 in which two separate mechanisms contribute to the critical attenuation. The first mechanism occurs only below T_λ and, as proposed by Pokrovskii and Khalatnikov[2], is caused by a relaxation process (rel.) attributed to coupling between first and second sound. The second mechanism comes from critical order parameter fluctuations (fl.). Attenuation in pure He^4 due to this mechanism seems to occur[1] with equal strength above and below T_λ, and in our analysis we assume the same to be true in mixtures.

255

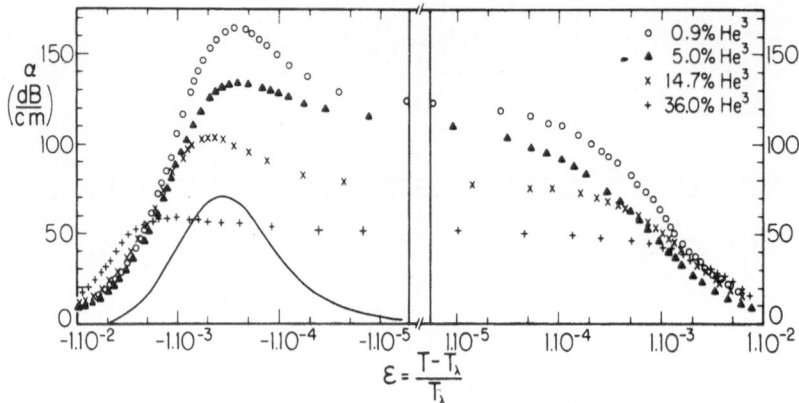

Fig. 1 Critical attenuation (dB/cm) versus reduced temperature
ε at 36 MHz below (left) and above T_λ. The relaxation peak α_{rel}
for 0.9% is also shown by the solid curve.

Hence, subtracting the measured critical attenuation $\alpha_{fl}(\varepsilon)$ above
T_λ from the total attenuation $\alpha(-\varepsilon)$ below T_λ will give $\alpha_{rel}(-\varepsilon)$
where $\varepsilon = (T - T_\lambda)/T_\lambda$.

The "relaxation" model[2] predicts that the maximum attenua-
tion $\alpha_{rel}^{max} \cdot \alpha \omega^{1.0}$. The "relaxation" peak is due to a process having
a single relaxation time $\tau = \xi/U_2$ where ξ is the coherence length
and U_2 is the second sound velocity. In pure He[4], it is predicted[2]
that $\tau \alpha \varepsilon^{-1.0}$ which is confirmed by experiments up to several MHz.
However, this model does not explain the very large attenuation peak
found at 1000 MHz.[3] Measurements at frequencies below 0.6 MHz in
a 19.7% mixture yielded $\alpha_{rel}^{max} \cdot \alpha \omega^{1.05}$ and $\tau \alpha \varepsilon^{-1.02}$.[4] Our data in the
36% mixture give $\alpha_{rel}^{max} = 3.8 \times 10^{-9} \omega^{1.17 \pm 0.1}$ dB/cm and $\tau = 6.5 \times 10^{-13} \times$
$\varepsilon^{-1.45 \pm 0.1}$ sec. For the "fluctuation" attenuation α_{fl}, we attempt a
scaling representation of the form[5]

$$\alpha_{fl}(\omega, \varepsilon) = \omega^y F(\varepsilon/\omega^z) \qquad (1)$$

which is a consequence of the dynamic scaling hypothesis and which,
in the limits of ε small and ε large, should reduce to the predictions of Kawasaki[6]. A test of this equation is shown in Fig.
2. In the 36% mixture, we find y=1.33±0.1, z=0.77±0.07, and α_{fl}^{max}
=3.7x10^{-10} ωy dB/cm. An analysis of acoustic results for He[4]
yields y=1.3±0.1 and z=0.9±0.1. We feel that this scaling hypothesis is valid for mixtures. There is some indication that the
width of $F(\varepsilon/\omega^z)$ becomes narrower as X_3 is increased.

We now turn to velocity measurements. For the 0.9%, 5% and
14.7% mixtures at 12 MHz we observed broad minima in the velocity
below T_λ similar to those observed[7] in pure He[4]. In the 36% mix-

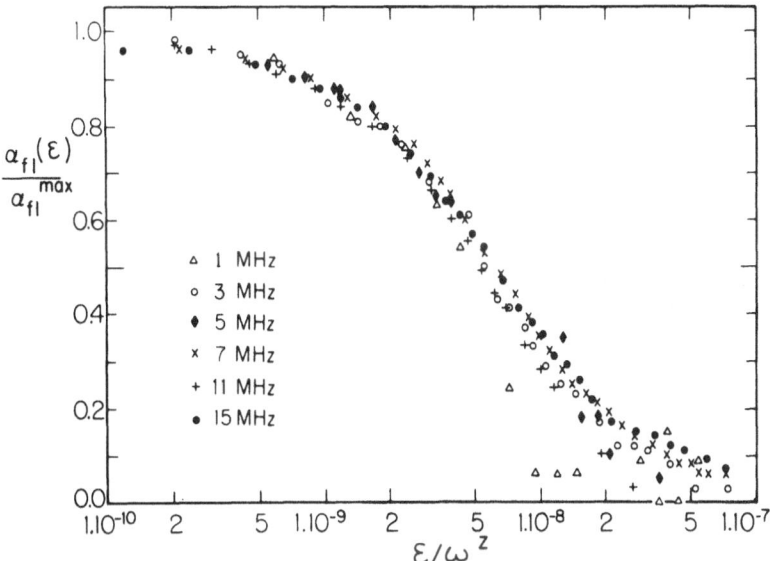

Fig. 2   Normalized critical attenuation α_{f1}. versus scaled temperature $\varepsilon/\omega^{0.77}$.   Six frequencies are shown at 36% He^3.   If
equation 1 holds, all points should lie on the same curve.

ture, there was a small cusp in the velocity at 1 MHz which dis-
appeared at 15 MHz. We have plotted the difference between the 1
MHz data and that at higher frequencies. Although the 1 MHz data
is a poor approximation to the limiting zero frequency velocity,
we find the qualitative behavior of the dispersion $U(\omega)-U(1\ MHz)$
to be similar to that of the attenuation. The asymmetric dispersion peak is comprised of a relaxation peak below T_λ and an order-
parameter scattering peak symmetric about T_λ.

[†]Supported by a grant from the National Science Foundation.
[*]Permanent address: Institute for Solid State Physics, University
    of Tokyo.
1. R. Williams and I. Rudnick, Phys. Rev. Lett. 25, 276 (1970).
2. V. Pokrovskii and I. Khalatnikov, JETP Lett. 9, 149 (1969).
3. J. Imai and I. Rudnick, Phys. Rev. Lett. 22, 694 (1969).
   D. Commins and I. Rudnick, Low Temp. Phys.-LT13, 396 and
   references therein to light scattering work.
4. C. Buchal, F. Pobell and W. Thomlinson, Phys. Lett. 51A, 19 (1975)
5. K. Kawasaki and J. D. Gunton, private communication
6. K. Kawasaki, Phys. Lett. 31A, 165 (1970).
7. Thomlinson and F. Pobell, Phys. Rev. Lett. 31, 283 (1973).

THERMAL CONDUCTIVITY OF RbCaF$_3$

J. J. Martin

Oklahoma State University

Stillwater, Oklahoma   74074   U.S.A.

Modine et al. (1) have reported that the perovskite compound RbCaF$_3$ undergoes a cubic to tetragonal structure change at 198 K and that additional structure changes take place at 43 K and 7 K. Ho and Unruh (2) have also found specific heat anomalies near these temperatures. Since the phase transitions are similar to those observed in other perovskite compounds which involve "soft" phonon modes, the thermal conductivity might be expected to show an anomalous behavior near the transitions. Steigmeir (3) found that phonon scattering from a "soft" zone center mode was important in both SrTiO$_3$ and KaTaO$_3$. Suemune and Ikawa (4) have observed anomalous thermal conductivities in magnetic fluoride perovskites such as KMnF$_3$ which also have structural phase transitions.

The thermal conductivity of single crystal RbCaF$_3$ has been measured from 5 to 300 K. The results are shown in Fig. 1 along with the thermal conductivities of KMgF$_3$, KZnF$_3$ and KMnF$_3$ which were measured in this laboratory. The thermal conductivity of RbCaF$_3$ is much smaller than the thermal conductivities of both KMgF$_3$ and KZnF$_3$. Above approximately 45 K the thermal conductivity increases with T in contrast to the usual T^{-1} dependence. Neither the 7 K nor the 198 K transitions significantly influence the thermal conductivity. We interpret the decreasing thermal conductivity as T is lowered towards 45 K as due to phonon scattering from a "soft" mode associated with the 45 K transition. While a complete understanding of the results requires detailed knowledge of the phonon dispersion curves which are not yet available, we have attempted a fit to the data in spirit of the Debye-Callaway model. (5) For the calculation, which neglects both polarization and dispersion, the combined phonon relaxation time contained the following: boundary scattering with the theoretical Casimir length, point defect scattering due to isotopes,

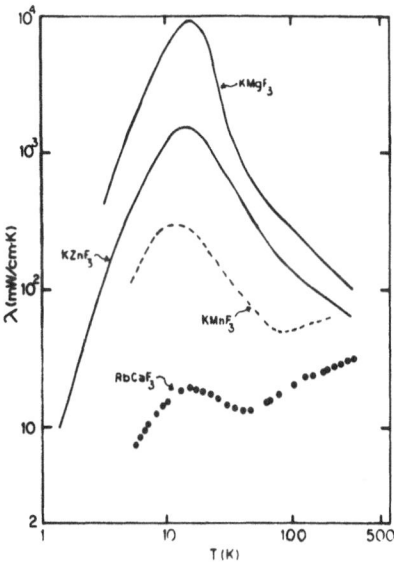

Figure 1.   The thermal conductivities of RbCaF$_3$ and three other fluoride perovskites are shown.

a strain field dislocation term, and a phonon–phonon scattering term.
All of the phonon–phonon scattering was assumed to come from a T dependent "soft" mode associated with the 45 K transition.   Below 45 K
the phonon–phonon scattering was described by

$$\tau_{pp}^{-1} = B\,e^{-\frac{\theta}{aT}}\,\omega^2\,T \tag{1}$$

where B is the strength of the interaction, θ is Debye temperature,
a is the phonon angular frequency.   Above 45 the lifting of the
"soft" mode would decrease the strength of the phonon–phonon scattering.   Several models were tried; the most successful model was
to multiply Eq. 1 by the factor $\exp\left[-C\left(\frac{T-45}{T}\right)^2\right]$, where C is a con-

stant, the other parameters had the same values used in Eq. 1.   In
order to fit the low temperature data it was necessary to include a
strong dislocation scattering term.   Figure 2 compares the calculated thermal conductivity with the measured values.   At the higher
temperatures, this soft mode should lose its T dependence and the
phonon–phonon scattering should return to the conventional type.
There may also be additional scattering near the 198 K transition.

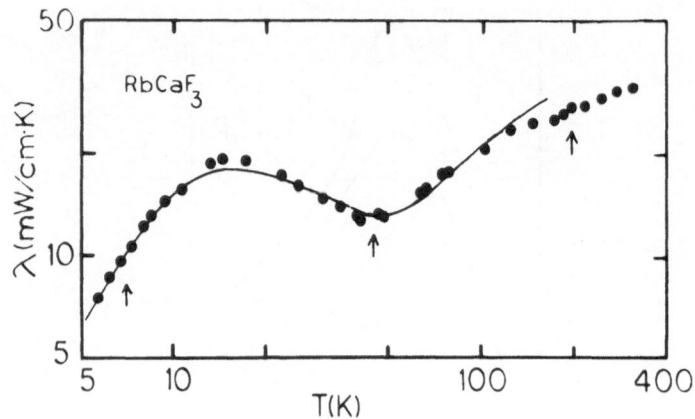

Figure 2.   The calculated thermal conductivity assuming phonon scat-
            tering by a soft mode is shown.

        The author wishes to thank E. Sonder of ORNL for supplying the
RbCaF$_3$ crystals, J. Ho and W. Unruh for their specific heat results
prior to publication and G. S. Dixon for several helpful discussions.

REFERENCES

1.   F. A. Modine, et al., Phys. Rev. B, 10, 1623 (1974).
2.   J. C. Ho and W. P. Unruh, paper to be presented at the 30th
     Calorimetry Conference, Seattle, Wash., July 16–19, 1975.
3.   E. F. Steigmeier, Phys. Rev. 168, 523 (1968).
4.   Y. Suemune and H. Ikawa,   J. Phys. Soc. Japan, 19, 1686 (1964).
5.   J. Callaway, Phys. Rev. 113, 1946 (1959).

PHONON ECHOES AND PHASE TRANSITIONS

Ch Frenois, J Joffrin and A Levelut

Laboratoire d'Ultrasons*, Universite P et M Curie

Tour 13, 4 Place Jussieu, 75230 Paris-Cedex 05, France

INTRODUCTION

The phonon echo method consists, in essence, of the following; two electromagnetic pulses separated by a time interval τ are applied to a sample and at a time 2τ (measured from the start of the first pulse) a signal (the echo) is received. Included in the information that can be obtained by this method, an important item appears to be the life-time of the phonons in the crystal: that is, the relaxation time T_2 of the echo.

The aim of the experiments presented in this paper was to examine the effect of the phase transition in KDP (potassium dihydrogen phosphate) on the time T_2. The frequency range employed ($\nu < 300MHz$) is justified by the high temperature of the transition ($\sim 120K$). In fact, to eliminate parasitic signals caused by ultrasonic reflections from the sample surfaces, powders of small size ($10\mu < d < 50\mu$) were used; it is known that there is no problem in using them with the phonon echo technique. Nevertheless it should be noticed that the acoustic wavelength used here ($\lambda \sim 40\mu$) is of the same order of magnitude as the dimensions d of the grains of powder. Hence the situation is different from that studied earlier at 1GHz when $\lambda \ll d$ (1). From a conceptual point of view, it is then only necessary to replace the idea of phonons with that of modes of vibration.

The results were obtained for the high temperature (paraelectric) phase where the frequency dependence of T_2 and its critical behaviour were examined.

---

*Associated with Centre National de la Recherche Scientifique.

NOTES ON THE PROPERTIES OF KDP

At a temperature $T_c \simeq 122K$, KDP undergoes a transition from a phase of symmetry $\bar{4}2m$, which is paraelectric and piezoelectric, to a phase of symmetry mm2, which is ferroelectric. At that temperature the elastic constant $c_{66}{}^E$ disappears; that is the result of the coupling (via the piezoelectricity) with a mode whose frequency is such that $\omega^2(T) = A(T-T_0)$ with $T_0 \simeq 117.7K$ (2). This transition has been extensively studied by ultrasonic methods (3,4,5); comparisons are thus made easy.

VARIATION OF T_2 WITH FREQUENCY

The experimental results show that the dependence of T_2 on frequency is straightforward. The reciprocal of the relaxation time is the sum of a constant and a term proportional to the frequency squared (fig 1). That is:

$$\Gamma_2 = a + b\nu^2$$

This qualitative result is easily explained. In fact, of the relaxation processes which add to give Γ_2, certain ones are independent of frequency. These are loss due to piezoelectric conversion, loss by elastic mode conversion and by phonon emission into the external medium (1); nevertheless, the last process is not important in this case because the experiments were made under vacuum. The processes which depend on frequency give a contribution proportional to ν^2; these are critical attenuation and attenuation by thermal phonons in this temperature range.

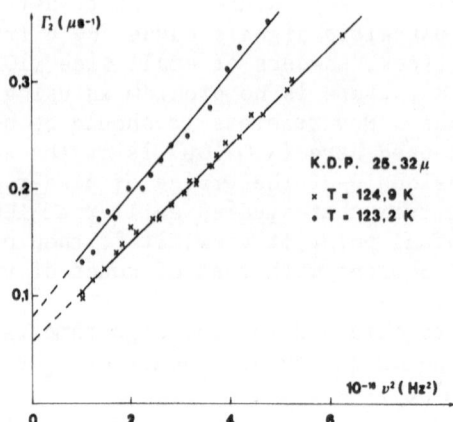

Figure 1: Reciprocal of the relaxation time vs the square of the frequency. The experimental points lie along the lines $\Gamma_2 = a + b\nu^2$. As the transition temperature $T_c = 121.65K$ is approached critical effects, which vary as ν^2, increase.

The coefficients a and b are temperature dependent. In the case of b, it is clear that this is because of the contribution from critical effects. In the case of a, the lowering of one of the acoustic velocities certainly plays a part.

CRITICAL EFFECTS ON T_2

In the neighbourhood of the transition temperature T_c, the relaxation time T_2 is considerably shortened. This fact, which is not very surprising, must nevertheless be emphasised. In fact, it contradicts the remarkable results obtained by Popov and Krainik (6) in a rather lower frequency range (from 2 to 70MHz) and where these authors observed a lengthening of T_2.

One of our results is shown in figure 2. Above 130K, T_2 is effectively constant up to at least 150K; in the following this value will be called $T_2 p^l$ (plateau). Below 130K, T_2 decreases and becomesso short in the region of the transition that the echo disappears completely. In order to make use of the experimental results the reciprocal of the relaxation time is written in the form:

$$\Gamma_2(T) = \Gamma_2{}^{pl} + \Gamma_2{}^{crit}(T)$$

It is straightforward to compare the following quantity with theory:

$$T_2{}^{crit}(T) = |\Gamma_2{}^{crit}(T)|^{-1}$$

Indeed, it can be shown (2) that the life-time θ of phonons of the deformation mode ε_{12} is such that:

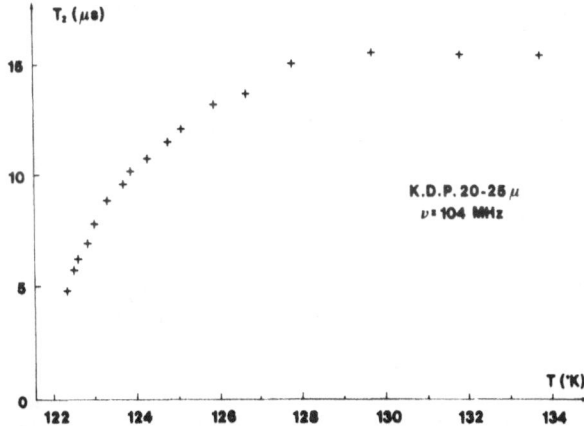

Figure 2: Relaxation time T_2 as a function of temperature. The phase transition is shown by a shortening of T_2.

$$\theta = 2c_{66}^P (T - T_o)(T - T_c)(\omega^2 D_o A)^{-1}$$

With the numerical values $A = 24 \times 10^{-12}$ sK^{-1}, $D_o = 0.3 \times 10^{12}$ dyne K cm^{-2}, $c_{66}^P = 7 \times 10^{10}$ dyne cm^{-2}, $\omega = 2\pi \times 10^8$ rad s^{-1}, it is found that:

$$\theta = 0.05 (T - T_o)(T - T_c) \qquad (\mu s)$$

The experimental results can be represented effectively by the relation:

$$T_2^{crit} = M(T - T_c)(T - T_o)$$

where it has been assumed that $T_c - T_o = 4.3K$ (3). The value of T_c is found to be 121.65K. The coefficient M is a decreasing function of d. For a powder with dimensions between 20 and 25μ, we take $M = 2.0$ (when T_2 is measured in μs). The temperature dependence is as expected, but the order of magnitude is not correct: the phonons which contribute to the echo have a critical life time nearly two orders of magnitude longer than that of the ultrasonic waves of the same frequency and acoustic mode ε_6.

A possible explanation is the occurence of mode conversion which continuously mixes the different polarisations excited in the microcrystals.

The dependence of M on d is an indication of this.

CONCLUSION

In order to check the use of the phonon echo technique by studying phase transitions, we have taken KDP as an example. In favour of the method, it has been shown that it is very sensitive to phase transitions whose effects are strongly coupled to the phonons and this is so even if one only has very small crystals of the material. On the other hand the most important disadvantage is the overall aspect of the method: firstly it is impossible to separate the roles of different phonon modes in the crystal; and secondly it is difficult to distinguish attenuation effects from variations in velocity.

(1) Ch Frenois, J Joffrin, A Levelut, J Physique Lett 35 (1974) L-221
(2) E M Brody, M Z Cummins, Phys Rev Lett 21, (1968) 1263
(3) C W Garland, D B Novotny, Phys Rev 177, (1969) 971
(4) E Litov, C W Garland, Phys Rev B2, (1970) 4597
(5) C W Garland, Physical Acoustics Vol 7, Chap 2 (Academic, 1970)
(6) S N Popov, N N Krainik, Sov Phys Solid State 14, (1973) 2408

CRITICAL ATTENUATION AND VELOCITY OF SURFACE WAVES AT STRUCTURAL

PHASE TRANSITIONS

L Bjerkan and K Fossheim

Department of Physics, The Norwegian Institute of

Technology, 7034 Trondheim, Norway

The structural phase transitions in $SrTiO_3$ and $KMnF_3$ occurring at T_c=103K and 187K respectively, have been studied using ultrasonic surface waves. Surface waves (in the present case Rayleigh waves) penetrate to a depth of the order of one wavelength, and since the wavelength, λ, may be varied so may also the penetration depth. The general motivation for this work is to find
   1. whether surface waves show critical behaviour similar to bulk waves,
   2. and if so, whether a detailed knowledge of the properties of bulk waves always will allow us to make correct predictions about critical properties of surface waves.

In this way specific surface effects may be revealed if present. The answer to the first question is, as will be shown below, affirmative for $T>T_c$. The second question is a very complex one, and requires better and more detailed experimental information on bulk waves in these materials before it can be answered completely. What makes it interesting to us in the present context, can be summarised as follows:
   i) In what ways will surface waves be affected when the coherence length near T_c becomes greater than the penetration depth $\sim\lambda$?
   ii) What role do the soft mode and the central mode play in surface wave characteristics near T_c?
With the techniques used here, the region of point i) above is not accessible. With regard to point ii) we expect that the critical behaviour of surface waves will be dominated by these modes in a way similar to bulk waves as long as the coherence length is very short.

$SrTiO_3$ and $KMnF_3$ have cubic symmetry in the high temperature

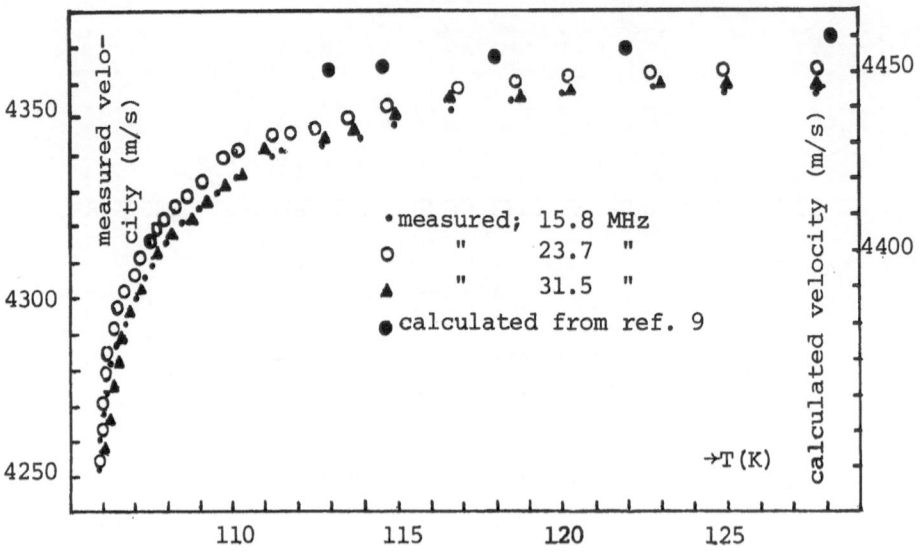

Fig 1: Measured Rayleigh wave velocities for the [100] direction in SrTiO₃ at three frequencies in the critical region. A few data calculated from ref 2 are also indicated, and can be read on the right-side axis.

phase and for waves in the [100] direction on the (001) surface, the following expression [1];

$$(\frac{C_{44}}{C_{11}})^2 \left[1 - \frac{C_{44}}{C_{11}}\right] x^3 - 2 \frac{C_{44}}{C_{11}} \left[1 - (\frac{C_{12}}{C_{11}})^2\right] x^2 +$$

$$\{2 \frac{C_{44}}{C_{11}} \left[1 - (\frac{C_{12}}{C_{11}})^2\right] + \left[1 - (\frac{C_{12}}{C_{11}})^2\right]^2\} x - \left[1 - (\frac{C_{12}}{C_{11}})^2\right]^2 = 0$$

(1)

relates the Rayleigh wave velocity v to the (bulk) elastic constants in an exact manner, where C_{11}, C_{12} and C_{44} are the three independent elastic constants for cubic crystals and $x = \rho v^2 / C_{44}$ where ρ is the density. Thus, the possible bulk effects due to soft modes and central mode are carried over to Rayleigh wave properties in an unambiguous way when the coherence length is short. Such a clear-cut analysis is not available for comparison of attenuation.

On SrTiO₃ the Rayleigh wave velocity was measured at 15.8, 23.7 and 31.5MHz or wavelengths approximately 275, 180 and 140μm respectively (fig 1). The measurements show qualitatively the same critical behaviour as that of bulk waves [2-4] with a marked slowing down of the velocity for $T_c < T \lesssim T_c + 30K$. Our technique did not allow accurate measurements of the absolute values of velocity, but on a relat-

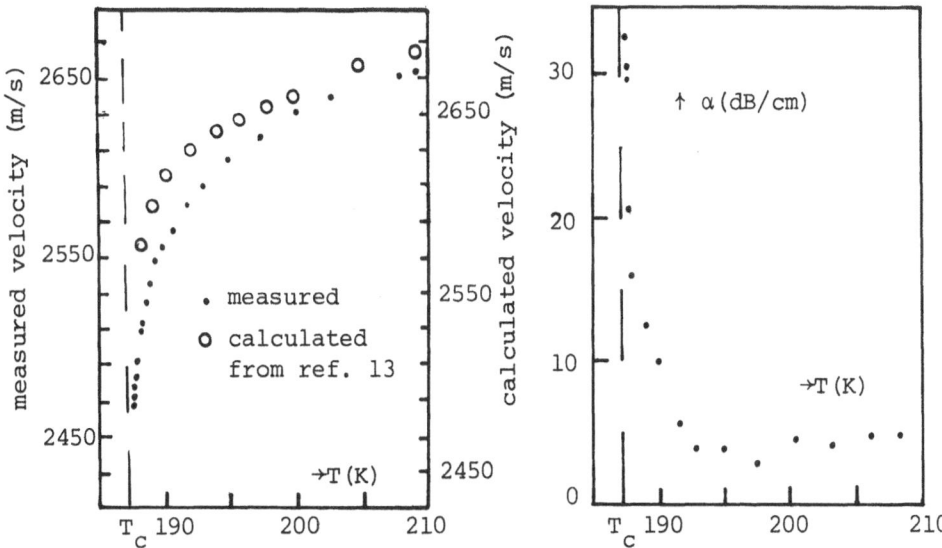

Fig 2: Measurements on Rayleigh wave velocity v and attenuation α
on KMnF$_3$. Calculations are based on data in ref 6.

ive scale the uncertainty is ∿±3m/s. We have inserted bulk wave data
from ref 2 in (1) and plotted the calculated values of velocity on
the shifted scale shown to the right of fig 1 so that measured and
calculated data coincide at room temperature. This comparison is
incomplete near T_c due to insufficient bulk wave data. The measurements could be extended down to ∿ 106K, but below this temperature
the signals were much too distorted to give reliable results. We
found T_c≈103K by bulk wave measurements. Preliminary attenuation
measurements indicate that critical attenuation sets in at about 10K
above T_c, roughly in a temperature region where bulk attenuation
also becomes important, [3-5].

On KMnF$_3$ critical attenuation and slowing down occurs in a similar way when approaching T_c. Fig 2 shows measured and calculated
velocities for a 10MHz Rayleigh wave in the [100] direction on the
(001) surface. Bulk data are from ref 6. In none of the measurements we performed could surface waves be observed below T_c which
can be understood in view of the domain formation in the tetragonal
phase.

1. R Stoneley, Proc Roy Soc 232A, 447 (1955).
2. R O Bell, G Rupprecht, Phys Rev 129, 90 (1968).
3. W Rehwald, Solid State Comm 8, 607 (1970).
4. K Fossheim, B Berre, Phys Rev B5, 3292 (1972).
5. B Berre, K Fossheim, K A Muller, Phys Rev Lett 23, 589 (1969)
6. K Fossheim, D Martinsen, A Linz, Proc NATO Advanced Study
   Institute on "Anharmonic lattices, structural transitions and
   melting", Ustaoset, Norway, ed by T Riste, Noordhoff (1974).

ULTRASONIC EVIDENCE OF COMMENSURATE DOMAIN STRUCTURE IN HOLMIUM

D.T.Vigren

Institut für Theoretische Physik, Freie Universität

Arnim Allee 3, Berlin 33, West Germany

Anomalous peaks in the attenuation of 15Mc/sec shear sound have been observed in Ho's spiral phase[1]. The sound is propagated along the $\hat{c}$ axis of the hcp structure and is polarized along a $\hat{b}$ axis. The peaks are suppressed by a magnetic field along $\hat{a}$ or $\hat{c}$, and although a marked shift of the peak position toward higher temperatures is observed with increasing field along $\hat{a}$, the peak position is quite insensitive to $\vec{H}$ along $\hat{c}$. The zero field peaks occur at 24K.

It was proposed [1] that these peaks mark the on-set of a tilted-spiral phase,first predicted by Sherrington [2,3] to exist as an intermediate phase between spiral and cone.Recently,however, it was shown that a sufficiently large exchange anisotropy will inhibit this phase completely[4], and careful neutron diffraction studies[5] give no evidence of it.Here,we wish to propose an alternative explanation of the peaks.

The magnetoelastic interaction[6] which couples the transverse strain field to the local spins, gives rise to an effective field in the $\hat{c}$ direction[1]. This field produces quasistatic oscillations of the spins $\perp$ to their equilibrium direction. Since the intrinsic relaxation time, ζ, of the spin system is very much shorter than $\omega_{\bar{p}}^{-1}$, the inverse sound frequency, these oscillations are in phase with the driving field and absorb no power. For $q{\neq}0$ they correspond to a distortive twisting of the spiral which results in a modulation of the exchange energy, that is linear in the shear amplitude, ϵ_o:

$$\Delta E_{ex} = 2S \sum_{i,j} J_{ij}\, S_y^i\, \sin(\phi_i - \phi_j). \tag{1}$$

Here J_{ij} is the exchange integral between ith and jth spin-planes, ϕ_i measures the angular displacement of the ith spin-plane with respect to some initial phase, α_o, and S_y^i is the transverse component of spin oscillation in the basal plane. One sees that this first order modulation vanishes in the case of a pure spiral: $\phi_i \rightarrow \phi_i^\circ = \vec{Q} \cdot \vec{R}_i + \alpha_0$, where $\vec{R}_i$ is the position vector of the ith spin and $\vec{Q}$ is the wavevector of the spiral. This result is in agreement with the calculations of Tachiki et al.[7]. In the presence of planar anisotropy or a basal plane field, however, ΔE_{ex} will be finite due to the higher harmonic content of ϕ_i[8,9]. Using a Bloch-Bloembergen description[10] of the spin dynamics one finds:

$$S_y(\vec{q}) = -4(N\hbar^2\omega_{\vec{q}}^2)^{-1} C_{44} \lambda^{\epsilon 2} \epsilon_c \sin(\omega_p t) \sum_j e^{i\vec{q}\cdot\vec{R}_j} \cos\phi_j \qquad (2)$$

Here N is the number of spin-planes along the c axis, $\hbar\omega_{\vec{q}}$ is the magnon energy of wavevector $\vec{q}$, and c_{44} and $\lambda^{\epsilon 2}$ are the relevant elastic and magnetoelastic constants, respectively. Substitution of (2) in (1) and use of the expression for ϕ_i, given by Cooper et al.[8] yields:

$$\Delta E_{e\lambda} = \sum_m f_m(\alpha,\beta,\gamma) \, g_m(\vec{Q}) \, \sin(m\alpha_c) \, \delta_{\vec{q}, \frac{2\pi s}{mc}\hat{c}} \qquad (3)$$

Here f_m is a function of the small quantities α, β, γ, which are essentially the ratios to the exchange energy, of the basal plane field energy, planar anisotropy energy and mixed field and anisotropy energies, respectively. The quantity g_m is a function of the Fourier-transformed exchange integral, $J(\vec{q})$, with various multiples of $\vec{Q}$ in the argument, s is a positive integer, c is the distance between spin-planes, and m takes on the integer values 1 through 8,11 and 13.

Neutron diffraction[5] shows that the spiral turn angle, ψ, ranges from $51°$ at 133K to $30°$ at 20K, so that the only contributions to the sum (3) occur at the commensurate angles $\psi = 45°$ (repeat distance 8 layers) and $\psi = 32.7°$ (repeat distance 11 layers). A study of the temperature dependence of the turn angle indicates that ψ passes through $32.7°$ almost precisely at 24K. Although we restrict our analysis to the low temperature absorption, it is interesting to note that anomalous peaks of the same character have been observed at 95.5K[11], very near the temperature at which ψ passes through $45°$.

At incommensurate turn angles the spin layers in a macroscopic length along the $\hat{c}$ axis point in all possible basal plane directions. Very near the temperature, T_c, at which the turn angle passes through its commensurate value, the spin-planes tend to "lock-on" to the lattice periodicity. This occurs with a slight loss of exchange energy, in order to gain the planar anisotropy energy, experienced when the spins align in discrete planar directions. Neutron diffraction[5] indicates that as the crystal is cooled from 25K to 20K, the planar anisotropy increases roughly by a factor 3. Because of this extremely rapid variation, one expects large spacial inhomogeneities in the planar crystal field, so that some parts of the spiral lock-on to the commensurate angle before others. One sees that a mixed phase, consisting of commensurate and incommensurate spiral domains, forms in a very narrow temperature interval,centered at T_c. Nagamiya et al.[12] have shown that for an applied basal plane field, $\psi(\vec{H}) < \psi(0)$. Since ψ is an increasing function of temperature, one sees that T_c shifts to higher temperatures with increasing field along $\hat{a}$. No such shift is expected for $\vec{H}$ along $\hat{c}$.

There is in general, a rather large angular separation between the last spin layer in a locked-on domain and the first layer of an adjacent incommensurate one, so that domain-walls are expected to form. These walls are necessarily thin, lie $\perp$ to the $\hat{c}$ axis, and possess a net magnetic moment, which rotates as the wall moves. At any temperature in the mixed phase, walls are pinned to crystal imperfections, which present local barriers to their motion. In the presence of the shear wave, commensurate domains experience a free energy density change, ΔE_{ex}, which is linear in the strain, so that they tend to grow and contract under the influence of the sound vibrations. It is clear that ΔE_{ex} is an effective pressure on the Bloch walls, which is capable of driving them harmonically in the potential valley of the pinning sites in which they sit. Local fluctuations of the magnetization, associated with the vibrating walls, create eddy currents, which provide the dissipative mechanism for resonant absorption.

The power dissipation may be calculated using a damped-oscillator model for the domain-wall motion[13] in a manner exactly analogous to that described in a previous work[14]. Apparently, the applied fields act to bring the sound vibrations further out of resonance with the characteristic frequency of the pinned walls, giving rise to a suppression of the attenuation with increasing field.

1.M. Tachiki et al., Solid State Comm. 15, 1071 (1974)
2.D. Sherrington, Phys. Rev. Lett. 28,364 (1972)
3.D. Sherrington, J. Phys.C: Solid State Phys. 6, 1037 (1973)
4.B.W.Southern and D. Sherrington, J. Phys.F: Metal Phys. 4,1755(1974)
5.W.C.Koehler et al.,Phys. Rev. 151, 414 (1966)
6.E. Callen abd H.B.Callen, Phys. Rev. 139, A 455 (1965)
7.M. Tachiki et al., Phys.Rev. Lett. 21, 1193 (1968)
8.B.R..Cooper et al., Phys. Rev. 127, 57 (1962)
9.K. Yosida, inProgress in Low Temperature PhysicsIV(North-Holland, Amsterdam, 1964), p.265
10. N. Bloembergen, Phys. Rev. 78,572 (1950)
11. M.C.Lee et al., to be published
12. T. Nagamiya et al., Progr. Theor. Phys. (Kyoto) 27,1253 (1962)
13. R.S.Tebble and D.J.Craik, in Magnetic Materials (Wiley,N.Y.,1969)
14. D.T.Vigren, Phys. Rev. Lett. 32, 1254 (1974)

DISCUSSION ON COOPERATIVE MAGNETISM AND CRITICAL PHENOMENA

Phonons in Cooperative Systems   G A Gehring   Page 231

J Joffrin: Is there any clear evidence that the central peak has been experimentally observed in vanadate crystals. If we remember that a large anisotropy of dispersion of the modes lead to similar but spurious effects?

G A Gehring: Drs Hutchings, Scherm, Smith and Smith are confident that they have seen a central peak: their paper has been submitted for publication.

****

Ultrasonic Attenuation in Thulium Vanadate at 10 GHz at Liquid Helium Temperatures   J H Page and H M Rosenberg   Page 237

D G Walmsley: If you change the frequency of the ultrasonic invest- igation do you expect to be able to distinguish resonant absorption from the background random strain effects?

J H Page: The attenuation arising from random strains should depend on frequency as ω^4(Rayleigh scattering!) so that at low frequencies the effect should be negligible. At these low frequencies, one would not expect to observe resonant absorption because the ultrasonic energy is too small, but one might expect to observe critical effects. This might be inferrred from the elastic constant measurements of Melcher, Pytte and Scott by looking at the region in which they lose their signal, but as yet I do not believe that a direct measurement of the attenuation at this lower frequency has been published.

****

The Thermal Conductivity of Holmium Phosphate, a Strong Ising Antiferromagnet with a Large Hyperfine Interaction
M W S Parsons   Page 240

P G Klemens: Why does the thermal conductivity increase above the T^3 line as the temperature is lowered below the transition tempera-

271

ature?

M W S Parsons: The scattering in the ground state is very much
reduced below the Neel temperature, as the ground state degeneracy
is raised.

****

Thermal Conductivity of GdCl$_3$, a Representative Magnetic Insulator
G S Dixon   Page 246

A F G Wyatt: The dipole-dipole interactions can be calculated and
so its a pity to use them as adjustable parameters. On the other
hand the exchange coupling cannot be easily measured in the bulk
GdCl$_3$, so I wonder how good your model fit to the data would be if
you only adjusted the exchange couplings.

G S Dixon: The derivatives of the dipole and exchange interactions
averaged over all phonon modes have been treated as parameters.
The dipole sums themselves have been calculated by the Ewald
technique and the exchange constants measured by Clover and Wolf
have been used to calculate the magnon dispersion curves.

H Weinstock: Can you explain the lack of critical scattering which
I have noticed in another magnetic system?

G S Dixon: We had expected to observe large critical scattering.
However, the critical scattering depends on the same coupling
constant as the exchange modulation process. There is a cancellation between the first and second nearest neighbour contributions
to the coupling constant giving a value almost two orders of
magnitude less than might have been expected.

M W S Parsons: Have you looked for any evidence of magnon
conductivity at intermediate fields?

G S Dixon: Yes, the magnon contribution to the conductivity is
negligible throughout the range of fields and temperatures of
these experiments.

****

Spin-Phonon Interactions in Exchange Induced Van Vleck Paramagnets
C M Care and J W Tucker   Page 249

G A Gehring: Your results would be applicable to two level
systems but I think that their application to singlet-triplet
systems is dubious - particularly as the soft mode behaviour is
probably different.

J W Tucker: Yes, I agree that our results only apply to a singlet-singlet system. Recent discussions in the literature suggest that both the singlet-singlet and singlet-triplet models are inappropriate and that the full ground multiplet of the ion has to be included.

****

Magnetic Field Dependent Thermal Conductivity in $FeCl_2$ and $FeBr_2$
D Petitgrand and G Laurence   Page 252

G A Gehring: Did the large changes observed in the thermal conductivity as a function of magnetic field imply that all the phonon branches were scattered by magnons?

D Petitgrand: No it does not - because of the layered-type structure of $FeCl_2$ and $FeBr_2$ there is a transverse phonon with vector parallel to the c‾axis (and thus parallel to the heat flux) which has a very low sound velocity and thus carries most of the heat flux - so it is sufficient that the interaction concerns this mode.

G S Dixon: Could you comment on the persistence of the scattering above the Neel point?

D Petitgrand: Just above the Neel point, we have surely an effect of resonant scattering between the $Jz = \pm 1$ and $Jz = 0$ levels of the paramagnetic ferrous ions. For higher temperatures (30K - 50K) we should also consider phonon scattering by the magnetic excitons associated with the $J = 2$ levels.

****

Thermal Conductivity of $RbCaF_3$ J J Martin   Page 258

J P Harrison: Is there evidence of any other phase transitions in $RbCaF_3$ below 5K that could be influencing the low temperature thermal conductivity?

J J Martin: No, there is no evidence of low temperature transitions to date.

J Daubert: As it seems desirable to measure at least part of the phonon dispersion with neutron inelastic scattering, do you think single crystals of sufficient size (> 1 cm^3) will be available?

J J Martin: Yes, crystals of that size have been grown and I believe that workers at CRNL are planning the experiment.

Phonon Echoes and Phase Transitions   Ch Frenois, J Joffrin and
A Levelut   Page 261

G A Gehring: Can your method be applied to systems with very large
attenuation? If so it would be very interesting to apply it to the
cooperative Jahn Teller phase transitions.

J Joffrin: Certainly it could be used in such conditions. KDP is
one such crystal where the attenuation of the soft acoustic mode
is large. But the method is only possible for crystals having no
centre of symmetry.

A Ikushima: Could you explain in some detail the difference
between the usual ultrasonic result and yours?

J Joffrin: The advantage of the phonon echo method is to apparently reduce the attenuation of the soft phonon mode, due to the
mode conversion effects on the surfaces of each grain of the powder.

****

Critical Attenuation and Velocity of Surface Waves at Structural
Phase Transitions   L Bjerkan and K Fossheim   Page 265

J D N Cheeke: 1. What is the efficiency of your quartz transducer Rayleigh wave generation technique? 2. Presumably there is
a critical length parameter which determines the difference between
surface and bulk effects. Can you give an estimate of its order
of magnitude?

L Bjerkan: 1. Not very high for this purpose, it is much less
efficient than generation by means of a good piezoelectric film.
It is limited by the quality of the grating array (General
reference: R F Humphreys, E A Ash, Electronics Lett $\underline{5}$ 175 (1969).
2. In $SrTiO_3$ it is probably only 10-100Å. We hope to find other
systems where this length is much larger.

QUANTUM OSCILLATION IN TRANSMISSION OF BALLISTIC HEAT PULSE IN BISMUTH

T. ISHIGURO, K. KAJIMURA, S. KAGOSHIMA, and H. TOKUMOTO

Electrotechnical Laboratory

Mukodai, Tanashi, Tokyo 188, Japan

J. KONDO

Electrotechnical Laboratory, and

Department of Applied Physics, University of Tokyo

Bunkyo, Tokyo 113, Japan

Spike-like quantum oscillation is found in the transmission of longitudinal (L-) heat pulse propagating along the binary axis of Bi under magnetic fields.

A sample studied was cut out of a crystal with the resistance ratio of 450. As a generator and a detector of heat pulses, we used a Au film and a CdS film [1], respectively. The sample was kept in vacuum. The sample temperature was 1.68 K and a supplied power to the generator was 1.1 W/mm^2. The spectrum of emitted phonons is estimated by the Planck radiation law [2]; in the present case the characteristic temperature T_p is estimated to be 8.1 K.

The height of the detected ballistic L-phonon propagating along the binary axis changed oscillatorily as sweeping the magnetic field up to 14 kG as shown in Fig. 1. The behavior is characterized by the periodicity in the reciprocal magnetic field plot of the absorption peaks. By comparing the period with that of the de Haas-van Alphen effect [3], it is found that the absorption peaks have concern with passing of the Landau levels through the Fermi level of the electron gas in the III electron-band (see inset of Fig. 1).[4]

By using the wave function of the electron in an ellipsoidal

energy band under magnetic field and the electron-phonon coupling
via the deformation potential, phonon absorption rates are calculated by the standard time-dependent perturbation theory in the
lowest order.[4] The requirement for energy and momentum conservation as well as Pauli's exclusion principle restrict the phonon
wave number which is absorbed. In addition to this, it is found
that absorption of any phonon in the heat pulse occurs only when
the change in the perpendicular (to the magnetic field) component
of the electron wave number accompanying the absorption is smaller
than the inverse of the cyclotron radius. The combination of these
restrictions explains why the absorption is strong only for small
ranges of field strength where the bottom of the highest occupied
Landau level is close to the Fermi level, in spite of the fact that
the heat pulse contains phonons of a broad spectrum.

We have calculated the heat pulse absorption as traversing
sample length of 5.9 mm by using the physical constants of Bi: the
electron effective masses [5], the deformation potential constants
and the sound velocities [6]. We first represent the spectrum with
the Planck distribution of T_p=8.1 K. The theoretical curve thus
obtained looks very similar to the curve of Fig. 1, except that the
magnitude of the oscillation is about one tenth of what is actually
observed. This seems to imply that the actual spectrum of the ballistic heat pulse is different from the Planck distribution of 8.1
K because of scatterings of high-frequency phonons. It should be
noted here that the thermal conductivity κ reaches maximum at 3.5
K in Bi [7] reflecting phonon scattering for phonons higher than
this temperature. Since Bi is isotopically pure, we consider that
normal phonon-phonon scattering process dominates at low temperature. According to Herring [8], the phonon-phonon scattering rates
for L- and T-(transverse) phonons are given by

$$\tau_L^{-1} = B_L \omega^3 T^2 \qquad \text{and} \qquad \tau_T^{-1} = B_T \omega T^4 \qquad , \qquad (1)$$

respectively, where B_L and B_T are constants, ω the angular frequency, and T the temperature. In Fig. 2 is shown a theoretical curve
which is obtained by taking account of cutoff in the ballistic phonon distribution caused by the phonon-phonon scattering (τ_L^{-1}) with
B_L=4×10^{-32}sec^{2}deg^{-2}. In the inset of Fig. 2 is shown κ calculated
after Holland [9] by using both τ_L^{-1} and τ_T^{-1} with B_T=1.5×10^{-9}deg^{-4}.

In Fig. 2 dotted lines represent the absorption lines obtained
by considering only the intra-Landau-level transitions of electrons.
We find that the absorption peaks come mainly from them, while the
background absorption which appears at low field is associated with
the inter-Landau-level transitions.

Concerning with the fast T-mode, the observed amplitude of the
quantum oscillation was smaller than that of the L-mode by one
order of magnitude. This observation is explained in terms of the
absorbed fraction in the phonon spectrum of ballistic heat pulse:

The phonon frequency of the T-mode is smaller than that of the L-mode due to its smaller sound velocity provided that the relevant wave number is the same, whereas the cutoff of the phonon spectrum due to the phonon-phonon scattering is dull against the frequency for the T-mode as shown by eq.(1). This situation reduces the absorbed fraction of the T-mode remarkably compared to the L-mode.

References
[1] Ishiguro T and Morita S 1974 *Appl. Phys. Lett.* 25 533.
[2] Weiss O 1969 *Z. Angew. Phys.* 26 325.
[3] Bhargava B N 1967 *Phys. Rev.*156 785.
[4] Ishiguro T, Kajimura K, Kagoshima S, Tokumoto H and Kondo J
     unpublished.
[5] Kao Y 1963 *Phys. Rev.* 129 1122.
[6] Walther K 1968 *Phys. Rev.* 174 782.
[7] Kopylov V N and Mezhov-Deglin L P 1971 *Pis'ma Zh. Eksper. Teor.
     Fiz.* 14 32.
[8] Herring C 1954 *Phys. Rev.* 95 954.
[9] Holland M G 1963 *Phys. Rev.* 132 2416.

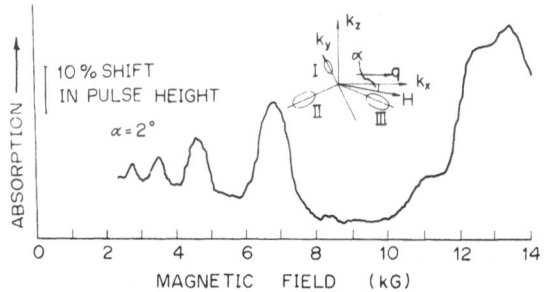

Fig. 1 Magnetic field dependence of the absorption of longitudinal mode. The directions of the phonon propagation and the magnetic field are shown in the inset together with the electron bands. The absorption is shown by fraction in %.

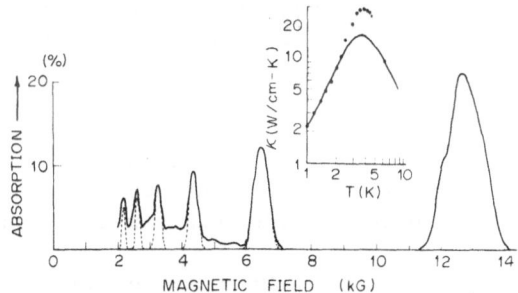

Fig. 2 Theoretical curve of the absorption corresponding to Fig. 1. The inset shows the temperature dependence of the thermal conductivity of Bi: Dots are the experimental data from ref. 7 and solid line is calculated one.

PHONON CONTRIBUTION TO THE HALL RESISTIVITY IN CADMIUM

A. N. Gerritsen

Physics Department, Purdue University

West Lafayette, Indiana, U.S.A.   47907

Earlier published[1] resistance and Hall data on single crystals of Cd, alloyed with less than 200 ppm Ag or In, with residual resistance values (at 4.2 K) between 1.7 x 10^4 and 100, have been subjected to further analysis.

The results can be summarized as follows:
1. The electrical resistivity ($\rho \| \langle 11\bar{2}0 \rangle$) decreases down to appr. 6 K according to a Grüneisen-Bloch expression with a characteristic temperature $\Theta_R \simeq 110$ K, in variance with the Debye temperature of Cadmium: $\Theta_D = 210$ K.
2. The coefficient A_1 of the term in first order in the development of the Hall resistivity in odd powers of H ($H \| \langle 0001 \rangle$) is positive and passes through a maximum between 10 K and 20 K, at a lower temperature for samples of higher purity.
3. The difference between the measured Hall resistivity and the term linear in H is positive for T > 10 K, zero for T $\approx$ 11 K and positive or negative (intersheet scattering) for T < 11 K.
4. The product $A_1\rho$ varies as exp(-43/T) between 200 K and the residual resistivity range, including temperatures below the occurrence of the maximum in A_1.

It was further observed that the temperature of appr. 11 K that seems to be characteristic for Cd, is within 10 to 20 per cent of a value that, by scaling down, corresponds to the onset[2] of a very high phonon distribution peak in the similar element Zn.

These results and considerations have led to an investigation by Hsu and Falicov. Starting from an electron and hole distribution model based on the known Fermi surface, they introduce in the collision term of the Boltzmann equation an additional term due to scattering by a flat, almost Einstein-like phonon mode, which is a hybrid of the TA and TO modes. This mode is considered responsible for the peak in the phonon distribution and it can scatter electrons from anywhere in the Fermi surface to any other point. Further they assume an anisotropic scattering, namely that the Einstein mode is dominantly effective in interhole scattering but inefficient in intrahole, intraelectron and electron-hole scattering. With a proper choice of an anisotropy parameter they can describe semi-quantitatively the decrease in the value (to $\approx$ .5 θ_D) of the characteristic temperature in the Grüneisen-Bloch relation, the maximum in A_1 and the exponential behavior of $A_1\rho$ (with $\theta_{Einstein}$ equal to 54 K or 60 K).

Details of the analysis and the theoretical considerations will be published[3,4] in two consecutive papers in the Physical Review.

References

1. D. A. Lilly and A. N. Gerritsen, Phys. Rev. B9, 2497 (1974).
2. L. Y. Raubenheimer and G. Gilat, Phys. Rev. 157, 586 (1967).
3. A. N. Gerritsen, to be published in Phys. Rev.
4. W. Y. Hsu and L. Falicov, to be published in Phys. Rev.

Part of this research was made possible by Grants GP 12633A1 and 7302651 from the National Science Foundation.

PHONON SCATTERING AND THE THERMOPOWER OF DILUTE ALLOYS

J. Kopp

Department of Physics, University of the Witwatersrand
Johannesburg, South Africa

1. Introduction

The presence of minute quantities of magnetic impurities
profoundly influences the electronic properties of noble metals,
especially the thermopower (Kondo, 1965; Berman & Kopp, 1971;
Kopp, 1975).   The theory of s-d scattering yields a single-impurity
thermopower in Au-Fe of about - 20 μV/K (Maki, 1969) but this is
never observed because of (a) other scattering mechanisms, which
reduce the expected thermopower by the Nordheim-Gorter factor, and
(b) impurity interactions, which lead to reductions at the lowest
temperatures.   In this paper, measurements of the thermopower of
very dilute Au-Fe alloys are compared with theory, and the phonon
scattering contribution is obtained explicitly.

2. Experimental

Fine wires of Au containing initially 93 ppm of Fe were treated
with chlorine at high temperatures to selectivity remove the Fe
(Kopp, 1975).   The table shows the properties at 4 K of the samples:

Sample	Fe concentration	Resistivity	Thermopower
1	93 ppm	80.0 μΩ.cm	−9.0 μV/K
2	13 "	14.4 "	−7.2 "
3	1 "	4.8 "	−1.7 "

The thermopower curves are shown in Fig. 1, which also includes the
theoretical, single-impurity thermopower applicable to Au-Fe (Maki,
1969; Berman & Kopp, 1971).   Interaction effects somewhat reduce

280

the thermopower of samples 1 and 2 at the lowest temperatures.
The dashed lines indicate the thermopower when corrections for
effect have been made.

3.   Theory

The observed thermopower is given by

$$S = S_m R_m / R$$

where S_m is the theoretical value, R_m is the magnetic and R the
total scattering.   This is often written as

$$S = S_m \rho_m / \rho$$

where ρ refers to the electrical resistivity (Nordheim-Gorter rule).
Strictly, the thermal resistivity W should take the place of ρ
(Macdonald, 1962).   If the Wiedemann-Franz law holds, both
approaches are equally valid, but whereas $\rho \propto T^5$ in the range
5 - 15K, $W \propto T^2$. Our results will show that the phonon part of the
scattering varies roughly as T^3. We shall assume that the various
scattering strengths can be added (Matthiessen's rule), so that

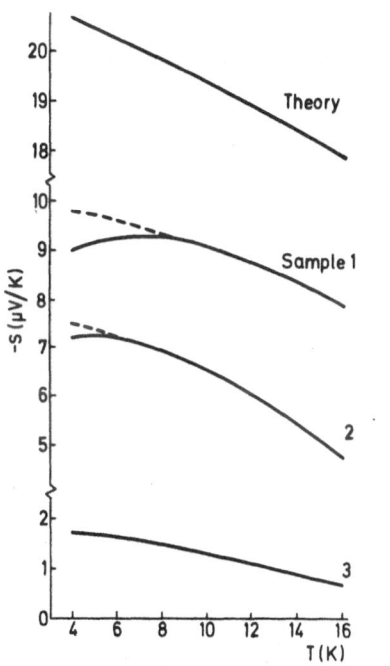

Fig. 1: Observed thermopowers.

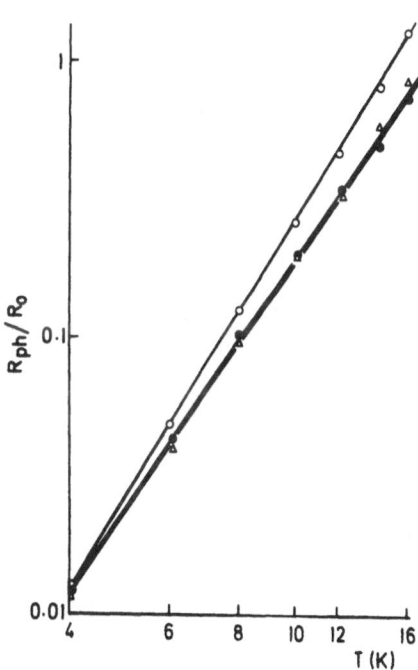

Fig. 2: Phonon scattering
        contribution.

$$R = R_{Fe} + R_o + R_{ph}$$

where R_{Fe} is due to the Fe impurities, R_o due to other (non-magnetic) impurities and physical defects and R_{ph} is the phonon contribution.   We shall neglect the slight temperature dependence of R_{Fe} (Kondo effect) and extract the ratio R_{ph}/R_o from the curves in Fig. 1.   This ratio is shown in Fig. 2.   It has the temperature dependence T^n, where n varies from 3.1 to 3.4.

4.   Discussion

If we had used the thermal resistivity, we should have expected an exponent of 3.   It is doubtful whether the slight deviations from this value are meaningful, since we have made the following approximations:
a)   additive scattering strengths, which may be invalid when the various mechanisms are comparable;
b)   application of nearly-free electron theory to the highly anomalous s - d scattering;
c)   neglect of phonon-drag scattering.   This means to be valid for temperatures less than 15 K in gold, provided the observed thermopower is large enough.   The slight deviations in sample 3 at 14 and 16 K may be indicative of phonon-drag.

Despite these difficulties, we have shown that phonon scattering reduces the thermopower as T^n where n is close to the expected value 3.

This work was supported by an equipment grant from the C.S.I.R. Pretoria.

References

R. Berman and J. Kopp (1971), J. Phys. F.1, 457.

J. Kopp (1975) J. Phys. F.5, 1211.

D.K.C. MacDonald (1962), "Thermoelectricity" (Wiley, N.Y.)

K. Maki (1969), Prog. theor. Phys. 41, 586.

USE OF A PHONON PROBE TO DETERMINE THE MAGNETIC DEPENDENCE

OF THE SUPERCONDUCTING GAP

D.Huet, B. Pannetier, and J.P. Maneval

Physique des Solides, Ecole Normale Supérieure

24, rue Lhomond, 75231, Paris 5 , France

Superconducting thin films in a magnetic field have been explored in a number of experimental ways including tunnel effect, thermal conductivity, and microwave absorption. Most of these effects, especially at finite temperature, involve the whole of the excitation spectrum, and the minimum, or gap, is usually deduced by assuming a BCS-type density of states dependent upon the magnetic field only through the order parameter (1). In fact the density-of -states functions found experimentally (2) do not agree with this simple model, so that an independent measurement seems useful.

We present here a heat-pulse determination of the gap, noted $E_g(T,H)$, valid for arbitrary conditions of magnetic field and of temperature, according to a method first proposed by Huet et al.(3). A pulse of ballistic phonons, at the onset of dispersion (Fig.1) is emitted from one face of a crystal (InSb), and a time-of-flight measurement of $(hf = E_g)$ phonons is performed with a superconducting tunnel junction on the opposite face. One then deduces the gap from the measured velocity through the dispersion relations.

Fig.1:Dispersion relation for acoustic waves. L:longitudinal; FT:fast transverse; ST:slow transverse. The threshold frequency for superconducting detection(280 Ghz in tin) determines the apparent velocity of heat pulses.

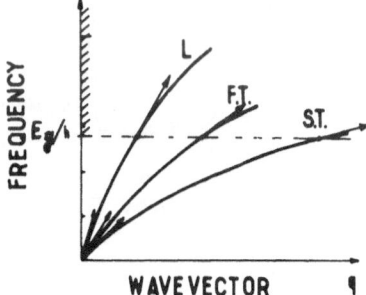

The frequency-dependent group velocity of low-energy phonons can be approximated in each branch (3) by:

$$V(f) = 2\pi df/dq = v_s (1 - Af^2) \tag{1}$$

where v_s is the relevant velocity of sound and A a constant. For normal dispersion (A>0), $V(E_g/h)$ is the heat-pulse velocity apparent in superconducting detection. Experimentally, one measures the delay Δt which separates the junction signal (J_o trace in Fig. 2) from the signal detected by a bolometer (B trace), used to mark the velocity of sound. Owing to the quadratic dispersion (Eq. 1), the field (and temperature) -dependent gap is related to the zero-field gap at T_o (the lowest accessible temperature: 1.30 K) by:

$$E_g(T,H) = E_g(T_0,0) \left[\Delta t(T,H)/\Delta t(T_0,0)\right]^{1/2} \tag{2}$$

The tunnel junctions,of area 1mm×1mm, were made of pure tin(5N). The oxide layer was formed in dry oxygen (pressure: 600mm Hg) at ambient temperature. For bolometric detection, we used one of the tin layers at the critical field. Typical signals are shown in Fig2. The long-term equivalent drift of the signals proved to be less than 5 nsec. It is remarkable that the limited response of the system (risetime of the PAR Boxcar Integrator:37 nsec) and the finite width of the heat pulses do not affect the definition of the arrival time (see Fig. 2b). The magnetic field was generated by a supercon-ducting Helmoltz pair having a homogeneity of 3% within 1×1×1cm³. Any lack of parallelism of the junctions to the field was detected by noting in very low field (10 Oe) the rounding of the I-V charac-teristics for voltages near the gap. To avoid flux trapping, the

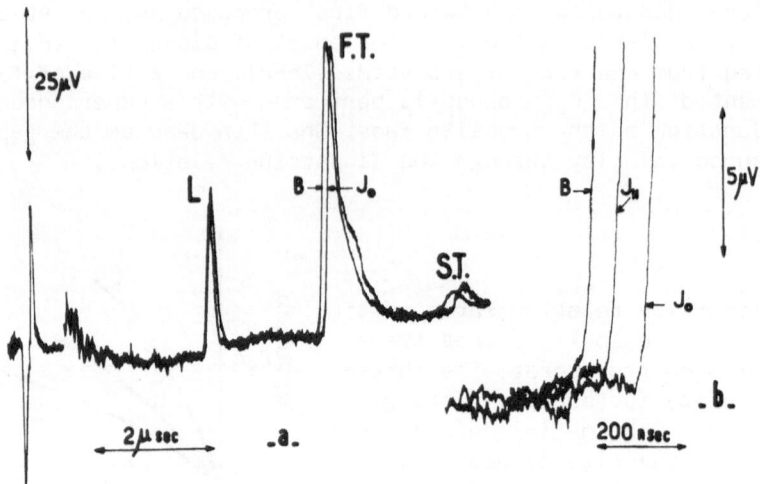

Fig.2: a/ Heat pulses detected by a bolometer(B) and by a tin diode(J_0). b/ Magnified onset of the FT signal (J_H is for the diode in a field). Direction(110) . Sample length: 11.42 mm.

sample-holder was lifted after each magnetic cycle above the liquid
helium level,in order to raise the temperature of tin above critical.

The value of the gap of a 2000 Å thick junction in parallel
field was plotted on the DC current-voltage characteristics (Fig.3).
One .observes that it is given by the onset of the steep current
rise rather than by the mid-height voltage (5). From measurements
of the resistivity and critical fields, the films parameters can be
estimated as follows: mean free path (800 Å), penetration depth
(500 Å),and de Gennes' coherence length (1400 Å) at T_0=1.30 K.

Fig.3:Tunnel characteristics
of a tin diode(2000 Å).Empty
circle :reference gap. Full
dots: gap deduced from dispersion of the FT signal.
 Empty square: from ST.
$H_{c//}$=465 Oe ; $H_{c\perp}$ =175 Oe

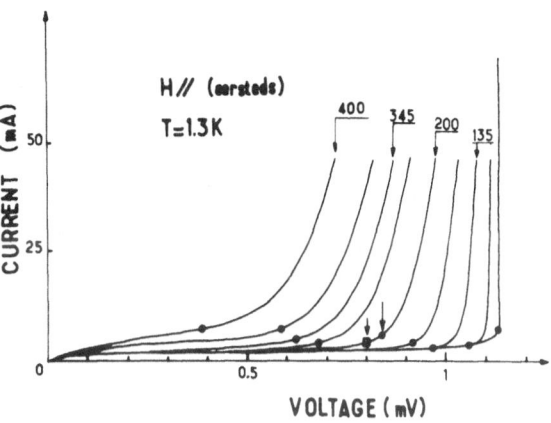

Another important experimental parameter is the bias applied to
the junction. The voltage was varied between 0.2 and 0.8 times the
gap voltage, without change of the apparent heat-pulse velocity, and
therefore of the associated gap. This is a strong indication that
detection occurs by pair-breaking only, and it sets an upper limit
to the "voltage-assisted" tunnel mechanisms one could imagine.

From another point of view, when considered as phonon detectors,
superconducting junctions are tunable by a magnetic field down to
about one fourth of the zero-field gap before the sensitivity decreases to an exceedingly low level.

REFERENCES :
1- R.S.Collier and R.A.Kamper, Phys.Rev. 143,323 (1966).
2- J.Millstein and M.Tinkham, Phys.Rev.158, 325 (1967).
3- Huet,Maneval and Zylbersztejn,Phys.Rev.Letters,29,1092 (1972).
4- I-V characteristics in H=0 provide an accurate value of $E_g(T_0,0)$.
5- According to the Rowell-Mac Millan construction, quoted by
   R.C.Dynes and V.Narayanamurti,Phys.Rev.B6, 143 (1972).

TEMPERATURE DEPENDENCE OF ELECTRON MEAN FREE PATH IN MOLYBDENUM

FROM ULTRASONIC MEASUREMENTS*

D. P. Almond, D. A. Detwiler and J. A. Rayne

Carnegie-Mellon University

Pittsburgh, Pennsylvania 15213, U.S.A.

The electronic contribution to the attenuation of sound waves is dependent on the product a = $q\ell$, q being the phonon wave number and ℓ the electron mean free path.[1] For $q\ell \gg 1$, the attenuation parameter $\frac{\alpha}{\nu}$, ν being the phonon frequency, is independent of $q\ell$. In general, the attenuation at arbitrary $q\ell$ can be expressed in the form

$$\frac{\alpha}{\nu} = \lim_{q\ell \to \infty} \left(\frac{\alpha}{\nu}\right) \cdot P(q\ell) , \tag{1}$$

where $P(q\ell)$ tends to unity for large $q\ell$.

Electronic attenuation data for longitudinal waves propagating along [100] in molybdenum have been fitted to Equation(1) to obtain the temperature dependence of the mean free path ℓ. Experiments were performed using the conventional pulse comparison method by measuring changes in attenuation between 77 K and 4.2 K. Measurements were made on specimens with specimens having residual resistance ratios of 400 and 17,000 to one, respectively.

Typical results taken at 150 MHz on the high-purity specimen are shown in Figure 1(a). The mean free path and hence the attenuation increase with decreasing temperature, the limiting values at 4.2 K being determined by impurity scattering and defects. Figure 1(b) shows the resulting impurity limited attenuation for both specimens plotted as α/ν versus ν. The two sets of data have been displaced horizontally relative to each other to give the best fit in the region of overlap. Clearly the resulting curve is equivalent to a plot of α/ν versus $q\ell$ corresponding to Equation(1)

*Work supported by National Science Foundation

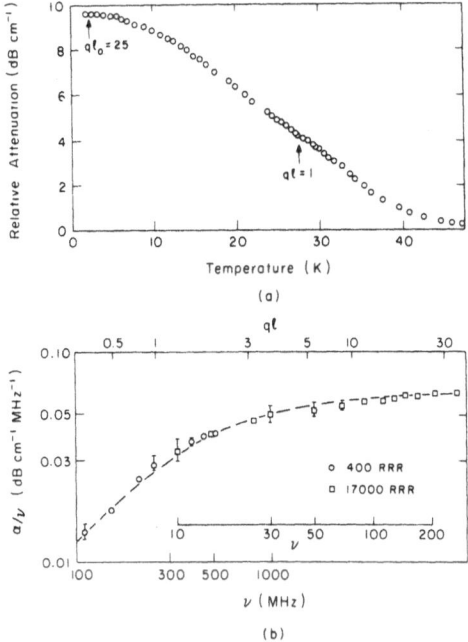

Figure 1:   (a) Temperature dependence of relative attenuation of 150 MHz longitudinal waves propagating along [100] in high-purity molybdenum. (b) Plot of α/ν versus ν for longitudinal waves propagating along [100] in molybdenum. Upper and lower frequency scales refer to the high and low purity samples respectively. Dashed curve is the free electron prediction.

with a fixed ratio of ℓ for the two samples. This ratio is 33 ± 2 to 1, which is in reasonable agreement with the value 43 ± 9 to 1 obtained from the residual resistivity measurements. The limiting value of α/ν for high $q\ell$ is in good agreement with that obtained from previous work.[2]

From the dashed curve in Fig. 1(b) it can be seen that the free electron form of $P(q\ell)$ gives a remarkably good representation of the data. This fit fixes the scale of $q\ell$ as shown and corresponds to values of 6 micron and 200 micron, respectively, for the limiting mean free path ℓ_0 in each specimen. Similar estimates of ℓ_0 are obtained from the residual resistivity data.

With the established dependence of α/ν on $q\ell$ it is easy to obtain the value of mean free paths at any temperature. Figure 2 shows the resulting plots of $(1/\ell - 1/\ell_0)/T^2$ versus T, assuming the validity of Mattheissen's rule. For all frequencies the relaxation rate follows a T^5 power law at higher temperatures. Deviations from this law occur when $q\ell \sim 1$, a T^2 dependence being observed at lower temperatures where $q\ell > 1$.

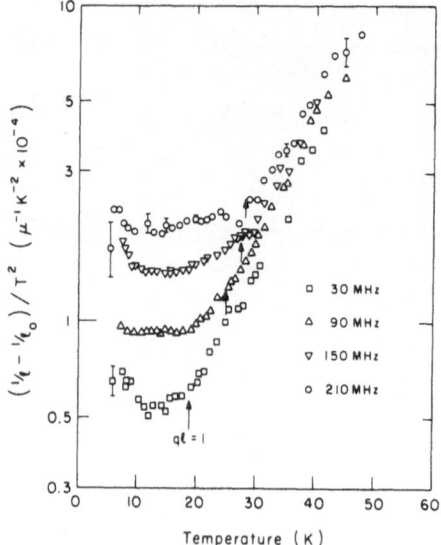

Figure 2: Semilog plot of $(1/\ell - 1/\ell_0)/T^2$ versus T for long-tudinal waves propagating along [100] in high-purity molybdenum.

The high temperature T^5 behaviour may be explained in terms of small angle scattering involving thermal phonons. At temperatures where $q\ell \gg 1$, each phonon would scatter electrons out of the effective zone and hence should give a T^3 dependence of the scattering rate. However, no clear evidence for this behaviour has been found here. The observed T^2 dependence is clearly related to the value of $q\ell$ and is believed to be caused by electron-electron scattering. A simple analysis, involving momentum and energy conservation, indicates that the Fermi surface geometry of molybdenum makes it probable that an electron is scattered out of the effective zone in a collision with another electron. Also, from more general arguments[3] it is expected that ultrasonic attenuation, like thermal conductivity, is more sensitive to electron-electron scattering than electrical resistivity. This increased sensitivity could account for the extended temperature range over which the T^2 power law is observed. Additional measurements are being made for other propagation directions to investigate the variation of the scattering over the Fermi surface.

References

1. A. B. Pippard, Proc. Roy. Soc. (London) A257, 165 (1960).
2. C. K. Jones and J. A. Rayne, Physics Letters 13, 282 (1964).
3. J. M. Ziman, Electrons and Phonons, Oxford University Press 1960, p. 417.

INTERACTION AND SCATTERING OF GREATER THAN 2Δ PHONONS WITH ELECTRONS IN SUPERCONDUCTING Sn FILMS[*]

I.L. Singer and W.E. Bron

Department of Physics, Indiana University

Bloomington, Indiana 47401, U.S.A.

Contrary to earlier results by others[1,2] phonons, with energy greater than the superconducting gap, escape Sn tunnel junctions in sufficient number to be detected by Pb tunnel junction detectors. An attempt is made here to describe an analytical basis for the observed spectral distribution and the temperature dependence of such phonons. The analysis is based on the electron-phonon interactions proposed by Tewordt[3] and Privorotskii.[4]

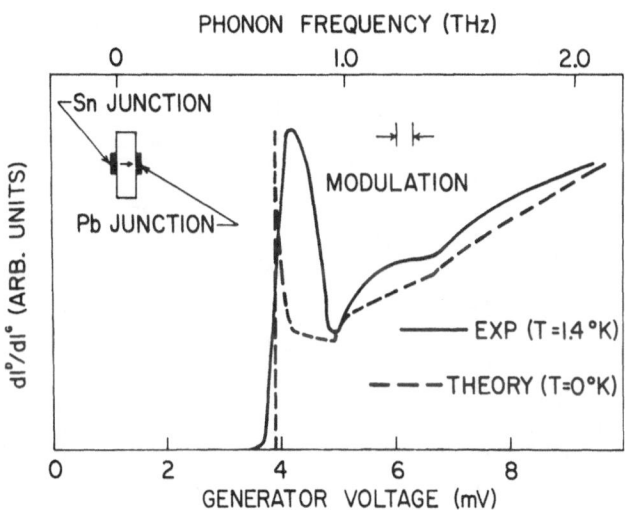

Fig. 1. Phonon transconductance as a function of generator bias voltage. The upper scale refers to the frequency of emitted "primary" phonons.

In a Sn tunnel junction, biased to a potential eV >> 2Δ(Sn),
an injected quasiparticle population is generated with a distribution of energies ε, such that Δ ≤ ε ≤ eV-Δ. These quasiparticles
relax to states at, and above, the gap emitting thereby a primary
distribution of phonons with energies 0 ≤ ℏω ≤ eV-2Δ(Sn).[5] In
addition phonons with ℏω=2Δ(Sn) are created by the recombination
of quasiparticles at the gap to form Cooper pairs. These latter
phonons are ignored in what follows. Since Sn is one of the
stronger coupled superconductors,[6] it is valid to assume that at
T=0°K all relaxation phonons with ℏω > 2Δ(Sn) are reabsorbed to
yield an additional distribution of quasiparticles with
Δ(Sn) ≤ ε ≤ eV-3Δ(Sn) and a secondary phonon distribution with
0 ≤ ℏω ≤ eV-4Δ(Sn). At 0°K < T < T_c, however, some relaxation
phonons cannot be reabsorbed simply because the excited quasiparticles states are filled through thermal excitation. (The
linear limit is assumed, i.e. the number of injected quasiparticles is small compared to thermally excited ones). It can
be shown that the number of such phonons varies with energy and
temperature as

$$N(\omega,T) \propto (kT\Delta(Sn)/(1 - 2\Delta(Sn)/\hbar\omega))^{1/2} \exp(-\Delta(Sn)/kT);$$

i.e. the number varies weakly with frequency but strongly with
temperature. Those phonons which are not reabsorbed may escape
the Sn junction and reach a Pb tunnel junction biased at
eV < 2Δ(Pb) and placed either directly behind the Sn junction or
after sapphire crystals of various thicknesses. Phonons with
ℏω ≥ 2Δ(Pb) generate quasiparticles in a manner analogous to that

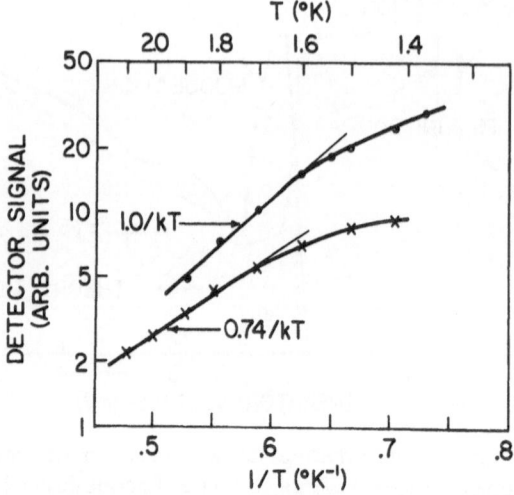

Fig. 2. Temperature dependence of the detector signal for two
different experimental cases. The exponent is in units of meV.

in Sn. The resultant quasiparticle current is then a measure of
the phonon transconductance between the generator and detector.
This current, for a given generator bias, may be obtained analytically from the detailed balance equations for quasiparticles and
phonons in the detector. The relevant integrations over continuous
sets of primary and secondary quasiparticle and phonon states are
carried out numerically. The analytically obtained incremental
change in the detector signal δI^D per incremental change in generator bias current δI^G as a function of the generator bias voltage,
at fixed temperature, is displayed in Fig. 1 together with the experimental result obtained through a standard, low frequency,
modulation technique.

Since the gap of Pb(2.70meV) exceeds that of Sn(1.14meV), if
no phonons with energy $\hbar\omega \geq 2\Delta(\text{Pb})$ escape from the Sn generator
then no incremental signal would be observed in the Pb detector.
Instead a sharp onset is observed at a generator bias of $2\Delta(\text{Pb})$
above the gap of the generator ($V^G \approx 3.8\text{mV}$) which corresponds to a
similar onset in the analytical curve arising from the production
of relaxation phonons in the generator with $\hbar\omega = 2\Delta(\text{Pb})$. A further
onset is observed at a bias of $2\Delta(\text{Pb}) + 2\Delta(\text{Sn})$ above the gap of
Sn($V^G \approx 5.0\text{mV}$) which corresponds to the production of secondary relaxation phonons with $\hbar\omega = 2\Delta(\text{Pb})$. A yet further onset is observed at
a bias $4\Delta(\text{Pb})$ above the gap of Sn($V^G \approx 6.5\text{mV}$) due to the additional
response of the detector to incoming phonons of $\hbar\omega \geq 4\Delta(\text{Pb})$. The
combined temperature dependence of the phonon flux escaping from
an ideal generator and the temperature dependence of the response
of an ideal detector leads to a predicted detector current which
varies as $\exp[(\Delta(\text{Pb}) - \Delta(\text{Sn}))/kT] \equiv \exp(b/kT)$. Exceeding the linear
limit, or the presence of Josephson currents, trapped flux, or
other non-thermal sources of quasiparticles in either the generator
or the detector leads to deviations from this temperature dependence.   Nearly ideal conditions have rarely been obtained. However, a strong temperature dependence is always observed. For high
quality Pb junctions, in the range of temperatures within the linear
limit, an exponential temperature dependence is indeed observed
with $1.05 > b > 0.75\text{meV}$, compared with the predicted value of
0.78meV.

References
*Work supported by the U.S. Army Research Office.
1.   M. Welte, K. Lassmann, and W. Eisenmenger, J. Phys. (Paris),
     Suppl. 10, G4 (1972).
2.   R.C. Dynes, V. Narayanamurti and M. Chin, Phys. Rev. Lett. 26,
     181 (1971).
3.   L. Tewordt, Phys. Rev. 127, 371 (1962).
4.   A. Privorotskii, Soviet Phys. JETP 16, 945 (1963).
5.   H. Kinder, Z. Physik 262, 295 (1973).
6.   R.C. Dynes and V. Narayanamurti, Phys. Rev. B 6, 143 (1972).

EXPERIMENTAL EVIDENCE OF PHONON FLUX FLOW INTERACTION

M. Martin, J.Y. Desmons, D.G. Thomas, E. Bridoux

M. Moriamez and M. Le Ray

59326 - CENTRE UNIVERSITAIRE DE VALENCIENNES (France)

We present in this paper some experimental results on longitudinal phonon detection by a very large constricted film. These results are related to the existence of a normal state region
in the middle of the constriction, surrounded by a flux flow state region. We think that the detection mechanisms involve a bolometric effect and a synchronisation phenomenon with vortex motion.

The detector is a tin film evaporated on a Xcut quartz rod
of diameter 6 mm and length 30 mm. The shape of the film is simi-
lar to that of a Dayem bridge |1| , the different films we have
tested having a width comprised between 10 and 30 μm, the aperture of the angles on each sides of the constriction being 90°. The
thickness of the films is 0.1 μm. Phonons are generated by the
classical piezoelectric method |2| at the frequency of 2.9 GHz.

The main characteristic of the amplitude of the detected voltage pulse H for one echo versus bias voltage V_{dc} , is the presen-
ce of voltage pulse plateaus |3| . In the experiment corresponding
to this curve, the voltage levels of the different echoes exhibit
similar behaviour. The height ΔH of the steps is 6 μV or submulti-
ples 4 μV or 2 μV. One can immediately remark that in the Josephson relation $2e\Delta H = h\nu$, 6 μV corresponds to the phonons frequency. But here, we have not the classical Josephson effect, since
in this latter, the voltage involved is the bias voltage across
the whole junction. The existence of a normal region at the cons-
triction has been considered in |4| and most recently in |5| where
its existence has been related to a hot-spot model. For the curve
$H(V_{dc})$ the current in the constriction is higher than the critical
current. Thus a normal region appears at the constriction which explains the large value of V_{dc}. Moreover, the envelope of $H(V_{dc})$ de-

creases beyond 6 mV. This is probably due to a bolometric effect
related to the growth of the normal part around the constriction.
The magnetic field of the current induces vortices of opposite sen-
ses near the sides of the layer and these vortices move towards the
middle of the constriction where they disappear. There will be a
stop of the normal part as long as the increase of the current is
not sufficient to create and move at least two additional vortices.
This mechanism provokes the appearence in the characteristic resis-
tance-current R(I) |6| of stable resistance plateaus. Coming back to
the plateaus on the curve $H(V_{dc})$, one can think that the plateaus correspond to a synchronisation between two frequencies associated respectively with the process of creation and destruction of the vortices in the lateral layer, and with the hypersounds, that is to say to
a "lateral Josephson effect".

    One of the properties of the Dayem bridge is the existence of
two critical temperatures |4| , one corresponding to the constriction
and the other one to the wide regions on each side of the constriction. We can here also define two critical temperatures on the curve
R(T) |7| . The curve of detected voltage versus temperature at cons-
tant current exhibits an important maximum corresponding to the critical temperature of the large regions on each side of the constriction and some secondary maxima. The voltage pulse detected by a bolometer is proportionnal to dR/dT. To each secondary maxima will correspond a dR/dT, too small to appear on the curve R(T) but suggested by the discrete growth of the normal part. It seems for this
curve that the bolometric effect is preponderant.

    On the other hand, a bolometer is a quadratic detector. The
detected voltage pulse is proportionnal to the absorbed power. The

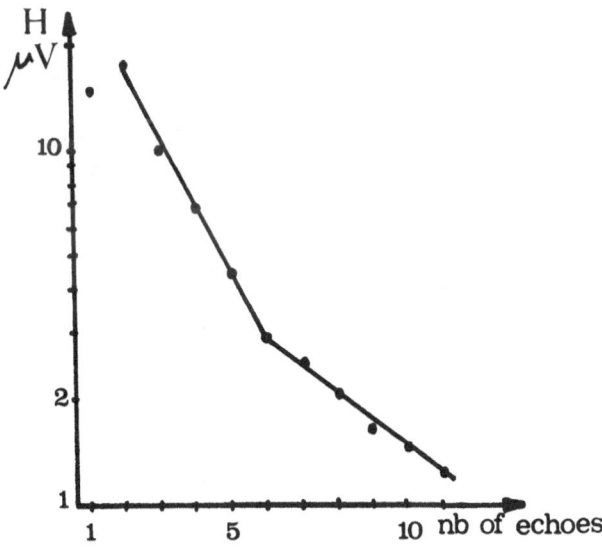

Fig. 1.   Envelope of the detected echoes for given I and T

envelope of different echoes in logarithmic coordinate (fig.1) and
the height of one echo versus incident power in log-log coordinates exhibit linear parts with different slopes. At high power, the
detected pulse is proportional to the incident acoustic power, which
may be related to a bolometric effect. But, at low power, the detected pulse is nearly proportional to the square root of the incident power. To each linear regions can correspond either different
effects (for instance bolometric effect, phonon flux flow interaction, including perhaps the role of a.c entropy transport by vortex
cores) or different systems of vortices for low and high power.

Another feature of some films is an inversion of the detected
pulses height. For instance, for one film at T = 3.852°K, for increasing bias current, we see the first echoes H = +2 to 3 μV at
I = 0.1 mA. With increasing current, the echoes increase, then decrease and disappear for I = 18 mA. These echoes reappear for
I = 28 mA but with reversed sign H = -2 to 3 μV. This feature can
perhaps be related to a negative slope in the characteristic
R(T) or to the difference between (I-V) with phonon on and (I-V)with
phonon off, the sign of which can change. Experimentally, it is difficult to conclude, the curve (I-V) with phonon on is obtained with
a discrete presence of phonons.

We are dealing with two main processes. The bolometric response due to the heating of the normal part, and the phenomenon of
synchronization of the vortex motion near the boundary of the normal region with phonons. We think to get a better understanding in
a simpler experimental situation, without a normal region at the
constriction. The first experiments on this case exhibit a linear
variation of the critical current versus temperature, like in some
Josephson junctions and multiple hysteresis loops in the (I-V)
characteristics |8| .

1 P.H.ANDERSON and A.H. DAYEM, Phys.Rev.Lett. 13 (1964) 195
2 H.E.BOMMEL and K.DRANSFELD, Phys.Rev.137 (1960) 1245
3 M.MARTIN and J.Y.DESMONS, Appl.Phys.Lett. 25(1974) 421
4 A.F.G. WYATT, Tunneling phenomena in solids,1969,ed.E.BURSTEIN
  and S.LUNDQUIST (Plenum)
5 W.J.SKOCPOL,M.R.BEASLEY and M.TINKHAM,Rev.Phys.Appl.9(1974)19
6 J.Y.DESMONS,M.MARTIN and D.G.THOMAS,Phys.Lett.51A(1975)123
7 M.MARTIN and J.Y.DESMONS,Phys.Lett.46A(1973)267
8 J.Y.DESMONS, M.MARTIN, D.G.THOMAS, M.MORIAMEZ, M.LE RAY,
  Proceedings of LT14, Helsinki (1975).

PHONON INDUCED STRUCTURE ON DAMPED MAGNETO-ABSORPTION OF WEAKLY POLAR SEMICONDUCTORS

J P Vigneron*, R Evrard, E Kartheuser

E S I S, Université de Liege

B-4000 Sart Tilman/Liège 1, Belgium

The purpose of this communication is to present a theory of the intraband magnetoabsorption of polarons at zero temperature and for weak electron-phonon coupling, taking into account the broadening of the Landau levels due to scattering by acoustic phonons and impurity centers. The spectrum will be calculated in the Faraday configuration, often used in experimentation.

Some features of the absorption spectrum have already been obtained by several authors considering a system for which the Landau levels are perfectly sharp[1,2,3,5].

At low electron concentration, the absorption coefficient $\Gamma(\omega)$ is related to the transition probability $P(\omega)$ by the following relation:

$$\Gamma(\omega) = \frac{8\pi n_e}{n(\omega)} \left(\frac{\alpha_r}{\omega}\right) P(\omega) \tag{1}$$

Units are such that $\hbar = 1$, $\omega_{LO} = 1$, $2m = 1$. In this expression n_e is the electron concentration in the conduction band, $n(\omega)$ the refractive index and α_r the atomic fine structure constant. The absorption probability is assumed to be well described by the ordinary golden rule. It is then further assumed that the considered band is simply parabolic and that the electron-phonon interaction can be described by Frohlich's polaron interaction. Under these assumptions the hamiltonian of the absorbing system is written

$$H = (A_+^+ A_+ + \tfrac{1}{2})\lambda^2 + P_z^2 + \sum_{\vec{k}} b_{\vec{k}}^+ b_{\vec{k}} + \sum_{\vec{k}} \sqrt{\frac{4\pi\alpha}{v}} \frac{1}{k}(b_{\vec{k}} e^{i\vec{k}\vec{r}} + b_{\vec{k}}^+ e^{-i\vec{k}\vec{r}}) \tag{3}$$

---

* Aspirant F N R S, Belgium

where $\lambda^2 = \omega_c/\omega_{LO}$ and α is the Frohlich coupling constant. Creation and annihilation operators for excitations in a magnetic field have been described in cylindrical coordinate by Feldman and Kahn[4].

In the Faraday configuration, the radiation active in absorption has a polarisation vector $\vec{\eta} = (1/\sqrt{2})(1,i,0)$ and the electron-photon interaction becomes simply

$$(\vec{p} - \frac{e}{c}\vec{A})\cdot\vec{\eta} = i(\lambda/\sqrt{2})\ A_+^+ \tag{6}$$

Supposing that all Landau states are all broadened in the same way due to scattering mechanisms other than electron-LO phonon interaction, an inverse lifetime γ may be introduced for each level and the golden rule may then be written

$$P(\omega) = 2\lambda^2\ \text{Re}\ i\ G(\omega) \tag{7}$$

$$G(\omega) = \langle i|A_+ \frac{1}{\omega+i\ \gamma/2 - H + E_i}\ A_+^+\ |i\rangle \tag{8}$$

Here, H, $|i\rangle$, E_i are the hamiltonian, the ground state and the ground state energy of the polaron. This expression will be calculated to first order in the coupling constant α. A tedious expansion may be avoided if one writes the equivalent expression:

$$(\omega + i\frac{\gamma}{2})\ G(\omega) = \langle i|\ [A_+,H]\ \frac{1}{\omega + i\gamma/2 - H + E_i}\ A_+^+|i\rangle\ +$$

$$\langle i|A_+A_+^+|i\rangle \tag{9}$$

in which the commutator is easily evaluated as

$$[A_+,H] = A_+\ \lambda^2 + \sqrt{\alpha}\ P \tag{10}$$

with

$$P = \frac{i}{\lambda}\ \Sigma_{\vec{k}}\sqrt{\frac{4\pi}{v}}\ (\frac{k_x - ik_y}{k})\ (b_{\vec{k}}e^{i\vec{k}\vec{r}} - b_{\vec{k}}^+e^{-i\vec{k}\vec{r}}) \tag{11}$$

Introduction of (10) into (9) makes the expression (8) for $G(\omega)$ appear in the right hand side of relation (9). This term can be added to the left hand side of (9) to give:

$$(\omega + i\frac{\gamma}{2} - \lambda^2)G(\omega) = \langle i|A_+A_+^+|i\rangle\ +$$

$$\sqrt{\alpha}\langle i|P\ \frac{1}{\omega + i\gamma/2 - H + E_i}\ A_+^+|i\rangle \tag{12}$$

Now we can apply to the frequency dependent matrix element in (12) the same kind of reduction used to transform expression (8) into expression (12). This leads to the final expression:

$$G(\omega) = \frac{\langle i|A_+A_+^+|i\rangle}{\omega-\lambda^2+i\gamma/2} + \sqrt{\alpha}\,\frac{\langle i|PA_+^+|i\rangle}{\left[\omega-\lambda^2+i\gamma/2\right]^2} +$$

$$\alpha\,\frac{\langle i|P\left[\dfrac{1}{H-E_i-\,-i\,\,/2}\right]P^+|i\rangle}{\left[\omega-\lambda^2+i\,\,\gamma/2\right]^2} \qquad (13)$$

In this expression, the last term may be computed using unperturbed energies and wave functions and the second one using first order expansions in $\sqrt{\alpha}$. The first term can be shown to be related to the polaron ground state, which has been calculated at first order in α (Ref 7). One can show that:

$$\langle i|A_+A_+^+|i\rangle = \tfrac{1}{2} + \frac{\partial E_i}{\partial\lambda^2} \qquad (14)$$

The calculation of the matrix element $\langle i|PA_+^+|i\rangle$ is straightforward, while the last term can be computed from:

$$\langle i|P\,\frac{1}{H-E_i-\omega-i\gamma/2}\,P^+|i\rangle = -\,\frac{\alpha}{\pi}\,\sum_{n=0}^{\infty}\int_o^\infty dk_\perp k_\perp\,\frac{(k_\perp^2/\lambda^2)^{n+1}}{n!}\,\exp(-\frac{k_\perp^2}{\lambda^2})$$

$$\int_{-\infty}^{\infty}\frac{dk_z}{(\omega-n\lambda^2-k_z^2-1+i\gamma/2)\,(k_\perp^2+k_z^2)} \qquad (15)$$

Figure 1 shows the numerical results in the region $\omega \sim \omega_c \sim \omega_{LO}$. The non crossing levels of the polaron induce a doublet structure in the absorption coefficient.

Besides this manifestation of the pinning effect it can be shown that the cyclotron line experiences a sudden broadening as ω_c passes through ω_{LO}. The linewidth is there mainly governed by the real part of expression (15) for $\omega \sim \lambda^2$. For the case $\gamma = 0$, the result obtained by Harper[5] is in perfect agreement with our results.

The oscillatory behaviour of (15) for $\omega > \omega_{LO}$ gives rise to the phonon sideband shown as a function of the incident frequency and as a function of the cyclotron frequency in figure 2.

The lineshape obtained in the present theory is in good agreement with the recent observations of Weiler, Aggarwal and Lax[6] of the phonon assisted transitions in InSb.

1. L I Korovin, S T Pavlov, Sov Phys JETP 26, 979 (1968).
2. D M Larsen, 'Polarons in Ionic Crystal and Polar Semiconductors',
   p 237 (1971).

3. Bass, Levinson, Sov Phys JETP 22, 635 (1966).
4. A Feldmann, A Kahn, Phys Rev B 1, 4584 (1970).
5. Harper, Proc Phys Soc 92, 793 (1967).
6. Weiler, Aggarwal, Lax, Sol St Commun 14, 299 (1974).

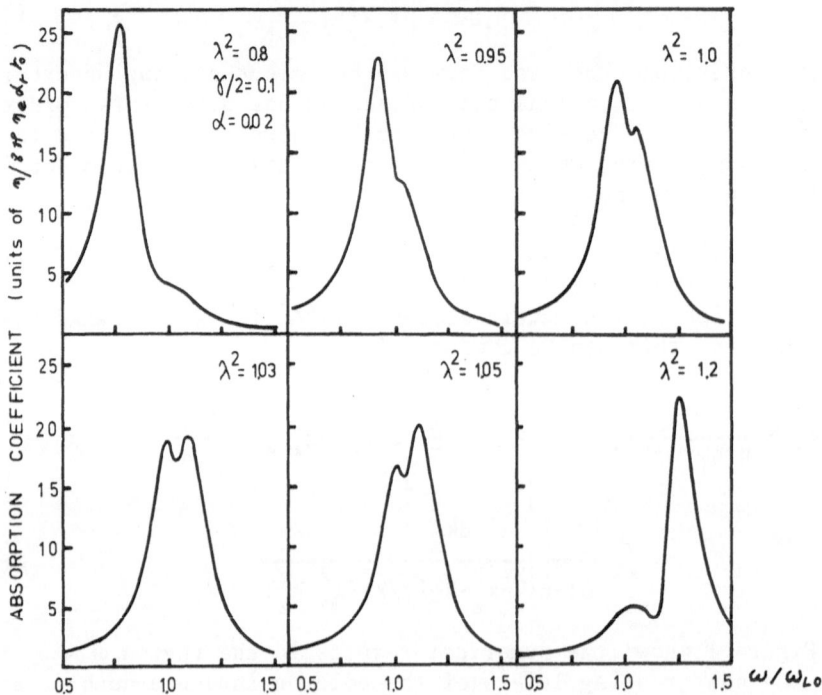

Fig 1: Evolution of the cyclotron lineshape with increasing magnetic fields for $\omega_c \sim \omega_{LO}$. r_0 is the polaron radius $(\hbar/2m\omega_{LO})^{\frac{1}{2}}$

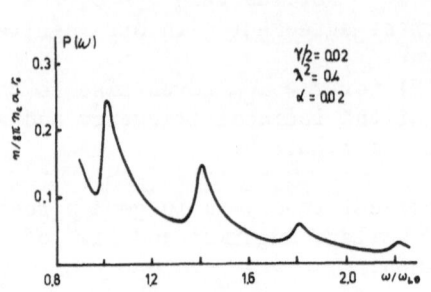 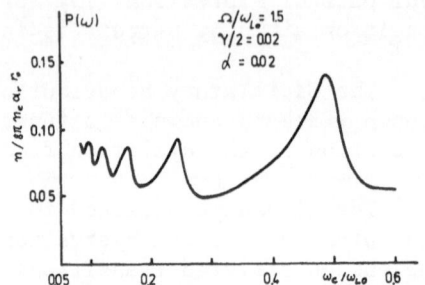

Fig 2: Phonon sideband. Abs. coeff. versus incident frequency.   Fig 3: Phonon sideband Abs. coeff. versus magnetic field.

ON THE ROLE OF NONELECTRONIC PHONON PROCESSES

IN MUTUAL ELECTRON-PHONON DRAG

P. Kocevar

Institut für Theoretische Physik

Universität Graz, Austria

Mutual electron-phonon drag and heating effects in semiconductors have been known and measured since the early fifties, but only recently detailed theoretical investigations, essentially numerical solutions of the Boltzmann equations for electrons and phonons, were published (1,2). Because of the great computational complexity of such calculations it seems desirable to have a first rough estimate of possible ph-induced corrections to measured carrier mobilities. We shall therefore use a simple electron-temperature model to derive a closed expression for the ohmic and nonohmic steady state mobility of nonpolar semiconductors. The main purpose of this paper is to show how strongly these ph-corrections depend on nonelectronic ph-dissipation mechanisms and to stimulate more experimental work on the determination of the corresponding ph-relaxation times as functions of temperature T and wave-vector q. In view of the many contributions of hot-electron experiments to our present understanding of the dynamics of charge-carriers in semiconductors the question arises, whether nonequilibrium ph-effects could have been overlooked in the analysis of certain measurements, especially at low temperatures. But up to now our lack of information about the details of the dominant ph-loss processes in most materials of interest has prevented quantitative work on this subject.

We start from the assumption, that carriers of standard energy-dispersion $\varepsilon = p^2/2m$ and density n_e, interacting only with longitudinal acoustic phonons and ionized impurities, obey a Maxwell-distribution displaced by a

mean drift velocity $\vec{v}_0$ and heated to an electron temperature T_e. The expressions for the total electronic energy
and momentum loss-rates to the phonons are then combined
with the steady-state solution of the phonon Boltzmann
equation. In general this solution deviates from the
usually assumed thermal Planck-distribution, because the
electrons try to enforce a Planck-distribution heated to
T_e and Doppler-shifted about $-\vec{q}\cdot\vec{v}_0$ (1). When the electronic
scattering rates of the phonons become comparable to the
nonelectronic rates τ^{-1}, which happens at sufficiently
low T, these ph-disturbances can lead to noticeable contributions to the overall energy- and momentum-balance
for the electrons (the necessary stationarity being provided by the thermal phonon sea, acting as a heat bath
for the electronically excited phonons (1,3)). After
linearizing with respect to $v_0/\tilde{v}$ and $g = s/\tilde{v}$, $\tilde{v}$ denoting
the mean thermal velocity of the electrons and s a proper
mean sound-velocity, the balance equations can be integrated in closed form for the following two standard
phonon relaxation mechanisms: 1.) Landau-Rumer loss (LR)
(4), where $\tau_{LR} \sim q^{-1} T^{-4}$ determines both energy- and
momentum-relaxation, and 2.) boundary losses, with momentum
relaxation $\tau_b^m \sim$ (crystal dimension)/s and energy relaxation
taken as $\tau_b^\varepsilon = \tau_b^m \cdot 1000/T$ because of acoustic mismatch at
the crystal boundaries (3,1,2). We introduce $r = v_0/s$ and
$t = T/T_e$ and, as usual (2), approximate the ratio of momentum losses to ionized impurities and phonons by $S \cdot t^3$,
where S should be fitted to the ohmic mobility. Restrict-
ing ourselves to LR-loss, the balance equations (whose
detailed form for LR- and B-losses will be given elsewhere
(7)) can be solved for

$$r^2 = 3(1-t)(1-\eta t^{1.5})/[St^3+t(1-\eta t^{1.5})+\eta t^{1.5}(1-2t)]$$

where $\eta = const \cdot n_e \cdot T^{-5.5}$, being proportional to τ, represents the effects of nonequilibrium phonons. The ph-drift
corrections enter through the lastterm of the denominator
of r^2; for $T_e < 2T$ they will tend to increase r ("phonon
drag") and counteract the damping effect of the remaining
η-terms ("phonon heating"). For pure ph-scattering (S = 0)
the borderline between dominant drag or damping would be
$T_e = 2T$, where $r^2 = 3$ independent of η. As numerical
example we show the relative change of mobility at 20K
(with a theoretical τ_{LR} from (4)) for three samples of
n-Ge, taken from the low-T analysis of Koenig et al. (6),
in which ph-disturbances had not been taken into account
explicitly. We find that noticeable phonon effects can be
expected for 15K < T < 25K only, because of carrier freeze-
out at lower temperatures (where n_e decreases more strongly
than $\tau^\varepsilon \sim \tau_b^\varepsilon$ increases, $\tau^m \sim \tau_b^m$ being constant) and due

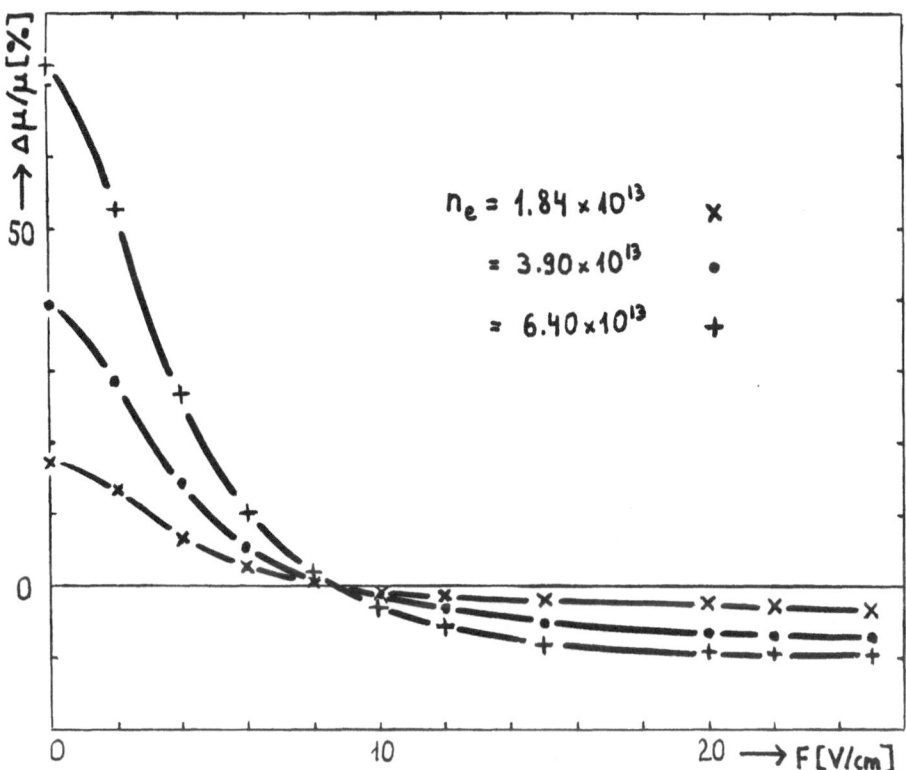

Relative change in mobility (in percent) vs. electric
field for n-Ge at 20K and three samples of different
electron concentration. Material parameters taken from (2).

to decreasing τ at higher temperatures ($\eta \sim T^{-5.5}$). Although very limited through the linearizations (e.g.:
$2\eta < 1$), our model nevertheless shows - in agreement with
the detailed theories (1,2) - the possibility of strong
ph-drift and -heating corrections to electron transport
at low temperatures and the need for precise numerical
expressions for nonelectronic phonon dissipation rates.

       References
1. P.Kocevar: J.Phys.C 5,3349(1972);
            Acta Phys.Austriaca 37,259,270(1973)
2. N.Perrin,H.Budd: Phys.Rev.B6,1359(1972),B9,3454(1974)
3. L.E.Gurevich,T.M.Gasymov: Sov.-Phys.-Solid St.9,78(1967)
4. L.Landau,G.Rumer: Phys.Z.Sowjet.U.11,18(1937)
5. M.Pomerantz: Phys.Rev.139,A 501(1965)
6. Koenig,Brown,Schillinger: Phys.Rev.128,1668(1962)
7. P.Kocevar: to be published

PHONON EXPERIMENTS WITH WEAKLY BOUNDED CHARGE CARRIERS AS DETECTORS

N. Riehl

Physik-Department der Technischen Universität München

Munich, Arcisstr. 21, Germany

Thermal release of charge carriers from traps or donors in phosphors or semiconductors is mostly a cooperative action of many phonons. But if a phosphor is excited at He-temperature the charge carriers can be stored in very shallow traps (trap depths being of the order 0,01 eV) so that a release by single phonons becomes possible and these traps can be used as single phonon detectors. The release is indicated by a light spike from recombination of the released charge carriers with luminous centers. Moreover the number and the depth of traps emptied by the phonon pulse is observable from a deficiency in the "glow curve" ("thermoluminescence curve") obtained subsequently by heating the phosphor. The glow curve shows the distribution of the charge carriers over the traps of different depths; a comparison of the glow curve obtained after the phonon action with a normal glow curve shows up to which depth the traps have been emptied by the phonon pulse; this maximal depth delivers the upper limit of phonon energies occuring in the applied phonon pulse. (This is correct only if no multiphonon processes are participating in the trap emptying i.e. if the phonon flux density is sufficient small; in our experiments this condition was quite fulfilled only if the phonon pulses were produced by bombardment of the crystal surface with gas-atoms.) - Shallow donors, acceptors or traps in semiconductors can also be used as detectors for single phonons if multiphonon processes are excluded (the charge carrier release being signaled by a current spike).

Three different methods of phonon generation were used in our experiments:

1) Bombardment of the crystal surface with Ne-atoms accelerated by means of the "skimmer-technique" which allows to obtain gas-atoms with kinetic

energies up to 1 eV and with a nonthermal, Gaussian energy distribution.
2) Short heat pulses (as used by many other authors).
3) Bombardment of the crystal surface with α-particles.

The density and the spectrum of the phonon pulses depend drastically
on the generation method used. It my be emphasized that the supplied
power and the phonon density in the phonon pulses produced by gas-
bombardment is by many orders smaller than in the case of heat pulses
but the mean energy of the produced single phonons is much greater and
their spectral distribution is quite different from any thermal equilibrium
(or Planck radiation law) preferring the high-energy phonons which are
able to release charge carriers; at the same power supply the releasing
effect of phonons produced by gas-bombardment is namely much greater
than of those produced by heat pulses!

For phonon range and propagation velocity experiments the phosphor
crystal (ZnO-phosphor with green luminescence) was excited only in a
thin layer near the surface opposite to the surface from which the phonons
started; this layer contained trapped charge carriers and signaled the
arrival of phonons. For the same measurements in a semiconductor (Ge)
the crystal was doped (with Sb) by diffusion only in a thin layer near its
surface which served as detector layer.

The upper limit of the range e.g. for 0,035 eV-LA-phonons in a
ZnO-crystal was found to be over 2,2 mm. Ranges of the same order were
found in monocrystals of different other materials for LA-phonons with
energies over 0,01 eV.

The propagation velocity obtained on Ge for 0,0096 eV-phonons is
essentilly smaller than expected from the known dispersion curve for LA-
phonons in Ge; this could be understood by the assumption that the
propagation is influenced (retarded) by Rayleigh-scattering (particularly
on isotopic inhomogenities) because of the very high scattering probability
for phonons of such small wavelength.

Experiments could be carried out also on crystals which are neither
luminescent nor semiconducting (quartz, boron, Al_2O_3, NaCl). In these
experiments a thin layer of polycristalline ZnO-phosphor on the crystal
surface served as phonon detector. The detected phonons had to pass
through the phase boundary crystal/ZnO so that reflexion and degradation
should be very probable at this boundary. But the experiments show that
surprisingly at least a part of the phonons is able to pass through the phase
boundary without degradation and to empty in a single phonon process
ZnO-traps with depths up to 0,035 eV (corresponding to the upper limit
of LA-phonon-energies in ZnO). The depth up to which traps are emptied
in the ZnO-detector by the phonons was compared with the maximal
LA-phonon energy of the given crystal. This comparison allows to
conclude whether the emptying is due to single phonons or to multiphonon processes and how the probability of multiphonon processes

depends on the method of phonon generation. If the phonons were produced by gas-bombardment the trap emptying was due only to the action of single phonons. Multiphonon processes (indicated by emptying of deeper traps) occur in the much more dense phonon fluxes generated by short heat pulses. - The highest probability for multiphonon processes was found if the phonon flux was generated by α-rays. This is understandable taking into account that a single α-particle produces a very great number of phonons (about 10^8) inside a very small volume and an extremely short time (about 10^{-12} sec). Because of such a short action time of α-particles the power supply from one single α-particle is enormous (of the order 1 W!). Therefore a phonon pulse from a single α-particle is to be considered as something like a shock wave (possibly as "soliton").

Some preliminary experiments concerning the propagation directions of phonons were carried out too. A "phonon-shadow" from a hole in a NaCl-crystal could be obtained on a ZnO-phosphor-screen on the crystal surface. An other observation seems to indicate a self-focussing of phonons: a thin excited (i.e. phonon-sensitive) layer at the surface of a ZnO-crystal showed some bright spots (with sizes 10-20 mμ) when the phonons came from an α-irradiated area (with a size 100 mμ) on the opposite crystal surface.

References: N. Riehl, A. Müller, Phys. Letters A 36, 487 (1971); phys. stat. sol. (a) 23, 345 (1974). N. Riehl, G. Baur, A. Müller, U. Puchner, phys. stat. sol. (a) 14, 453 (1972). N. Riehl, R. Wengert, phys. stat. sol. (a) 28, 503 (1975).

ANOMALY IN THE ELASTIC PROPERTIES OF A SEMICONDUCTOR UNDER PRESSURE

S.K. Ghatak

Groupe des Transitions de Phases
Centre National de la Recherche Scientifique
B.P. 166, 38 042, Grenoble, France

The monochalcogenides of Sm and Yb are semiconductors at normal pressure and they undergo the Semiconductor-Metal transition under high pressure. The transition is discontinuous in SmS whereas it is continuous for other compounds [1,2]. The lattice constant measurements show that the compressibility behaves anomalously in the transition pressure interval and is reduced very much after the transition [2]. These transitions are due to the partial delocalization of 4f-electrons into the 5d-6S conduction band states resulting in a 'mixed-valence' state for rare-earth ion. Many of the observed facts on these compounds can be explained on the basis of the Falicov-Kimball model [3,4]. In this model the transition is driven by interaction between the holes in a localized state and the electron in conduction band. In rare-earth compounds these states are respectively the localized 4f-state and 5d-6S conduction band state separated by an energy gap which decreases with pressure. In the following we consider this model for the transition and calculate the ultrasonic attenuation and the compressibility. An increase in attenuation is expected from the following argument. The electron-phonon scattering amplitude depends upon the probability that a conduction electron is available for scattering. In an intrinsic semiconductor this probability goes as $e^{-\Delta/kT}$, Δ being the energy gap and hence is small for large Δ/kT. In the metallic state the probability is large and constant at low T and thereby an increase in attenuation will occur.

Assuming that quasi-particle interactions are short range, the Hamiltonian of the electron and hole is given by [3]

$$H = \sum_{k\sigma} \varepsilon_{k\sigma} c^{+}_{k\sigma} c_{k\sigma} + \sum_{i\sigma} E_{o} \, b^{+}_{i\sigma} b_{i\sigma} - G \sum_{i\sigma} b^{+}_{i\sigma} c^{+}_{i\sigma'} c_{i\sigma'} b_{i\sigma} \qquad (1)$$

The first term represents the electron in the band states with
energy $\mathcal{E}_{k\sigma}$ and the second term is hole energy. The last term is
the electron-hole interaction of strength G. The operator C_k^+ (bi)
is the creation operator for an electron (hole) in the band state
k (in localized state at i-th site). The energy gap between the
localized state and the band state is given by Δ = Min $\mathcal{E}_k$ +E_0.
In the Hartree-Fock approximation the density of states of the
conduction band and the localized state are respectively

$$\rho_c(\mathcal{E}) = \rho_c^\circ(\mathcal{E}+Gn_h) \quad \text{and} \quad \rho_\ell(\mathcal{E}) = \delta(\mathcal{E}-Gn_c)$$ taking E_0 =0 as reference. $\rho^\circ(\mathcal{E})$ is the unperturbed density of state for the con-
duction band which is assumed to be parabolic, n_h and n_c are, respectively, the average number of holes and conduction electron
and they are given by

$$n_h = \int_{-\infty}^{+\infty} \rho_\ell(\mathcal{E}) \frac{d\mathcal{E}}{1+e^{(\mathcal{E}_F-\mathcal{E})\beta}} = \frac{1}{1+e^{(\mathcal{E}_F-Gn_c)\beta}} \cdots (3) \& \quad n_c = \int_{-\infty}^{+\infty} \frac{\rho_c^\circ(\mathcal{E}+Gn_h)d\mathcal{E}}{1+e^{(\mathcal{E}-\mathcal{E}_F)\beta}} = 2\left(\frac{1}{\Gamma\beta}\right)^{3/2} F_{1/2}(\eta) (4)$$

were $\Gamma = \left(\pi^2\sqrt{2}/V\right)^{2/3}/m^*$; $\beta=(1/k_BT)$. F_m (n) is the Fermi-Dirac
integral and $\eta = (\mathcal{E}_F-\Delta+Gn_h)\beta \cdots (5)$. The Fermi energy $\mathcal{E}_F$ is
determined from the condition n_h = n_c and therefore equ (3) and(4)
constitute the self-consistent equation for n_c (or n_h).

Attenuation coefficient : We consider the contribution to the
attenuation only due to electrons because the contribution due to holes will be small as they are strongly localized at lattice sites.
In the semiconductor the electron-phonon interaction can be well
described by the deformation potential approximation[5] and the
Hamiltonian for it can be written as $H_{ep}=\sum_{q,k} V_q(a_q^+ + a_{-q})C_{k+q}^+ C_k \cdots (6)$
with $V_q = iDq\cdot(1/2\omega_q\rho)^{1/2}$; ρ is the density of system, D
is the deformation potential and a_q is phonon creation operator
with frequency ω_q. The attenuation coefficient α is then given
by [6] $$\alpha = \left(\pi V_q^2/\upsilon\right)\sum_k (f_k - f_{k+q})\, \delta(\mathcal{E}_{k+q} - \mathcal{E}_k - \omega_q). \cdots (7)$$
υ is the velocity of sound, f_k is the F-D distribution func-
tion and the energy of the electron in H-F approximation $\mathcal{E}_k = k^2/2m^* - Gn_h$.
Integrating over K-values we get $\alpha = \alpha_0/(1+e^{-\eta}) \cdots (8)$
with $\alpha_c = \left(D^2\omega_q m^*/4\pi\rho\upsilon^2\right)$ and η is given by equ.(5).
    When η is large negative $\alpha \approx \alpha_0 e^{-|\eta|}$ and for η large positive
$\alpha \approx \alpha_0$. For G = 0 and large $\Delta/_{kT}$ (i.e. semiconducting state) η is
large and negative and the attenuation due to the electrons is
very small. We also note from equ.(5) that α is large or small depending on whether $\mathcal{E}_F$ lies in the band or in the gap. In Fig.1,
n_h and α as obtained from the equs.(3),(4) and (8) are shown as
a function of energy gap for different values of G/Γ and T $\approx$ 40°K.
For discontinuous transition from $n_h \approx$ 0 to $n_h \approx$ 1,   changes discontinuously from 0 to α_0. But when the discontinuity in n_h is
small (curve 3) α has a smooth behaviour. Again for continuous
transition α changes smoothly but attains the maximum value α_0 much
before n_h becomes large (curve 4). The reason for this behaviour is
that for a discontinuous transition with large change in n_h the

position of ε_F jumps from the gap to the band at transition pressure, whereas for the transition with small discontinuity or with no discontinuity ε_F enters the band before the transition occurs and so even for small n_h ($<< 1$) α is large. A numerical estimate of α_o is in order. Taking $D \sim 5$ ev^2 ; $m^* \sim m$, $v \sim 5 \times 10^5$ cm/sec, $\rho \sim 4$ gm cm^{-3}, and $\omega_q \sim 10^8$ sec^{-1} $\alpha_o \sim 10^0 \sim 10^1$ cm^{-1} which is large and can be observed experimentally.

<u>Compressibility</u> : The free energy of the system is

$$F = Gn_h^2 + k_BT \ln(1-n_h) - \tfrac{4}{3}(k_BT)^{5/2}\Gamma^{-3/2}F_{3/2}(\eta) + 9av_c\left[\left(\tfrac{v_o}{v}\right)^{4/3} - 2\left(\tfrac{v_c}{v}\right)^{2/3} + 1\right]/8 \cdots (9)$$

The first three terms are the electronic contribution as obtained from equ (1). The last term is the elastic energy as obtained from the Birch equation [7] (with $\zeta = 0$) which is found to hold good for many ionic compounds. B_o and V_o are the bulk modulus and the volume at normal pressure. The compressibility is then given by

$$\frac{\chi_o}{\chi} = \left[\left(\tfrac{v_o}{v}\right)^{8/3} - \left(\tfrac{v_c}{v}\right)^{10/3}\right]/2 + n_h\chi_o D/v - \frac{\partial n_h}{\partial v}\left[2Gn_h - k_BT/(1-n_h) - D\right]\chi_o \cdots (10)$$

In writing eq (10) we assume the energy gap variation as $\Delta = \Delta_o - D\frac{v_o-v}{v_o}$. The first term is pure pressure effect which increases with the decrease in volume (i.e. increase in pressure). The second term follows the change in number of conduction electron and decreases the compressibility in the metallic phase. Since the deformation potential D is usually larger than G and T, the dominant contribution of the third term goes as $D\chi_o(\partial n_h/\partial v)$ and it is large and negative near the transition pressure. In Fig. 2 the overall behaviour of χ is depicted schematically and it shows that χ decreases at low pressure, increases anomalously near the transition and then decreases again. Also the compressibility in the semiconducting phase is larger than that in the metallic phase. This is in agreement with experimental results on SmS [8] SmSe and SmTe [2].

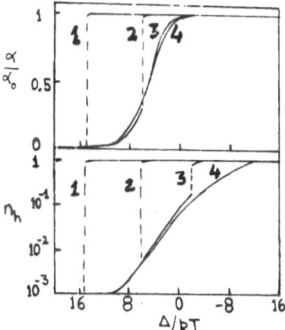

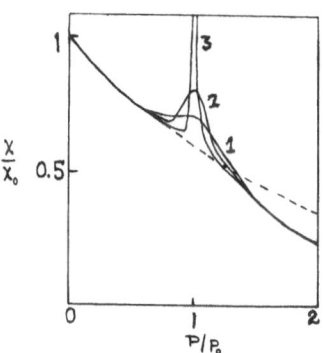

<u>Fig. 1</u> : Variation of α/α_o and n_h as a function of Δ/kT for different values of $G/\Gamma : G/\Gamma = 0.8$ (1), 0.65(2), 0.47(3) and 0.3(4). T is taken to be 40°K.

<u>Fig. 2</u> : A schematic plot of χ/χ_o as a function of p/p_o ; p_o being the transition pressure in arbitrary unit. The curve 1 & 2 — smooth transition, 3)— discontinuous transition and --- without electronic contribution.

I am grateful to Drs. M. Avignon and J.M.D. Coey for their comments and to Dr. B.K. Chakraverty for his hospitality.

References

1. A. Chatterjee, A.K. Sinh and A. Jayaraman, Phys. Rev. B 6, 2285, (1972).
2. A. Jayaraman, A.K. Sinh, A. Chatterjee and S. Usha Devi, Phys. Rev. B9, 2513 (1972).
3. L.M. Falicov and J.C. Kimball, Phys. Rev. Lett. 22, 997 (1969).
4. M. Avignon and S.K. Ghatak, Solid State Commun. 16, 1243 (1975). M. Avignon and S.K. Ghatak, International Conference on Electronic Properties of Solids under high pressure, Leuven, Belgium, Sept.1-5, 1975.
5. W.A. Harrison, Solid State Theory (McGraw Hill Co. 1970) p. 391.
6. C. Kittel, Quantum Theory of Solids (John Wiley Inc, 1963). p. 327.
7. F.J. Birch, Geophys. Res. 57, 227 (1952).
8. A. Jayaraman, V. Narayanamurti, E. Bucher and R.G. Mains, Phys. Rev. Lett. 25, 1430 (1970).

X-RAY SCATTERING STUDY OF AMPLIFIED PHONONS IN THE PRESENCE OF

PLASMONS

S. KAGOSHIMA and T. ISHIGURO

Electrotechnical Laboratory

Mukodai, Tanashi, Tokyo 188, Japan

Acoustoelectrically amplified phonons in n-GaAs are studied spectroscopically by the x-ray Brillouin scattering at 77 K, under the conditions of $q\mathit{l}>1$ and $\omega_p\tau>1$, where q is the phonon wave number, l the electron mean free path, ω_p the plasma frequency, and τ the electron collision time.

Samples used were epitaxially grown on the (100) plane of insulating GaAs substrates. Their characteristics are listed in Table I. For the x-ray scattering observation, a monochromator of GaAs and a sample were set to be the (+-) arrangement. The scattering intensity was measured by rotating the sample near the (004) reflection. Thus, the amplified phonons with q//[110] and e//[001] were investigated (e; the polarization vector).

The gain factor $(n_q+1/2)\hbar\omega/k_B T$ which is the ratio of the observed phonon population to the thermal equilibrium level (n; the average occupation number of q-phonons, ω; the angular frequency, T; the temperature) is calculted by

$$I_1/I_0 = (1/2\rho)(2\pi)^{-3}S^2(n_q+1/2)\hbar V_q/\omega \quad , \qquad (1)$$

where I_0 and I_1 are the intensities of the Bragg and the first Brillouin scatterings, ρ the crystal density, S the scattering vector and V_q is the volume in the reciprocal lattice space which is determined as an overlapped volume of the cone spreaded by amplified phonons with both the thickness of the Ewald sphere and the aperture subtended by a receiving slit. The size of the clipped part of the Ewald sphere by the cone subtended by the receiving slit with spreading angle ϕ is given by $k_{sc}\phi$ (k_{sc} is the scattered x-ray wave number, $\phi=2\times10^{-3}$rad), whereas the corresponding size by the phonon cone is given by $q\psi$ (ψ; the spreading angle of the phonon cone). In a thin layer only the phonons which propagate almost parallel to the layer are amplified. Then, ψ is estimated as less than 10^{-1}rad. Since k_{sc}/q is of order of $10^3 \sim 10^4$, $k_{sc}\phi > q\psi$. Thus, provided that the thickness of the Ewald sphere is constant, V_q is proportional to q^2. We should mention that Carlson et $al.$[1] assumed that V_q is independent of q. This corresponds to assume that $k_{sc}\phi < q\psi$, which may be satisfied if we do not take account of the geometrical restriction to the amplified phonons.

When an applied electric field exceeds a threshold, acoustic instabilities associated with current oscillations appeared in #14 -#16, whereas in #17 we could observe neither current oscillation nor non-ohmic behavior. In #14-#16, the amplified phonon spectra were measured at the phonon population which is of order of 10^8 times of the thermal equilibrium level. This means that the amplified phonons are still in the weak signal level: According to the studies of the phonon spectra in bulk specimen by the light scattering, the nonlinear effect dominates when the phonon level exceeds 10^9 times of the thermal equilibrium level.[2]

As one of the characteristic parameters of the phonon amplification we adopt the maximum gain frequency ω_m. We compare the observed ω_m with the calculated by using the gain factor α_{sp}[3] and the lattice loss α_L. In Fig. 1 by the curved line is shown the frequency dependence of α_{sp} calculated by adopting the experimental conditions. α_L is evaluated by using $\omega^{1.5}$ dependence[4] and the data by Pomerantz [5]. In Fig. 2 the observed ω_m are compared with the calculated.

Mosekilde calculated the acoustoelectric gain factor quantum mechanically for arbitrary degeneracy of the electron gas and proposed some deviations from α_{sp}.[6] One of the points to be noted is that α_{sp} should be multiplied by a factor of $\exp(-\lambda_e^2/4\lambda_{ac}^2)$ where $\lambda_e = \hbar/(2m^*k_BT)^{1/2}$ and $\lambda_{ac} = 2\pi/q$, for the electron gas with non-degenerate statistics which is applicable for #14-#16. The factor causes downshift in ω_m if the value is appreciable, but in the present cases the factor becomes 0.85 for $q=10^6$cm^{-1} and 0.98 for $q=10^7$ cm^{-1} and the effect to ω_m is negligible.

To explain the observed downshift in ω_m of #16, we consider that in the present case the effect of the collision-less plasma may appear since $\omega_p\tau > 1$. The gain factor α_p relevant to the plasma is given by Miranda and ter Haar [7]. In Fig. 1 the calculated α_p is shown, where we cut α_p at $q=k_D$ (k_D; the Debye wave number) to reflect the effect of Landau damping. In non-degenerate electron gas, the Landau damping may occur near but below k_D and it is expected that α_p or $\alpha_{sp}+\alpha_p+\alpha_L$ of #16 shows a maximum below k_D. By comparing Fig. 1 with Fig. 2, it is found that the deviation in ω_m is closely related to the contribution of the plasma effect.

Two reports on the amplified phonon spectra at 77 K [1][8] do not contradict the present work if we process the data in correct way. The results are shown in Fig. 2. Thus, we conclude that the systematic deviation in ω_m vs carrier concentration reflects the contribution of the plasma effects.

References
[1] Carlson D G et al. 1971 Appl. Phys. Lett. 18 330.
[2] Spears D L 1970 Phys. Rev. B2 1931.
[3] Spector H N 1966 Solid State Physics (New York:Academic) 19 291.
[4] Keller K R and Abeles B 1966 J. Appl. Phys. 37 1937.
[5] Pomerantz M 1965 Phys. Rev. 139 A501.
[6] Mosekilde E 1972 J. Appl. Phys. 43 4957.
[7] Miranda L C M and ter Haar D 1972 Phys. Lett. 39A 15.
[8] Ishibashi T et al. 1973 Phys. Lett. 44A 371.

Table I   Sample characteristics. n is the carrier concentration and μ is the mobility. $q\ell$ is estimated for ω_m calculated by α_{sp}.

#	$n(cm^{-3})$	$\mu(cm^2/V\ sec)$	$\omega_m(GHz)$	$q\ell$	$\omega_p\tau$
14	$1.5x10^{14}$	$1.0x10^5$	5.2	10	3.3
15	$7.9x10^{14}$	$6.7x10^4$	11	15	4.9
16	$1.1x10^{16}$	$1.7x10^4$	40	14	4.6
17	$1.5x10^{17}$	$4.8x10^3$	70	7	4.8

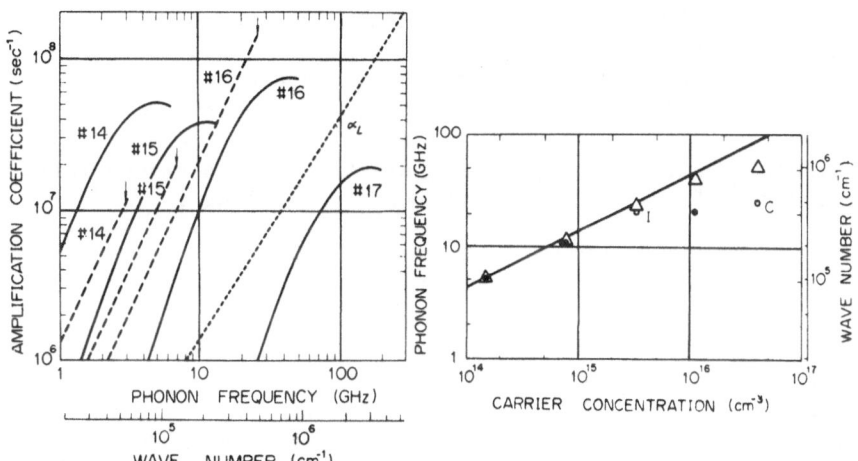

Fig. 1 (left)   Frequency dependences of α_{sp}(solid line) and α_p (broken) for #14-#17. α_L(dotted) should be read as negative.

Fig. 2 (right)   Carrier-concentration dependence of ω_m.   The solid line shows ω_m estimated by α_{sp}. $\triangle$ shows ω_m estimated by α_{sp}+ α_L. ● shows the observed ω_m. ○ marked with C or I shows the result from ref. 1 or ref. 8.

COUPLED PLASMON-PHONON EFFECTS IN IV-VI SEMICONDUCTORS

O H Hughes, P M Nikolic, J M Chamberlain, C J Doran and
M Merdan

Department of Physics, University of Nottingham
University Park, Nottingham NG7 2RD, U K

INTRODUCTION

Many solids with layer structures can be considered as essentially two-dimensional networks with weak interaction between the
layers. Such materials have lately attracted a certain amount of
interest; recent studies include directional photoemission, transport and optical measurements of the layer materials themselves and
their intercalated derivatives with a view of elucidating their band
structure[1,2,3]. Although some attention has been focussed on the
phonon behaviour[4] in this type of system, no investigations have
so far been carried out into the nature of the phonon-plasmon interaction in the layer compounds. We have grown single-crystal specimens of the IV-VI semiconductors SnS, GeS and GeSe and carried out
optical studies[5] in the far infra-red range to investigate this
behaviour. All of these materials crystallise into distorted rock-
salt-structure units arranged in layers for which a strong intra-
layer and weak (van der Waals) inter-layer coupling is found.

EXPERIMENTAL DETAILS

Single crystal specimens of SnS, GeS and GeSe were prepared
using the Bridgeman technique. They were subsequently examined by
X-ray analysis and confirmed to be single crystals. Far infra-red
reflection measurements were carried out at 77K and 300K at near-
normal incidence using conventional Michelson interferometric techniques. The range of frequencies studied was $10cm^{-1}$-$500cm^{-1}$. In
all cases the crystals were easily cleaved perpendicular to the c-
axis and this was done before all experimental runs to ensure high
quality surfaces for reflection studies. The results were subsequ-

312

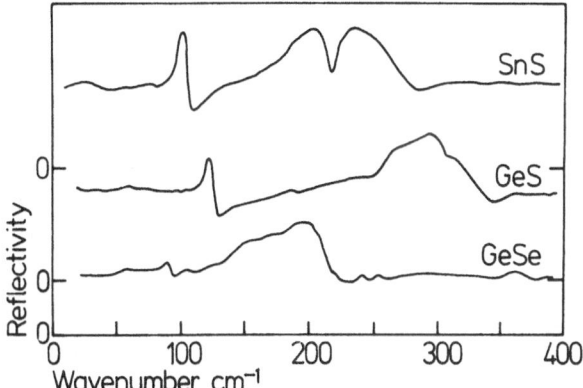

Fig 1: Infra-red reflectivity of SnS, GeS and GeSe.

ently analysed by Kramers-Kronig computer programs to obtain dielectric constants and refractive indices. Ancillary transport measurements were also undertaken in the temperature region of interest to
provide carrier concentration and mobility values. The specimens
were all p-type at room temperature (a consequence of the lack of
stoichiometry) and had typical carrier concentrations of $\rho \sim 10^{18}$/cc.

RESULTS

Some typical 300K results for the three materials are shown in
fig 1. It is of some interest to compare these results with reflec-
tances obtained from isotropic three-dimensional semiconductors,
e.g. GaAs. As is well-known, such a system displays a simple plasma
edge and restrahl pattern for $\omega_p < \omega_L$ and a plasma edge and dielectric anomaly for $\omega_p > \omega_L$. It is apparent that the results obtained
for the layer compounds conform to neither of these patterns. An
edge is observed at 280cm^{-1}, 230cm^{-1} and 340cm^{-1} in SnS, GeS and
GeSe respectively and a restrahl-like feature at 100cm^{-1}, 124cm^{-1}
and 90cm^{-1} for these materials respectively. A shift in the edge
position of 20cm^{-1} to lower energies is noted at 77K and this figure
is typical for the other two materials. No change of position or
shape with temperature is observed in the $\sim$ 100cm^{-1} maximum for all

of the materials. The 210cm^{-1} dip in SnS is similarly unaffected
by temperature, but the half-width varies from sample to sample.

DISCUSSION

The problem of coupled plasmon-phonon modes in isotropic three-
dimensional semiconductors has been discussed by a number of workers[6,7]. It is concluded that the reflectance curves will display
minima at frequencies (ω_+, ω_-) which, in the limit of zero coupling,
become the free plasma and TO phonon frequencies respectively. For
coupled systems, ω_+ and ω_- cease to be identifiable with the pure
modes of the system. Under these conditions ω_- will vary only slightly with carrier concentration, and hence temperature, whereas ω_+
will vary more rapidly. We have reason to believe that the layer
materials investigated here are exemplars of this behaviour. The
higher energy feature in each curve is thus identifiable as an ω_+
(plasma-like mode), and the restrahl-like feature as ω_-. Thus, although the carrier concentration for SnS diminishes by an order of
magnitude between 300K and 77K implying a shift for a 'pure plasma'
edge from about 300 to 100cm^{-1} a shift of an order of magnitude less
than this is observed. Such a small shift is consistent with the
magnetoplasma theory of Perkowitz[6] and Devreese[7] for such coupled systems. The restrahl-like feature is similarly accounted for
as the ω_- mode, with a negligible temperature dependence. This mode
may consist of two almost degenerate modes arising from the two in-
layer modes. Unfortunately, this theory is only applicable to systems with degenerate carrier densities, which is not the case for the
samples examined. This behaviour is attributable to the two-dimensional nature of the material; the hole plasma density in this material will be increased because of the confinement of the plasma to
the layers. This increase of the order (layer separation/ionic radius) would imply an order of magnitude increase in the plasma density for SnS, with similar results for the other materials. One further feature merits attention; namely the 216cm^{-1} dip observed in
SnS which is not apparent in the other materials. The fact that
this feature varies in width from sample to sample, but not in position, with either temperature or sample change indicates that it
may be a local mode in this non-stoichiometric material.

1. Yoffe A D, Proc XIIth Int Conf Semiconductor Physics; B G Teubner
   Stuttgart, 1974, p. 611.
2. Williams R H, Murray R B, Thomas J M, J Phys C 6, 3631 (1973).
3. Murray R B, Williams R H, Stuttgart Conference, p. 637.
4. Zallen R, Slade M L, Phys Rev B 9, 1627 (1974).
5. Nikolic P M, Vujatovic S S, Hughes O H, Doran C J and Chamberlain
   J M, Stuttgart Conference, p. 331.
6. Perkowitz S, Phys Rev B 9, 545 (1974).
7. Devreese J T, Lemmens L F, Stuttgart Conference, p. 483.

ANISOTROPY OF PHONON EMISSION FROM HOT ELECTRONS IN GERMANIUM

W. Reupert, K. Lassmann, and P. de Groot

Universität Stuttgart, Physikalisches Institut

Teilinstitut 1, 7 Stuttgart 80, Germany

We have measured quantitatively the anisotropy of the electron – acoustical phonon scattering in n-germanium in five directions of the $(1\overline{1}0)$ plane utilizing tunneling junctions for calibrated detection of the phonon radiation emitted from a small avalanche breakdown region in the germanium surface. For comparison and evaluation of phonon focusing a constantan heater and a tunneling junction were also used as phonon sources. A similar experiment has been reported by A. Zylbersztejn /1/, but he could only compare the ratio of the amplitudes of longitudinal and transverse phonon pulses in a given crystal direction to the calculation.

Our sample was doped with Sb $[3 \times 10^{14} \text{ cm}^{-3}]$, the schematic arrangement is shown in Fig.1. The avalanche heater consisted of two alloyed n^+n contacts 0.5 mm wide. The signals were generated by current pulses in the $[001]$ direction. This implies the same electron temperature T_e and the same electron density in each valley of the conduction band. The Constantan film as well as the $Sn-SnO_x-Sn$ tunneling junction had the same area as the avalanche heater. The sample was immersed into liquid helium at 2K. Current pulses of 0.4 us at a repetition rate of 4 kHz were applied. The signals of the avalanche heater received by the tunneling junctions are shown in Fig. 1.

E. Conwell /2/ calculated the anisotrpy of the steady state phonon distribution for an isotropic many valley semiconductor. In our calculations we took into account the anisotropy of the sound velocity and evaluated the electron phonon matrix elements for the special case of phonon propagation in the $(1\overline{1}0)$ plane from interpolation formulas given by C. Herring and E. Vogt /3/. The contributions of the two transverse modes were calculated separately. To calculate the phonon rate detected by the junction we may write

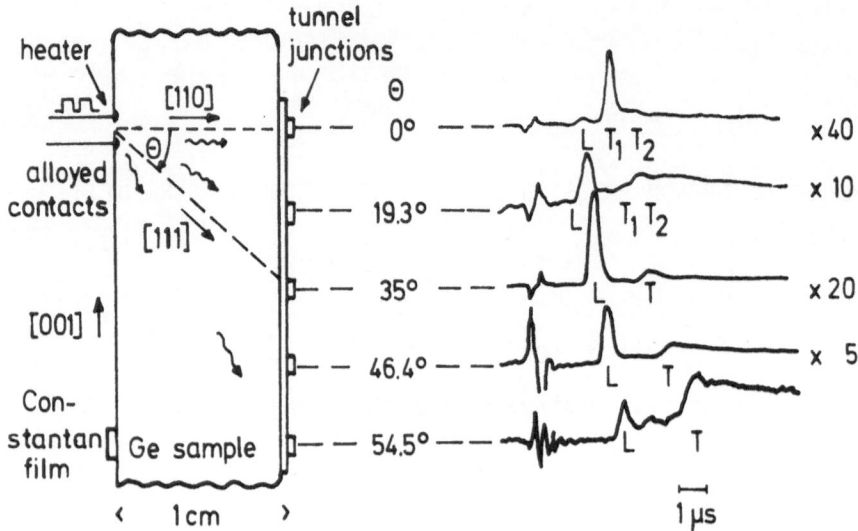

Fig.1 Schematic sample arrangement and signals in 5 directions of the (1$\overline{1}$0) plane at the same input power. Lattice temperature 2K.

$$\dot{N}_p = \int_{2\Delta}^{4\Delta} ((N_{qp} - \overline{N}_{qp})/\tau)Z(\varepsilon)d\varepsilon \ ,$$

where N_{qp} is the steady state number of phonons of polarization p and wave vector q and $\overline{N}_{qp}$ thermal equilibrium number small compared to N_{qp} at helium temperatures. $Z(\varepsilon)$ is the phonon density of states. The relaxation time τ was identified with the signal rise time of about .2 us (the response time of the receiving tunneling junctions is much faster at 2 K). The lower limit of integration is given by the spectral sensitivity of the tunneling junction (2Δ is the energy gap of tin). For electron temperatures below T_e = 100K we found that the number of phonons with energy $\varepsilon > 4\Delta$ is negligible. The tunneling junction operates as a phonon counter in this energy interval.

The normalized signals of the Constantan heater and the tunnel-ling junction were compared to the theoretical values for an isotro-pic phonon source taking into account the geometrical factors and phonon focusing values for LiF /4/, which should be valid for Ge within a few percent. We found good agrement and took the measured focusing factors to eliminate phonon focusing from the signals of the avalanche heater. In the saturation range of the avalanche breakdown, when the donors are ionized, the kinetic energy of the hot electrons can be estimated from the ionization energy of the Sb donors (9.7meV) corresponding to 75 K. Fig.2 shows the good agreement of the calcu-lated phonon distribution for T_e = 75 K with measured values in the

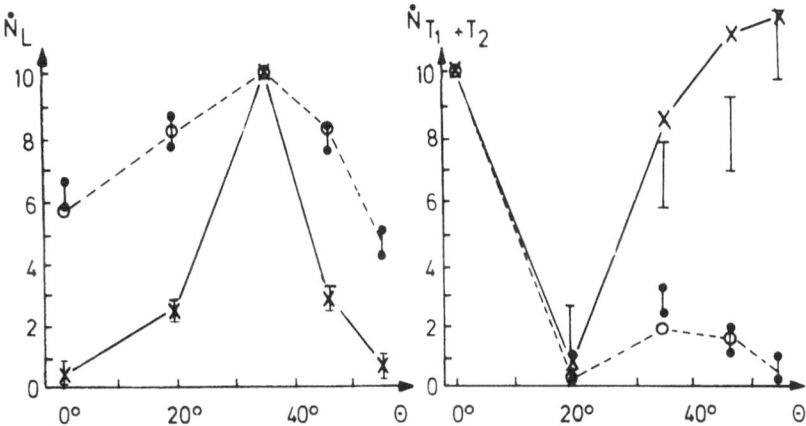

Fig.2 Relative rates of phonons received at the tunnel junctions.
The longitudinal (L) pulses are normalized in [111] ,the transverse
(T₁+T₂) in [110] . Heat pulse measured (⧧), electron-phonon inter-
action measured (I), and calculated (+).

saturation range. In accordance to theory the anisotropy of the phonon distribution increased with decreasing input power (that means
decreasing T_e). At the beginning of the avalanche breakdown we estimate from experiment T_e = 30 K. To estimate the electron temperature
we evaluated the longitudinal phonon signals because of a background
signal of the transverse phonons rapidly increasing with growing input power of the avalanche heater. This background was not seen at
corresponding powers of the Constantan heater and may be a property
of the avalanche heater.

References:

/1/ A.Zylbersztejn, Proc. 10th Int. Conf. Semiconductors p 134 (70)
/2/ E.Conwell, Phys. Rev. 135, A814 (64)
/3/ C.Herring and E. Vogt, Phys. Rev. 101, 944 (56)
/4/ B.Taylor, H.J.Maris, and C.Elbaum, Phys. Rev. B3, 1462 (71)

DISCUSSION ON FREE CARRIERS

Quantum Oscillation in Transmission of Ballistic Heat Pulse in
Bismuth   T Ishiguro, K Kajimura, S Kagoshima, H Tokumoto and
J Kondo   Page 275

G Bellessa: Does the effect always exist (with spike-like character)
when the magnetic field is perpendicular to the wave vector and the
magnetic field along the binary axis?

T Ishiguro: Let us think of the case when the magnetic field is
parallel to the binary axis and the direction of the wave propagation is parallel to the bisectrix axis. Then, the absorption by
the I band will not be spike-like but the absorptions by the II or
III bands will be spike-like.

G Bellessa: Why does the last peak disappear when the angle between
the magnetic field and the wave vector is large in your last slide?

T Ishiguro: It is simply due to the fact that the attainable field
of our magnet is limited to 14kG. If we raise the magnetic field
intensity, the absorption related to the lower Landau level will
appear.

L J Challis: Would you expect to see quantum oscillations in heat
pulse transmission in other systems? Also do you expect them to be
visible in thermal conductivity experiments?

T Ishiguro: It may be possible if the system has a sufficiently
anisotropic band structure. The difficulty in seeing oscillations
in thermal conductivity experiments is that the oscillations in
attenuation of transverse phonons are very small so that the overall effect is much smaller and has not been seen in Bi for example.

****

Phonon Contribution to the Hall Resistivity in Cadmium
A N Gerritsen   Page 278

K F Renk: Would it be possible to see this Einstein mode using
the Mossbauer technique?

A N Gerritsen: We think it might be possible in a Raman experiment.

P G Klemens: Would the Einstein oscillator also show up in the temperature dependence of the electrical resistivity?

A N Gerritsen: Yes. The apparent fit to a Bloch-Gruneisen relation with $\theta \underset{\sim}{} .5\theta_D$ is due to (a) lower cut off frequencies for electrons and holes than given by θ_D and (b) the large quasi-elastic scattering by the Einstein phonons.

J Rayne: Do you expect to find similar results in the other HCP metals Be, Mg and Zn?

A N Gerritsen: Possibly. But one should remember that Cd and Zn are the most "open" hexagonal crystals, i.e. with largest c/a ratios. This may reduce the possibility for Mg and Be.

****

Phonon Scattering and the Thermopower of Dilute Alloys   J Kopp
                                                         Page 280

J G Collins: I'd like to note that measurements of resistivity in our laboratory on very pure copper (magnetic impurities less than 10^{-9}) and silver show a temperature dependence which varies between $T^{2 \cdot 8}$ and $T^{3 \cdot 5}$ in the temperature range 1 to 8·5K. This is similar to yours in gold.

J E Parrott: Have there been any measurements of thermal resistance on systems of this kind?

J Kopp: I think only some unpublished work by Brock. Most of the work has been on specimens containing $\sim$ 100's ppm of magnetic impurities where it is now known than spin-spin interactions dominate all other effects.

****

Use of a Phonon Probe to Determine the Magnetic Dependence of the Superconducting Gap   D Huet, B Pannetier and J P Maneval
                                                         Page 283

W Eisenmenger: What happens if you apply a field perpendicular to the junction? Does the field penetrate? You could presumably test this by switching the field on and off since you would see a change in the I-V characteristic. We find a change if the field is greater than 19 Oe applied perpendicular to a junction of $1mm^2$.

B Pannetier: We see no change in the characteristic though it is
reasonable to assume that the flux should penetrate since the
junction is wide.

V Narayanamurti: We find for thinner films, 600-700Å evaporated
onto cold substrates, that if the field is not exactly in the
plane of the junction, the I-V characteristic shows slight
hysteresis.

N A Lockerbie: Can you put an upper limit to the "phonon assisted"
tunnelling in your detector junction as a percentage of the
tunnelling due to direct breaking of Cooper pairs?

B Pannetier: Voltage-dependent mechanisms contribute certainly to
less than 5% of the tunnelling due to pair breaking.

N A Lockerbie: Do you see a sharply defined onset in the phonon-
induced signal when the applied magnetic field is perpendicular to
your detector junctions?

B Pannetier: Fig 2b corresponds to the case of perpendicular field.

****

Temperature Dependence of Electron Mean Free Path in Molybdenum
from Ultrasonic Measurements   D P Almond, D A Detwiler and
J A Rayne   Page 286

J G Collins: Have you measured the electrical resistivity of
these samples?

J Rayne: We're hoping to do this shortly. The resistance ratio is
very large and one has to use a squid voltmeter.

P G Klemens: The thermal resistivity should give you a $WT \propto T^2$ term.
Has anyone measured this?

J Rayne: No, not on samples as pure as these although it would be
interesting to do so. There have been measurements on thermoelectric power of pure tungsten which vary as T^2.

****

Interaction and Scattering of Greater Than 2Δ Phonons with Electrons
in Superconducting Sn Films   I L Singer and W E Bron Page 289

H Kinder: I am not clear where the singularity at 2Δ comes from
in your theoretical phonon spectrum. I should expect only a small
overshoot as in ultrasonic attenuation.

W Bron: The singularity is not a generator phenomenon but is in
the detector. It is the onset of detection from the lead. The
analysis of this is quite complicated; perhaps we should discuss
it later.

****

Experimental Evidence of Phonon Flux Flow Interaction   M Martin,
J Y Desmons, D G Thomas, E Bridoux, M Moriamez and M Le Ray
Page 292

K Kajimura: Do you think the inversion of the detected pulse
heights is related to the change in the sign of the entropy flow?

M Martin: We think that it is a possible mechanism. Such change
in sign of the entropy flow has been also seen in low κ type II
superconductors.

****

Phonon Induced Structure on Damped Magneto-Absorption of Weakly
Polar Semiconductors   J P Vigneron, R Evrard and E Kartheuser

Page 295

R F Wallis: How does your work compare with that of Korovin's?

J P Vigneron: The main difference is that in Korovin's work, the
width of the cyclotron line in the absence of electron-phonon
interaction is zero. I have included phenomenologically other
scattering processes such as acoustic and impurity scattering.
These broaden the line even before the LO phonon is switched on.

****

On the Role of Nonelectronic Phonon Processes in Mutual Electron-
Phonon Drag   P Kocevar   Page 299

T Ishiguro: Have you compared your theory with any experimental
data?

P Kocevar: In the work of Koenig et al. I referred to, current
versus field curves are only given for lattice temperatures outside the range of maximal phonon-disturbances. Quite generally,
most low temperature work in germanium has concentrated on carrier
– avalanche and freeze-out processes at somewhat lower temperatures.
But for the highly non-linear case of strong phonon-amplification
by hot electrons at 4K, corresponding time variations of the
mobility were seen.

Phonon Experiments with Weakly Bounded Charge Carriers as Detectors
N Riehl (read by Professor Renk)   Page 302

J D N Cheeke: Is the α-particle experiment a pulse experiment and
if so what is the pulse width?

K Renk: It is not really a pulse experiment as they see individual
α-particles. A shutter is opened for maybe 10s and perhaps 10^5
α-particles arrive.

D Pohl: Could you explain how the heat pulse experiment is done?

K Renk: A short pulse, $\sim$ 100 ns, is obtained from a constantan
heater. The high frequency phonons excite the traps. One can then
look at this 'frozen in' information by changing the temperature
etc.

****

Coupled Plasmon-Phonon Effects in IV-VI Semiconductors
O H Hughes, P M Nikolic, J M Chanberlain, C J Doran and M Merdan
                                                              Page 312

K Renk: Do you know the nature of the defect responsible for the
local mode?

O H Hughes: No. However the material is highly non-stoichiometric
with defect concentrations up to $\sim$ 20% - indeed these are responsible for the p-type conduction - so it seems likely to be associated with these. Certainly there is no change in position
from one sample to the next to indicate that it is due to chemical
impurities.

****

Anisotropy of Phonon Emission from Hot Electrons in Germanium
W Reupert, K Lassmann and P de Groot   Page 315

T Ishiguro: Does the existence of anomalously high transmission
of phonons into the helium cause any problems in your analysis?

K Lassmann: We assume that the phonons striking the interface are
wholly transmitted into the helium which is of course an exaggera-
tion.

PHONON SCATTERING AND ULTRASONIC ABSORPTION BY ACCEPTORS

IN Si AND Ge

T. ISHIGURO and H. TOKUMOTO

Electrotechnical Laboratory

Mukodai, Tanashi, Tokyo 188, Japan

The lattice thermal conductivity of semiconductors below the low-temperature maximum is substantially depressed by localized shallow impurities. The phonon scattering due to donors has been explained in terms of a virtual excitation of electrons in an impurity atom.[1-5] The localized acceptors also scatter phonons effectively.[6-12] In this paper we discuss the interaction of the acoustic phonons with the localized acceptors as laying emphasis upon the relation to splittings in the acceptor-ground state. We do this as comparing the situations in Si and Ge.

The localized acceptor states are represented by a hydrogen-like electronic structure, whose first excited states are about 80 K and 330 K above the ground states in Ge and Si, respectively.[13] Thus, as far as the thermal conductivity below 10 K are concerned, we can neglect effects of the excited states. The ground state has fourfold degeneracy reflecting the properties of the valence band in the absence of static fields of lower symmetry. By using the acceptor-hole-lattice interaction Hamiltonian of the same form as that for free holes near the band edge, Suzuki and Mikoshiba (SM) have calculated the matrix elements of the hole-lattice interaction among the ground states.[14] Then, they obtained the single mode relaxation time in the second Born approximation by considering elastic scattering processes of phonons accompanying virtual excitations of the acceptor-holes. Assuming that the splitting in the quartet is negligible compared with $4kT$, dominant energies at the temperature T, they obtained theoretical curves which fit well the experimental data in the limited temperature range.[14] In the lower temperature region the theory gives less scattering. The deviation arises from the neglect of the splitting: The holes in the split states cause resonance and/or inelastic scatterings.

   Since uniaxial stresses and magnetic fields split the quartet
into two doublets and four singlets, respectively, these external
fields affect the phonon scattering by the acceptor-holes.[15]
The uniaxial stress dependence of heat-pulse transmission in p-Si
or p-Ge is semiquantitatively in agreement with the SM theory.[16]
The deviation appeared in low stress field was explained in terms
of the presence of randomly distributed splittings in the acceptor-
ground state. We should mention, however, that the observed piezo-
thermal conductivity does not agree with the theory.[17,15]

   Concerning with the magnetic field H dependence, the phonon
scattering by the acceptor-holes increases at moderate fields and
decreases rapidly at high field in Si, whereas it is almost independent of field below 55 kG in Ge.[18,19] The striking difference
comes from the H-dependences of the level splittings in Si and Ge
as shown in Fig. 1: In Ge the acceptor ground state is affected by
the second order Zeeman or diamagnetic energy shift, which is proportional to H^2, as well as the ordinary Zeeman splitting.[19]
Then, the thermal phonons are scattered staggeringly by causing
hole transitions between different pairs of levels as varying H in
Ge. In fact, the magnetothermal conductivity is explained semi-
quantitatively by extending the SM theory as taking account of the
above situation and the initial splittings in the ground state.[20]

   As mentioned above, it has turned out that the splittings in
the ground state are substantial for the scattering of the phonons
whose energy is less than a few degrees of K. To pick out the
small splittings effectively, ultrasonic absorptions are useful
since the energy is comparable with or less than the splittings.
In fact, the observed ultrasonic absorptions in p-Si[21-23] and p-
Ge[20] are not explained unless taking account of the internal
splittings.[24] SM calculated the absorption by using the formulae
derived by thermodynamic Green's function method by Kwok [25].[24]

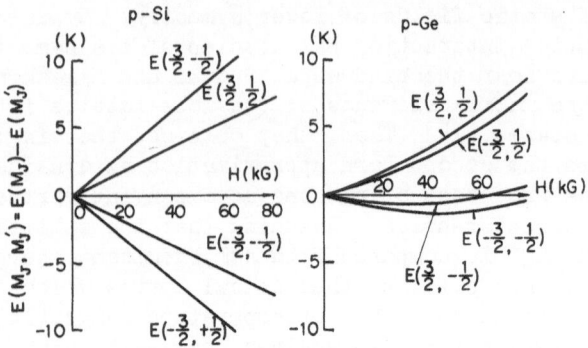

Fig. 1. Calculated splitting energy in acceptor-ground state in
Si and Ge as a function of the magnetic field along the [001] axis.
Here the initial splitting is taken to be zero. Physical constants
are qouted from refs. [27] and [19].

When uniaxial compressional stress is applied, the ultrasonic absorption decreases, and it is quenched with the stresses of 10^8 dyn/cm^2 for Si and 10^7 dyn/cm^2 for Ge at 4.2 K, since the splitting becomes too large to alter occupation number of upper level appreciably.[26,20] If we apply the magnetic field instead of the uniaxial stress, the ultrasonic absorption is also quenched by the magnetic field of 30 kG in p-Si.[27] In p-Ge we found that the resonance absorption associated with the level crossings as shown in Fig. 1.[19] In Fig. 2, the variation of the resonance behavior is shown as a function of the uniaxial compressional stress applied simultaneously.[20] By assigning the peaks to the relevant level crossings, we found that the $M_J=\pm 1/2$-like states stay above the $M_J=\pm 3/2$-like states, based on the analysis using the SM theory.[20]

The splittings are induced by the static field which breaks the tetrahedral symmetry at the acceptor sites. In the absence of the external fields, the splittings may be caused by internal local stresses which are brought about by randomly distributed crystal imperfections and dislocations as well as weak correlations among distributed impurities. In order to find physical origins of the internal splittings, the effects of the impurity content and the dislocation density to the ultrasonic absorption have been studied in p-Si [22] and the following facts are found. (1) The presence of the oxygen atoms with content of less than 10^{18}cm^{-3}, which are supposed to act as point imperfections, does not give appreciable effect to the absorption and hence to the splittings. This fact seems to be explained by the fact that the strain field around the point imperfection decreases rapidly with r^{-3} where r is the distance from the source. Actually the calculated strain field after Stoneham [28] is much smaller than the observed. (2) The dislocation density of 10^4cm^{-2} affects the temperature dependence of the

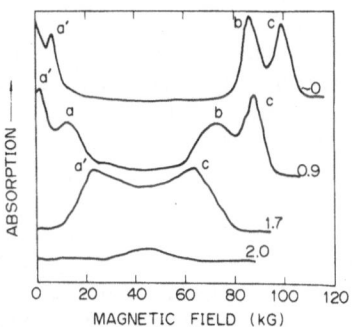

Fig. 2 Uniaxial compressional stress ($//$[001]) dependence of the resonance absorption of 1.11GHz fast-transverse acoustic wave propagating in the [110] axis in Ga-doped Ge with 3.5×10^{15}cm^{-3} at 4.2 K. The magnetic field was applied along the [001] axis. The numbers represent the stresses in unit of 10^7dyn/cm^2. The letters assign corresponding peaks.

ultrasonic absorption and hence the splittings. This may be related
to the fact that the strain field around the dislocation decreases
with r^{-1} and may reach far distance. (3) The splitting increases
with the acceptor concentration. The spread of the wave function ψ
is approximately represented by $\exp(-r/a^*)$ where a^* is the effective
Bohr radius and weak perturbations to the acceptor-ground state may
occur due to the overlapping of ψ's at tails even when the acceptors
are separated by several times of a^*.[29] The average interaction
strength will be stronger as increasing the content, since the pairs
of the atoms which get close increase.

To calculate the ultrasonic absorption, SM take the splittings
into account by assuming that normal stresses along the <111> axes
exist locally, whose magnitudes are distributed (e. g. expressed by
the Lorentzian).[24] Though this assumption makes the calculation
tractable, the selection of the stresses along the <111> axes is
not conventional. Then, if we take the stresses along the <100>
axes, the proportional constants of the relaxation and the resonance
absorptions [24] are changed. Thus, on fitting the experimental
data,these proportional constants are regarded as adjustable parameters which may be related to the detailed structure of the internal splittings.

So far we have discussed the observed ultrasonic absorption in
terms of the SM theory. Schad and Lassmann [30] criticized the SM
formula saying that in the classical limit it does not agree with
the Zener relaxation formula. In fact the formula does not take
account of the local equilibrium of the holes and, therefore, seems
to be valid for $\omega\tau>1$ where ω is the angular frequency of phonon and
τ the relaxation time of holes. To take the local equilibrium into
account, the density matrix formulation may be useful.[31] We point
out, however, that the SM formula is useful even when $\omega\tau\leq1$: When the
instantaneous equilibrium of the holes is taken into account after
Karplus and Schwinger [32], the resonance absorption near resonance
condition and the relaxation absorption are substantially unchanged.
[33] Since the contribution of the tail of the resonance absorption
appearing at off-resonance condition is not significant in our case,
the observed absorption seems to be well described by the SM theory.

We have revealed that the internal splitting is essential for
the scattering/absorption of low frequency phonons. We are in a
dilemma, since practically the theory cannot take account of the
randomly distributed internal splittings as they exist. Nevertheless, it is conceivable that the essential aspects of the acceptor-
hole-phonon interaction have been grasped. We conclude this paper
by pointing out that in real crystals the local splittings may be
closely related to habits of crystals and impurities and processes
of crystal growth; the aggregation of the acceptor atoms such as
precipitating on dislocations or clustering.

The authors are much obliged to Dr. K. Suzuki for helpful discussion and Professor N. Mikoshiba for useful communication.

References
1. Keyes R W 1961 *Phys. Rev.*112 1176
2. Griffin A and Carruthers P 1963 *Phys. Rev.* 131 1976
3. Pomerantz M 1967 *Phys. Lett.* 24A 81
4. Perlman N and Goff J F 1967 *Phys. Lett.*25A 480
5. Suzuki K and Mikoshiba N 1971 *J. Phys. Soc. Jpn* 31 186
6. Carruthers J A, Geballe T H, Rosenberg H M and Ziman J M 1957 *Proc. Roy. Soc.* A238 502
7. Carruthers J A, Cochran J F and Mendelssohn K 1962 *Cryogenics* 2 160
8. Holland M G and Neuringer L J 1962 *Proceedings of International Conference on Semiconductor Physics, Exeter* p.474
9. Challis L J, Cheeke J D N and Williams D J 1965 *Proceedings of 9th International Conf. on Low Temperature Phys.* p.1145
10. Challis L J, Cheeke J D N and Harness J B 1962 *Phil. Mag.*7 1941
11. Poujade A M and Albany H J 1969 *Phys. Rev.*182 802
12. Holland M G 1964 *Proceedings of International Conference on Semiconductor Physics, Paris* p.173
13. Kohn W 1955 *Solid State Physics* 5 ed. by Seitz F and Turnbull D (Academic, New York) p.257
14. Suzuki K and Mikoshiba N 1971 *Phys. Rev.* B3 2550
15. Suzuki K and Mikoshiba N 1971 *J. Phys. Soc. Japan* 31 44
16. Fjeldly T A, Ishiguro T and Elbaum C 1973 *Phys. Rev* B71 1392
17. Sladek R 1962 *Proceedings of International Conference on Semiconductor Physics, Exeter* p.35
18. Challis L J and Halbo L 1972 *Phys. Rev. Lett.*28 816
19. Tokumoto H, Ishiguro T, Kajimura K, Inaba R, Suzuki K and Mikoshiba N 1974 *Phys. Rev. Lett.* 32 717
20. Tokumoto H and Ishiguro T, unpublished
21. Mason W P and Bateman T B 1964 *Phys. Rev.*134 A1387
22. Ishiguro T 1973 *Phys. Rev.* B8 629
23. Ishiguro T and Tokumoto H 1974 *J. Phys. Soc. Japan* 37 1716
24. Suzuki K and Mikoshiba N 1972 *Phys. Rev. Lett.* 28 94
25. Kwok P C 1966 *Phys. Rev.* 149 666
26. Ishiguro T, Fjeldly T A and Elbaum C 1972 *Solid State Commun.* 10 1039
27. Ishiguro T, Suzuki K and Mikoshiba N 1970 *Proceedings of Inter. Conf. on Semiconductor Physics, Warsaw* p.1239
28. Stoneham A M 1969 *Rev. Mod. Phys.*41 82
29. Sonder E and Schweinler H C 1960 *Phys. Rev.*117 1216
30. Schad H and Lassmann K 1974 *Microwave Acoustics* ed. by Dobbs and Wigmore (Institute of Physics, London) p.191
31. Isawa Y and Mikoshiba N, private communication
32. Karplus R and Schwinger J 1948 *Phys. Rev.* 73 1020
33. Suzuki K, private communication

THE EFFECT OF MAGNETIC FIELD AND UNIAXIAL STRESS ON THE THERMAL

CONDUCTIVITY OF p-GERMANIUM

L J Challis, S C Haseler, M W S Parsons and J Rivallin

Department of Physics, University of Nottingham

University Park, Nottingham NG7 2RD, U K

Measurements have been made in the temperature range 1 to 4K of the thermal conductivity of p-germanium under uniaxial compressions up to 10^9 dynes cm^{-2} and in magnetic fields up to 130kOe, both applied along a <110> axis. Measurements were also made in zero field and stress in the range 0.1 to 20K. The samples were doped with gallium or indium in the range 10^{14} to 10^{16} cm^{-3} and had dislocation densities of $\sim 10^4$ cm^{-2}.

These shallow acceptors have a ground state energy level which is a four-fold degenerate Γ_8 quartet in an undistorted site. The quartet splits into two Kramers doublets under stress, and into four singlets in a magnetic field. Thermal phonons are scattered by transitions of acceptor holes within the ground state quartet.

Earlier measurements of the thermal conductivity[1], the magnetothermal conductivity[2], and data from heat pulse[3] and ultrasonic measurements[4], have indicated however that the ground state is more complex than this simple model predicts. The present work aims to investigate the nature and origin of this complexity.

The more highly doped samples have a zero field conductivity which has a T^3 dependence below 1K (Fig 1), and a rather more rapid ($\sim T^4$) dependence above (Fig 2). This indicates strong scattering by the acceptors which is increasing with decreasing frequency down to ~ 3 cm^{-1} and then flattens off. The results above 1K can be explained quantitatively[5] using a relaxation rate for scattering by the degenerate quartet proportional to the acceptor concentration. Above ~ 3 cm^{-1} this ω^2 term shows a rapid cut-off as the phonon wavelength becomes <a^*, the Bohr radius of the scattering centre. Below 1K the ω^2 term is too weak to explain the data, but by incl-

328

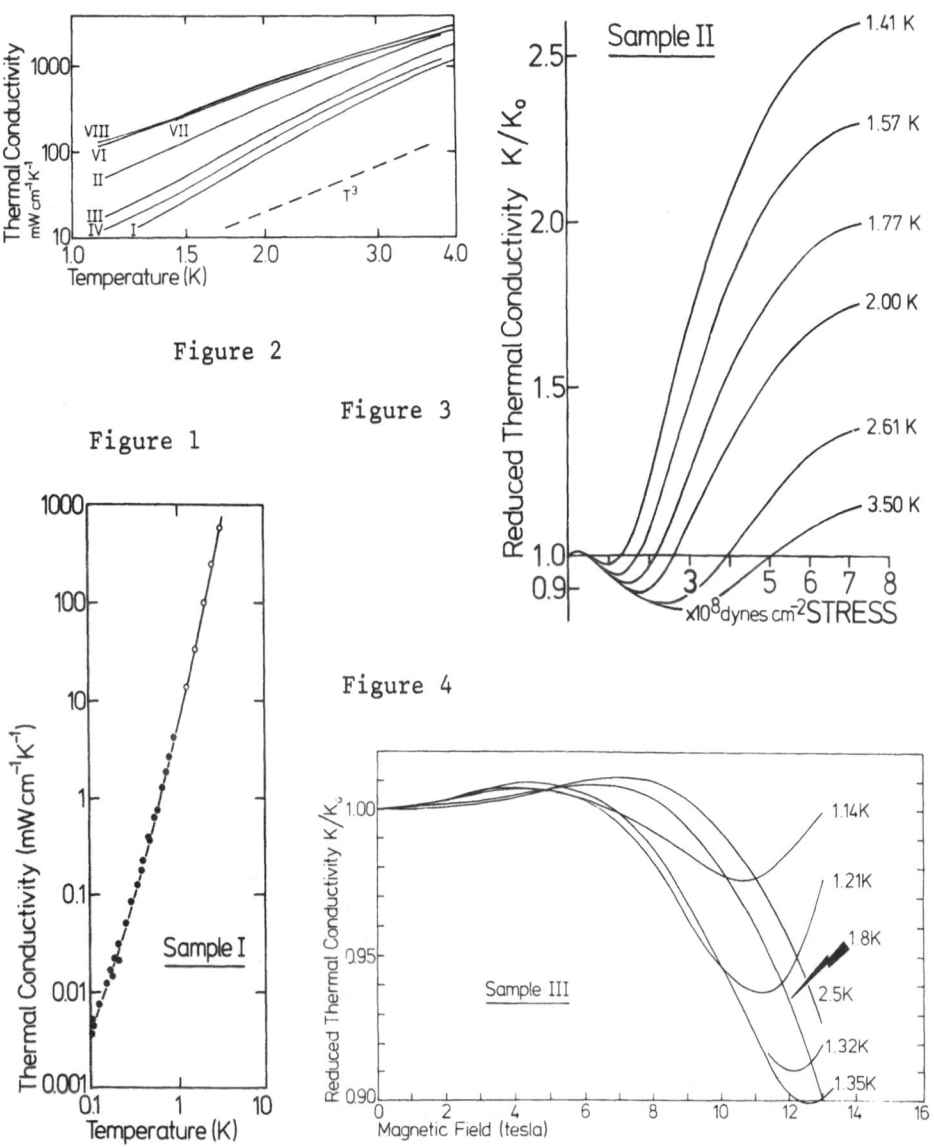

Figure 2

Figure 3

Figure 1

Figure 4

Impurity Concentrations: I:Ga 1.3×10^{16}; II:Ga 2×10^{15}; III:Ga 8×10^{15}; IV:In 1.2×10^{16}; VI:Ga 1×10^{14}; VII: undoped.

uding in the relaxation rate an additional term of square wave form in the frequency range 0 to 1.4 cm^{-1}, good agreement may be obtained. This term represents a square wave distribution of ground state splittings enabling resonant phonon scattering to occur.

The effects of a stress up to $\sim 10^8$ dyne cm^{-2} (Fig 3) and magnetic field up to 130kOe (Fig 4) are broadly similar. In the low
field region (<70kOe, <5 x 10^7 dyne cm^{-2}), the conductivity shows
a slight increase before decreasing to a minimum at higher fields
and stresses. The variation of the conductivity with temperature
at various fixed stresses has been fitted by a scattering term of
the form $D\omega^4/(\omega^2 - \omega_o^2)^2$, where ω_o is taken to be the splitting of
the two Kramers doublets and D the strength of the resonance (both
being used as adjustable parameters). ω_o varies linearly with
stress and extrapolation to zero stress indicates a value $\lesssim 1.5$ cm^{-1},
in agreement with the conclusions drawn from the zero-field conductivity data. In the highest stresses, where ω_o becomes so large
that resonant phonon scattering rapidly diminishes, the conductivity
begins to saturate and approaches the value for pure germanium.

The very small effect in low fields might possibly be understood in terms of a large ground state zero field splitting
(>7 cm^{-1})[2], so that the additional splitting is but a small perturbation. (A splitting of ~ 4 cm^{-1} was assumed to explain data from
heat pulse work[3]). Such values are considerably larger than those
determined from our high and zero field data. The effect might also
be explained in terms of a distribution of splittings, but we have
not yet tested this. (Tokumoto et al[4] assume a Gaussian distribution centred on 0.7 cm^{-1}; work by Laßmann et al[6] indicates a distribution of width 3.5 cm^{-1}). It is far from clear how these large
splittings originate, and there is no evidence for them in optical
data[7]. It can be shown that a dislocation density of 10^4 cm^{-2}
amongst randomly distributed impurities produces a ground state splitting distribution centred at 0.03 cm^{-1} of width 0.03 cm^{-1}, and a
density of ~ 3 x 10^7 cm^{-2} is needed to account for the effects seen
here. The splittings would be larger however if the impurities clus-
tered around the dislocations. The possibility[2] of a Jahn-Teller
effect generating additional low-lying acceptor levels cannot at
present be rejected.

We are very grateful to the Service Basses Temperatures, Centre
d'Etudes Nucleaires, Grenoble where the stress work was carried out.
We mention in particular the help of M B Salce. We are also very
grateful to M M Locatelli and Dr D Waldorf for their help.

1. Carruthers J A, Cochran J F, Mendelssohn K, Cryogenics 2, 160
   (1962).
2. Challis L J and Halbo L, Phys Rev Lett 28, 816 (1972).
3. Fjeldy T, Ishiguro T, Elbaum C, Phys Rev B7, 1392 (1972).
4. Tokumoto et al, Phys Rev Lett 32, 717 (1974).
5. Suzuki K and Mikoshiba N, Phys Rev B3, 2550 (1971).
6. Private Communication. See also these Proceedings.
7. Jones R L and Fisher P, Phys Rev B2, 2016 (1970).

HOLE-PHONON INTERACTIONS IN P-TYPE InSb BY HEAT-PULSES EXPERIMENTS

F.R. Ladan and J.P. Maneval

Physique des solides, Ecole Normale Supérieure

24, rue Lhomond, 75231, Paris 5, France

Heat-pulse experiments in n-type InSb (1) have clearly demonstrated such properties as the coupling of the longitudinal (L) phonons to the free electrons, and the absence of coupling of the transverse (T) phonons, which result from the isotropy of the conduction band. On the contrary, the valence band wavefunctions are of P-type symmetry instead of S-like, and quite different behaviour may be expected in p-type materials, which we consider now.

It is well known that acceptor impurities in excess of 10^{16} cm^{-3} form an impurity band, which degenerates into the valence band for a concentration of about 10^{17} cm^{-3} (degeneracy is not removed at low temperatures). If the phonon-to-free-carrier interaction is handled in the deformable ion model, the matrix element connecting two states $\underline{k}$ and k' is proportional to (2):

$$\int (\vec{q}.\vec{\nabla} u_k(\vec{r})).(\vec{e}.\vec{\nabla} u_{k'}(\vec{r}))d^3r \qquad (I)$$

where $\vec{q}$ and $\vec{e}$ are the wavevector and polarisation of the phonon. The u_k and $u_{k'}$ are the Bloch cell-functions of the initial and final hole states, related by the law of momentum conservation $(\vec{k'}=\vec{k}+\vec{q})$, and assumed to belong to the heavy-hole band. Near the center of the Brillouin zone, and with neglect of the $\vec{k}$-dependence this space can be spanned by linear combinations of $(X+iY)\uparrow$ and $(X-iY)\downarrow$. Then, coupling of the T modes to the holes is allowed due to the presence of non-vanishing integrals of the form:

$$\int \frac{\partial X}{\partial x}.\frac{\partial Y}{\partial y} d^3r \quad , \quad \int \frac{\partial X}{\partial y}.\frac{\partial Y}{\partial x} d^3r \ , \ etc... \qquad (2)$$

in the matrix element(eq.1). Coupling of the L phonons is allowed as well.

Experiments were performed at 1.3 K along the symmetry directions of p-type InSb crystals with dopings (Ge) in the range 10^{14} to 10^{18} holes per cm^{-3}. The heat generator was a gold film driven by current pulses (10^{-7} sec) and the bolometer a tin film biased at the critical magnetic field. Both were insulated from the crystal by evaporated layers of SiO (5000 Å). Fig.1 shows heat-pulse signals in the (110) direction. Only the doped sample (lower trace) corresponds to a degenerate hole distribution. We note a complete vanishing of the L signal, and a strong attenuation of the FT(fast) and ST(slow) phonons, since ballistic propagation lasts hardly longer than 3 millimeters. From the width of the diffusion bump, a mean free path of 0.15 mm can be deduced for T phonons, in good agreement with thermal conductivity data (3). An even shorter free path for L phonons is probable. This very strong attenuation, as compared to n-doped InSb (fig.2), is due to the ratio of the effective masses (0.4 m_o for heavy holes; 0.02 m_o for electrons). On the other hand, the enhancement of ST compared to FT in the p-doped sample is understandable, since FT phonons induce piezoelectric effects, in addition to the short-range interaction (Eq.1).

Detection by superconducting tunnel junctions provides additional information about the frequency composition of the transmitted heat pulse. The gap of tin stands at 1.145 meV= 13.3 K in phonon energy, which implies that phonons of lower energy are not detected. Let us consider Fig.3(lower part) where the signal (B) was taken by a bolometer, (J) by a tin junction.

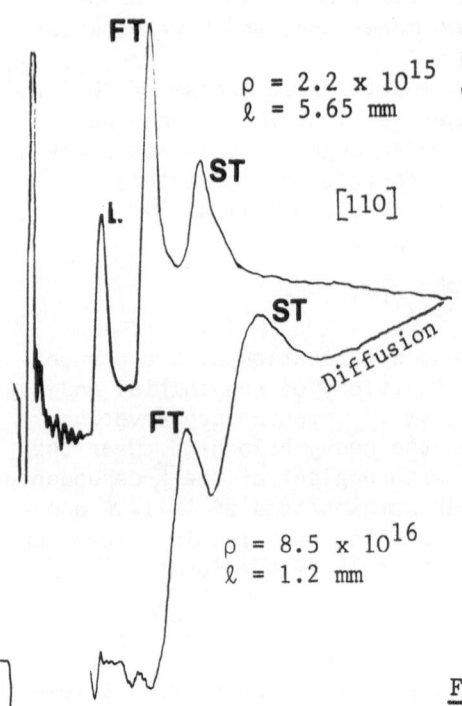

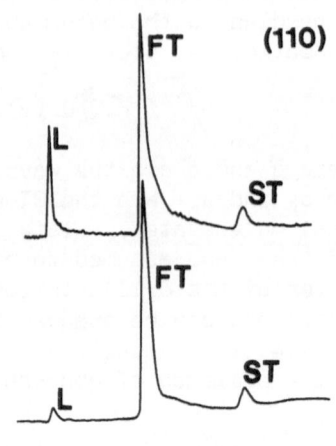

$\rho = 2.2 \times 10^{15}$
$\ell = 5.65$ mm

$\rho = 8.5 \times 10^{16}$
$\ell = 1.2$ mm

Fig.1 Heat-pulse signals versus time in p-doped InSb.

Fig 2 Heat pulses in n-type InSb n= 1.9×10^{14} cm^{-3} (top trace), and n= 6.5×10^{17} cm^{-3} (lower trace) Note the sample lengths ~18 mm.

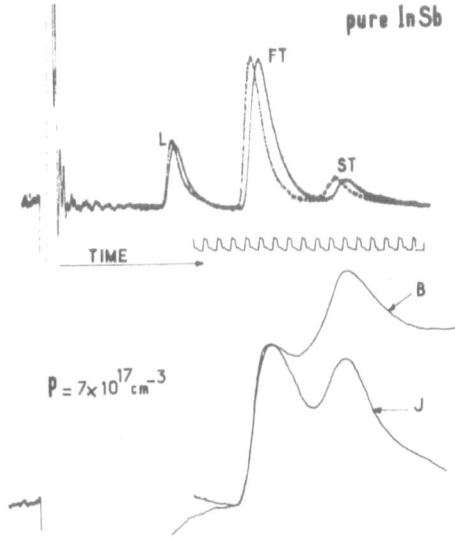

Fig.8 Top: heat pulses in pure InSb taken with a tin junction (full curve) and a bolometer (dotted curve).
Bottom: p-doped sample 1.54 mm long. The L pulse is absent; only FT and ST phonons are visible.

It is clear that the high energy (above 2Δ) phonons detected by the tin junction are less diffused than the low frequency phonons appearing in the bolometric detection. This is consistent with the cut-off of the phonon-free carrier interaction at q= 2k_F (phonon wavevector= diameter of the Fermi sphere), already observed in the heat conductivity of p-InSb (3), and in the heat-pulse transmission of n-InSb (1). Another confirmation comes from consideration of chromatic dispersion: it has been shown (4) that the frequency-dependent time-of-flight of thermal phonons can be resolved in heat-pulse experiments. Thus, (see top of Fig.3) the bolometer signal of a pure InSb sample is advanced with respect to the junction, as the junction detects only dispersed phonons. No such advance is observed in the highly doped sample, where all phonons of low frequency (q<2k_F) have been absorbed by the holes.

Further frequency and time-of-flight measurements with super-conducting tunnel junctions are planned. In the same manner, the use of the complete wavefunctions (5) should provide a more exact treatment of the hole-phonon interactions.

REFERENCES :
1- J.P. Maneval and D.Huet, Phys.Letters,48A,463 (1974).
2- W.Zawadzki and W.Szymańska, Phys.Stat.Sol.,45,415 (1971).
3- C.R.Crosby and C.G.Grenier, Phys.Rev. B4,1258 (1971).
4- D.Huet, B.Pannetier and J.P.Maneval, this conference.
5- G.L.Bir and G.E.Pikus, Soviet Physics-Solid State,3,2221 (1962).

ULTRASONIC ATTENUATION IN P-TYPE GERMANIUM

E. Ortlieb, Hp. Schad, and K. Lassmann

Universität Stuttgart, Physikalisches Inst., Teilinst. 1

7 Stuttgart 80, Germany

The ground state of shallow acceptors in cubic semiconductors is fourfold degenerate, the degeneracy being partly lifted by random local fields. Such a distribution $N(\delta)$ of level splittings δ may be probed by resonant interaction ($\hbar\omega = \delta$) of an ultrasonic wave. For an evaluation, the theory of the distributed-two-level-model for the ultrasonic attenuation of glasses should be applicable /1/. If the distribution $N(\delta)$ does not vary much within the linewidth of the resonance one obtains for the acoustic attenuation at low intensities

$$\alpha = (N(\delta)h/4\varrho v^3 k_B)\ D^2\ (\omega^2/T) \quad \text{with} \quad \int_{-\infty}^{+\infty} N(\delta)d\delta = N$$

Here, N is the concentration of neutral acceptors, ϱ the density of the crystal, v the sound velocity and D an appropriate combination of the two deformation potential constants of the acceptor accounting for an averaged interaction with the various "sorts" of resonance splitting. D is difficult to estimate but should be the same combination under analogous conditions. Thus, if $N(\delta)$ is constant over the frequency range measured the attenuation should vary as ω^2/T. At high acoustic intensities the attenuation should become amplitude dependent due to saturation of the levels in resonance with the ultrasonic wave /1/ /2/.

In germanium, there is some evidence that the width of the distribution may be rather large, even in clean crystals with low doping. We have, therefore, investigated the resonance attenuation of longitudinal sound waves in several p-type Ge crystals (see Table) with CdS transducers evaporated onto the samples.[+] Magnetic fields up to 7 kG have been applied perpendicular to the propagation direction. The intensity dependent part of the attenuation was determined in such a way as to eliminate nonlinearities (which at any

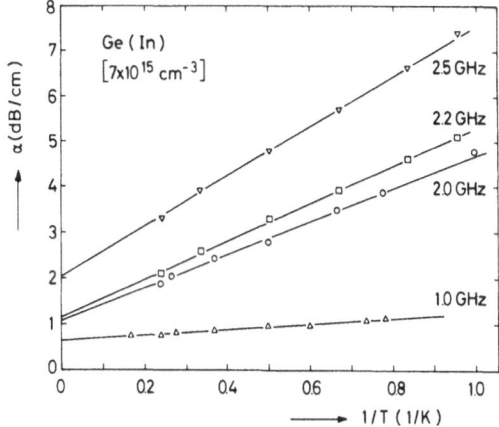

Fig.1 Dependence of the attenuation on 1/T for low intensities
      at several frequencies.

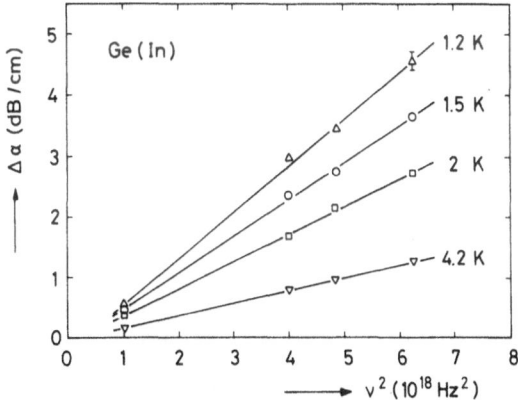

Fig.2 Dependence of the attenuation on ν^2 at several temperatures
      (apparent residual attenuation subtracted).

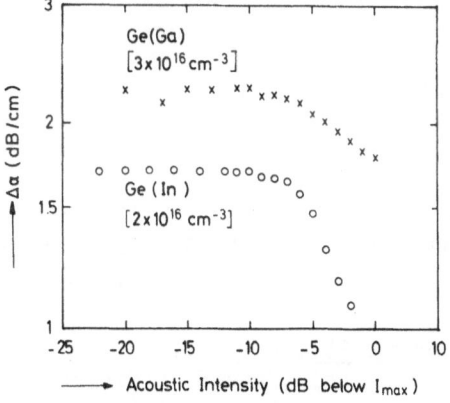

Fig.3 Power dependent part of the ultrasonic attenuation at 1 GHz.

rate are small) of the receiver by working at the same signal level
by means of a calibrated attenuator at the receiver input. The absolute value of the acoustic intensity could only roughly be estimated. The highest intensity at 1 GHz may be of the order of 1 mW/cm^2.

For all crystals the attenuation is $\propto 1/T$ at low temperatures
for low intensities. (Fig. 1). Extrapolating $1/T = 0$ we get an apparent residual attenuation due to geometrical effects. If this is subtracted, the attenuation at constant temperature is $\propto \omega^2$ (Fig. 2) .
This means that $N(\delta = \hbar\omega)$ is constant in the frequency range measured.
For a rough estimate of the width of the distribution from the absolute value of the attenuation we may assume $N(\delta)$ to be constant up to
a maximum value δ_o. The values δ_o for the different crystals obtained
with D = 3.5 eV are listed in the Table. They are consitently rather
large, but may be smaller, if $N(\delta)$ is peaked at some nonzero value
/3/ or if D has to be reduced. However, from an analogous estimate
for GaAs(Mn) we obtain δ_o = 0.04 meV and $N(\delta) \sim$ constant up to
$\delta' \sim 0.01$ meV. Thus, in that case a large reduction of δ_o is not possible because it should not be smaller than δ'.

Crystal	Ge 1	Ge 8	Ge 9	Ge 10
Orientation	[111]	[111]	[111]	[110]
Dopant	Ga	In	In	In
Concentration (10^{15} cm^{-3})	30	7	20	8
Dislocation density (cm^{-2})	700	1500	0	300
δ_o(meV)	0.6	0.5	0.3	0.3

(Ge 10 has been kindly supplied by S.C. Haseler, Univ. of Nottingham, a sample from his piezo-and magnetothermal measurements.)

At higher intensities for all crystals we find an intensity dependent attenuation due to saturation of the resonant levels /2/ reported here for the first time for the case of the acceptor ground
state (Fig. 3). Such a saturation has not been seen for GaAs(Mn) but
may be responsible for some variance in the low temperature part of
the attenuation published in literature.

Magnetic field dependence up to 7 kG was very small (just outside the experimental error) and varied from crystal to crystal. We
did not find a Raman relaxation peak of the attenuation expected near
20 K. From our estimate it should be rather broad due to the extended
acceptor wave function and in fact barely discernible from the T^4
rise of the attenuation due to phonon - phonon interaction.

+We are indepted to W. Arnold for his help with the transducers.

/1/ J. Jäckle, Z. Phys. 257, 212 (72)
/2/ S. Hunklinger, W. Arnold, and S. Stein, Phys. Lett. 45 A, 311(73)
/3/ B. R. Watts Journ. Phys. C 6, 1930 (72).

ULTRASONIC ATTENUATION DUE TO THE NEUTRAL ACCEPTOR INDIUM IN SILICON

Hp. Schad and K. Lassmann

Universität Stuttgart, Physikalisches Inst., Teilinst. 1

7 Stuttgart 80, Germany

We have measured the temperature and frequency dependencies of the u.s.-attenuation of a longitudinal u.s. wave in [111] excited by an evaporated CdS-transducer in the frequency range 0.38 to 4 GHz in silicon doped with the deep acceptor indium ($p = 5 \cdot 10^{15}$ cm^{-3}) for comparison with the results of the attenuation measurements in GaAs due to the deep acceptor Mn /1/ /2/. There we have seen that this deep acceptor ground state may be characterized by a distribution of splittings δ of the two Kramers levels due to random local fields and with a level 3 meV above, which may be connected to a dynamic Jahn Teller effect.

Fig. 1 shows the main features of our results for Si(In). For high temperatures we observe a T^4-rise due to the phonon – phonon interaction. For low temperatures there is a rise $\propto 1/T$, if the acoustic intensity is low. Such a behavior is expected from the resonance interaction between the u.s. wave and a distribution of split electronic levels at the acceptor ground state. For high acoustic intensities this rise vanishes following the relation $\alpha_{res} \propto I^{-1/2}$ (I = acoustic intensity), as was found in the u.s. attenuation in glasses /3/. This is the first time that such a saturation effect has been seen in the u.s. absorption due to the acceptor ground state.

In the intermediate temperature range there is an attenuation peak, as was found similarly in GaAs(Mn) and also in Si(B) /3/. The peak height is proportional to frequency ν and its maximum temperature shifts with frequency to higher temperatures. We interpret this peak as due to relaxation. For $\delta \lesssim 0.2$ meV the relaxation formula is approximated by /2/:

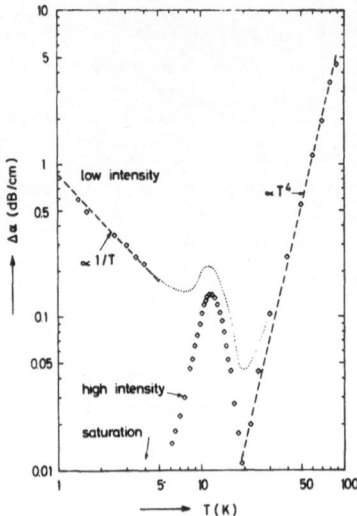

Fig. 1: Temperature Dependence of the Attenuation; 3 GHz, Si(In)

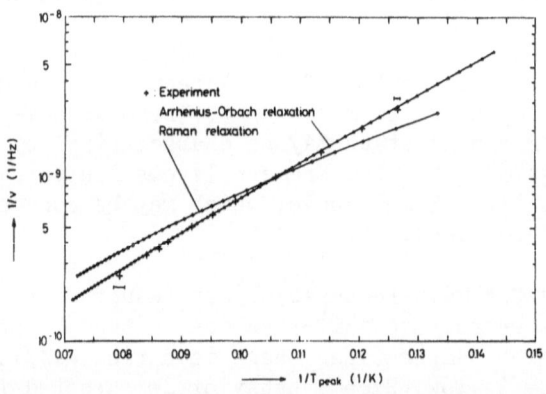

Fig. 2: Frequency Dependence of the Peak Maximum Temperature

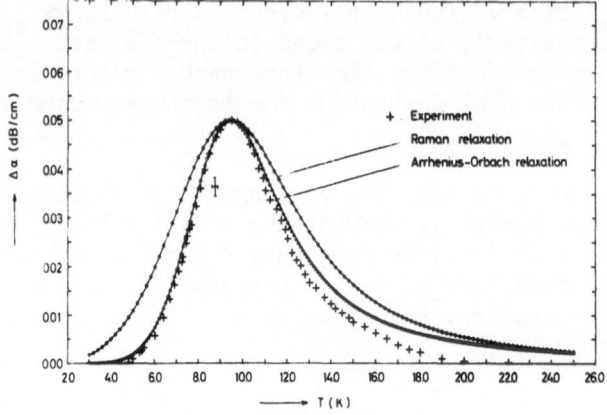

Fig. 3: Relaxation Peak: Comparison of Experiment with Calculations

$$\Delta \alpha_{rel} \propto D'^2/T \cdot \gamma^2 \tau(T)/(1 + \gamma^2 \tau^2(T))$$

where D' is an effective deformation potential and $\tau(T)$ the relaxation time of the corresponding relaxation process. The relaxation time can be determined from the peak maximum (approximate condition $\gamma \tau(T_{max}) = 1$). Measuring at high echo numbers, with the intensity still high enough to saturate and thus eliminate the additional resonance attenuation, we were able to determine the relative attenuation with high resolution recording several echos continuously with varying temperature. T_{max} was accurate to ± 0.05 K for the lower and ± 0.1 K for the higher temperatures. In our calculations (accounting for the 1/T dependence of the prefactor in the relaxation formula) the best fit of the experimental frequency dependence of T_{max} is given by an Arrhenius equation $\tau_A(T) \propto \exp[\Delta /k_B T]$ for the relaxation time (Fig. 2). This indicates an Orbach relaxation process with an excited level $\Delta = 4.2 \pm 0.3$ meV.

In Fig. 2 we have also plotted the frequency dependence of T_{max} to be expected for Raman relaxation with the relaxation time $\tau_R \propto D''^4 T^5 f(T)$ (f(T) = cutoff function, which describes the reduced interaction of the Raman phonons with shorter wavelength than the Bohr radius of 7.3 Å, and D'' = effective deformation potential fitted in Fig. 2 for coincidence with experiment at 1 GHz. Clearly, the Raman relaxation does not describe the experimental results. Also, comparing the whole peak region, a Raman relaxation would be too broad, while an Arrhenius peak fits rather well the observed attenuation (Fig. 3). Furthermore, taking for D'' the experimentally determined value of Si(In), the Raman peak would appear at higher temperatures and would be far too broad. A combination of both relaxations (Raman relaxation with experimental deformation potentials) modifies only to a small amount the Arrhenius-Orbach results.

In summary we find that for Si(In) there is a level 4.2 meV above the ground state as similarly found for GaAs(Mn); it seems that such a level is characteristic for deep acceptors. Preliminary measurements in the far infrared and with superconducting tunnel junctions did not show a line at 4.2 meV, which may be due to the low content of indium, but only the small absorption line of oxygen at 3.6 meV /5/ (content $\approx 10^{18}$ cm^{-3}, estimated from measurements in the near infrared).

References:

/1/ Hp. Schad and K. Laßmann, Conf. on Microwave Acoustics Lancaster (74), Proc. p. 191
/2/ K. Laßmann and Hp. Schad, to be published
/3/ e. g. S. Hunklinger et al., Phys. Lett. 45 A, 311 (73)
/4/ T. Ishiguro and H. Tokumoto, J. Phys. Soc. Japan 37, 1716 (74)
/5/ W. Hayes and D. R. Bosomworth, Phys. Rev. Lett. 23, 851 (69)

PHONON SCATTERING DUE TO DEEP ACCEPTORS IN SEMICONDUCTORS

A. de Combarieu[+] and K. Lassmann[++]

[+]S.B.T., CENG, BP.85, 38041 Grenoble, France

[++]Physik. Institut, Teil 1, 7 Stuttgart 80, Germany

We have measured the magnetothermal conductivity in GaAs(Mn) [3.8×10^{18} cm^{-3}] and Si(In) [5×10^{15} cm^{-3}] for temperatures between 1.4 K and 90 K at magnetic fields up to 8 T. In both cases the dopants are deep acceptors with binding energy much larger (110 meV and 165 meV respectively) than given by the effective mass theory (~35 meV). There is a double interest in such systems: First, an excited level 3 meV (4.2 meV) above the acceptor ground state has been concluded from ultrasonic measurements /1/ /2/. Such an excited state might be connected with a Jahn-Teller effect of these deeper acceptors and should be seen by resonant phonon scattering in thermal conductivity. Second, an anomalous behavior of the magnetothermal conductivity has been found for shallow acceptors in Ge (but not in Si) /3/ making comparison with systems with different g-factors desirable. The g-factors of acceptors in GaAs are roughly three times, the g-factor of Si(In) about 0.6 times that of Si(B).

In zero magnetic field in both cases we see a strong reduction of the thermal conductivity at low temperatures as compared to the pure material scaling well in concentration with the reduction found for shallow acceptors; furthermore, a distinct dip is seen at about 23 K for Si(In) (Fig. 1) and at about 12 K for GaAs(Mn) (Fig. 2). The same dip has been found for GaAs(Mn) by Holland /4/ at concentrations lower than ours. The maximum thermal conductivity for Si(In) is reduced due to about 10^{18} cm^3 oxygen contained in the sample /2//5/. The dependence on magnetic field is analogous to that found by Challis and Halbo /3/ for Si(B): The relative thermal conductivity $\varkappa_B/\varkappa_o$ for GaAs(Mn) first falls to a minimum value of about 0.6 (due to an enlarged resonant scattering) and then rises rapidly (Fig. 3). However, this rise tends to saturate at the highest fields to a value about ten times below that of the pure crystal, which may be due to

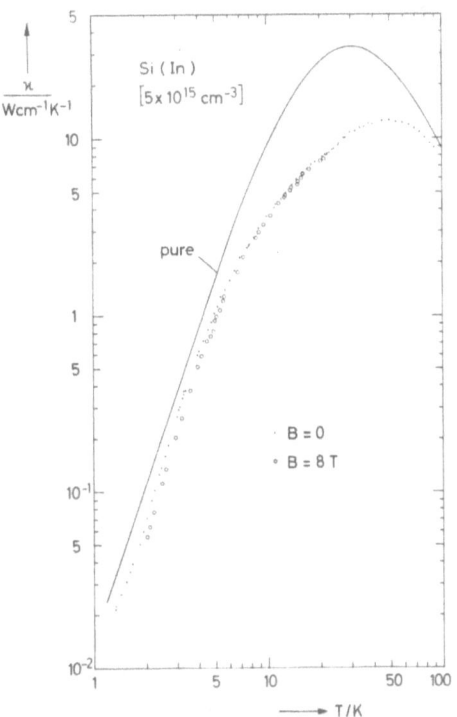

Fig.1 Thermal conductivity in Si(In).

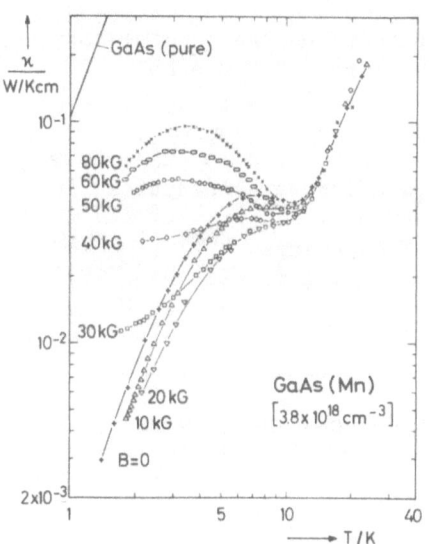

Fig.2 Thermal conductivity in
GaAs(Mn) at several magnetic
field strengths.

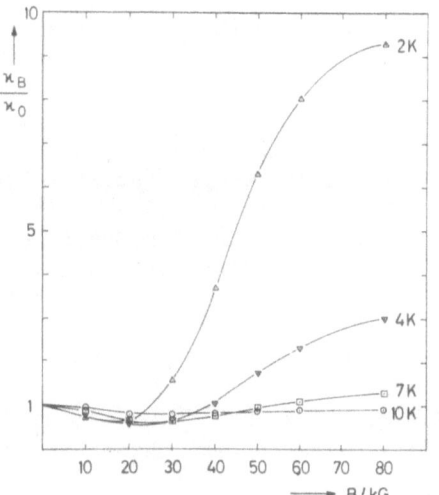

Fig.3 Relative thermal conductivity
$\varkappa_B/\varkappa_0$ in GaAs(Mn) as a function of magnetic field at
various temperatures.

the low frequency part of the resonance scattering associated with the dip at 12 K. As is expected from the small g-factor, the variation of $\varkappa_B/\varkappa_o$ for Si(In) is slow; that is, only a reduction is seen attaining a minimum value of 0.8 at 2 K and 8 T. Thus, in both cases the magnetic field dependence does not show the "anomaly" found for shallow acceptors in Ge. However, the variation of $\varkappa_B/\varkappa_o$ for Si(In) seems to be somewhat slower than can be accounted for by the g-factor and, in the case of GaAs(Mn), the minima of $\varkappa_B/\varkappa_o$ at different temperatures do not scale very well with B/T, but there is a part $\propto (B/T)^2$.

For an analysis of $\varkappa_o$ we applied a modified model for the acceptor ground state: A distribution of small splittings δ (determining the low temperature part of the scattering) and an excited level Δ above these split levels. The extended nature of the acceptor wave function, reducing the interaction with short wavelength phonons, was taken into account (Bohr-radius 10 Å for GaAs(Mn) and 7.4 Å for Si(In)). The formula for resonance fluorescence scattering in the form given by Suzuki and Mikoshiba /6/ was applied. The assumption $\delta \ll kT$ made for ease of calculation underestimates the scattering at the lowest temperatures. A good fit following the acute variation of $\varkappa_o$ in the dip region especially for GaAs(Mn) was not possible. In both cases it was necessary to take a smaller Bohr radius for the excited level (3 Å for GaAs(Mn)). The best values for $\varkappa_o$ thus obtained are 5 meV for Si(In) and 3 meV for GaAs(Mn). This has to be compared with the values analyzed from ultrasonic measurements. A more direct determination of these energies with quasimonochromatic phonons /7/ is desirable.

One of us (K. L.) gratefully appreciates the cordial hospitality of the group of Service Basses Températures, CEN Grenoble.

References:

/1/ Hp. Schad and K. Laßmann, Conf. on Microwave Acoustics, Lancaster (74) p. 191
/2/ Hp. Schad and K. Laßmann, these Proceedings
/3/ L. J. Challis and L. Halbo, Phys. Rev. Lett. 28, 816 (72)
/4/ M. G. Holland, Proc. 7th Int. Conf. Semiconductors, Paris (64), p. 713
/5/ M. G. Holland, Proc. Int. Conf. Semiconductors, Prague (60) p. 633
/6/ K. Suzuki and N. Mikoshiba, J. Phys. Soc. Jap. 31 44 (71)
/7/ W. Eisenmenger, these Proceedings.

SCATTERING OF THERMAL PHONONS IN P DOPED Si

D.FORTIER and K.SUZUKI[*]

DPh-EP,Laboratoire de Physique des Materiaux

C.E.N.SACLAY,91190 Gif-sur-Yvette,France

In this paper,we have investigated the thermal conductivity of P-doped Si in the concentration range
2.5×10^{14} to 10^{18} cm^{-3}.It is shown that,in the low concentration region,the phonon scattering by isolated donors
is weak and a theory can fairly explain the experimental
data of K(T),while in the concentration region higher to
4.7×10^{17} cm^{-3},the thermal resistance becomes larger than
that predicted by the theory.

EXPERIMENTAL and THEORY

The technique of the thermal conductivity measure-
ments and its accuracy are the same as those described in
Ref.1.The symbols,the donor concentrations n ($\times 10^{16}$ cm^{-3})
and Casimir's length L (cm) of samples are as follows:
N°1(▼)n=0.025,L=0.301 - N°2(O)n=0.37,L=0.303 - N°3(Δ)
n=0.65,L=0.284 - N°4(□)n=7.5,L=0.277 - N°5(▽)n=25,
L=0.302 - N°6(▲)n=47,L=0.308 - N°7(●)n=100,L=0.261.
The measured thermal conductivity is shown in Fig.1.

In the effective mass approximation[2],the ground
state of a shallow donor in Si is 6-fold degenerated,
reflecting the six equivalent conduction band minima.This
degeneracy of the ground state is partially split into a
singlet (A_1) a doublet (E) and a triplet (T_2) by the
valley-orbit interaction and the central cell correction.
The singlet is the lowest state and the triplet is higher
by 11.7meV and the doublet lies above the triplet by
1.35 meV[3] .

From the symmetry argument,it is shown that only the
matrix elements of donor-phonon interaction among the A_1
and E states have non zero values[4].And we have found
that,for P donor (also for Sb and As donors),the relaxation rates of phonons by the inelastic scattering and
"thermally assisted"phonon absorption contribute hardly
to the total relaxation rate.Then we need consider only
the elastic phonon scattering by P donors.The relaxation
rate by this process is given by Eq.(10) in Ref.1.
We have calculated K(T) by using the usual semiphenome-
nological[4] expression for the lattice thermal conductivity.
We have assumed the additivity of the relaxation rates
due to various scattering processes,i.e,boundary scattering,isotopic scattering,Umklapp and normal processes of
phonon-phonon interaction and electron-phonon interaction.

COMPARISON and DISCUSSION

        It is noted from Fig.1 that,in samples with donor
concentrations lower than 2.7×10^{17} cm^{-3},the phonon scattering at lower temperatures is very weak in contrast with
Group-V donors in Ge[4,5].This can be seen from ultrasonic
attenuation by neutral donors[6].The strong scattering of
thermal phonons observed in Li-doped Si is caused by the
degenerate lowest state and the small valley-orbit splitting of the ground state[1].The weak scattering of P donor
(also Sb and As donors) in Si is chiefly due to the large
valley-orbit splitting.
Fig.2 shows the comparisons of the calculated thermal
conductivity,K(T),with the experimental data for samples
N°4,5 and 6,respectively.The theory can fairly explain
the experiment in the concentration region lower than
2.7×10^{17} cm^{-3}.
In sample N°6 the phonon scattering is strong even at
low temperature and the calculated K(T) cannot reproduce
the experimental data.It seems,at this concentration,
that P donors cannot be considered to be isolated and
thus one may expect the existence of an interaction[7]
among donors.This should give rise to a modification of
the donor ground state and then to a phonon scattering
different from that by an isolated donor center.In addition phonon scattering by clusters might perhaps occur.
Furthermore,if some compensation exists,phonon scattering
by homopolar or polar pairs can be expected.At present
time,it is not clear if the scattering mecanisms mentioned above give the temperature dependence of K.

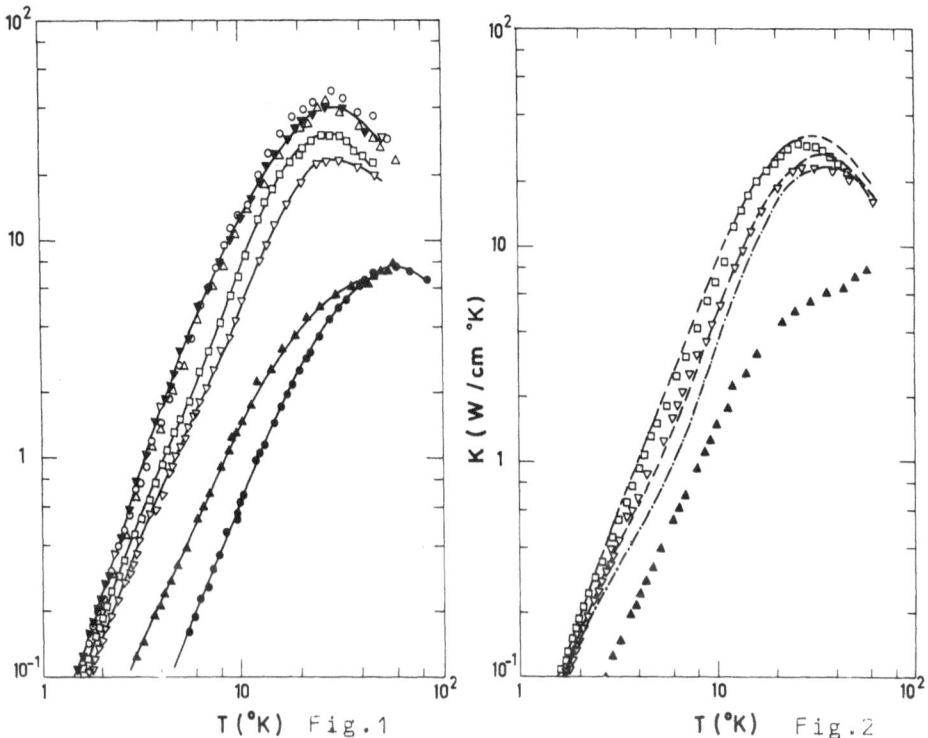

Fig.1.Thermal conductivity of P doped Si as a function
      of temperature.The samples were cut in the shape
      of rectangular parallelepipeds with a length of
      35mm along the <111> direction from cristals grown
      by Floating-zone technique

Fig.2.Experimental and calculated K for samples N°4,5 and
      6 (□━━/▽━━/▲━━/respectively).The values of phy-
      sical parameters used in the calculation of K(T)
      are the same as in Ref.1,withΞ_i=10eV and Δ=13meV.

REFERENCES

✸Permanent address:Dept of Electrical Engineering,Waseda
University,Shinjuku,Tokyo.
1) D.Fortier and K.Suzuki,Phys.Rev.B9,2530 (1974).
2) See W.Kohn,in Solid State Phys.edited by F.Seitz and
   D.Turnbull (Academic Press,New York,1957)vol.5,p.257.
3) R.L.Aggarwall and A.K.Ramdas,Phys.Rev.146,A1246(1965).
4) A.Griffin and P.Carruthers,Phys.Rev.131,1976 (1963).
5) See,for exemple,K.Suzuki and N.Mikoshiba,in Proc.Int
   Conf.on Phonon scattering in Solids,Paris,1972,ed.
   H.J.Albany,p.125.
6) M.Pomerantz,Phys.Rev.B1,4029 (1970).
7) S.Maekawa and N.K.Mikoshiba,J.Phys.Soc.Japan,20,1447
   (1965).

DISCUSSION ON THE LOW TEMPERATURE MAGNETOTHERMAL CONDUCTIVITY

IN LIGHTLY DOPED Ge(Sb) AND Ge(As)

Leif Halbo
Dept. of Elect. Eng., University of Nairobi, Kenya
After Sept. 1, 1975: Central Institute for Industrial
Research, Blindern, Oslo 3, Norway

Some time ago we reported[1] measurements of magnetothermal conductivity in n-type Ge in the temperature range $1.3 \lesssim T \lesssim 4.4$ K, in fields $B \lesssim 8$ Tesla, over the whole range of Sb and As doping concentrations. Only qualitative suggestions were given for interpretation, due to lack of adequate theory at the time. In the present paper we give an analysis by computer of the case of low doping concentration, where interaction between neighbouring donors may be neglected. Briefly, the experimental results for this case are as follows[1]: For Ge(Sb) with the field B ‖ <100> the relative change in conductivity, $\Delta K(B)/K_o$, is negative, increases in magnitude approximately linearly with B, and for a given B the magnitude is larger for the lower temperatures. For $B \gtrsim 4$ Tesla there is large anisotropy: $\Delta K/K_o$ becomes positive for higher fields when B‖ <111> or <110>. For Ge(As) $\Delta K/K_o$ is always positive, increasing approximately linearly with B. There is only a weak T- dependence, and small anisotropy.

Our analysis includes the scattering mechanisms considered by Suzuki and Mikoshiba[2] (SM): elastic and inelastic phonon scattering by donor electrons, isotope and boundary scattering. The SM theory can be adapted to include the presence of a magnetic field along <100> by introduction of the B-dependence of the donor ground state effective Bohr radius a* and the chemical shift 4Δ. Physically, the extension of the electron orbit will be reduced, as the electron gets more tightly bound to the donor atom in the presence of the field; the scattering cut-off phonon frequency then increases[1]. The chemical shift is due to the "central cell correction". Since the probability of finding the electron near the donor nucleus increases when B increases, 4Δ will increase with B. The field dependence of 4Δ thus increases the resonance frequency, where the scattering is strongest.

346

Variation calculations have been made[3],[4] to find quantitatively how a magnetic field affects the Ge donor ground state.
For the relatively weak fields in question here ($\gamma = \hbar\omega_c/2\text{Ryd} < 1.3$
for B along conduction band valley axis,) the envelope of each wave
function component is assumed to have the form[3]:

$$\psi = (\pi\, a_t^2\, a_l)^{-1/2}\, \exp\{-[(x^2+y^2)/a_t^2 + z^2/a_l^2]^{1/2}\} \qquad (1)$$

where a_t (B) and a_l (B) are determined by the variation calculation.
The effective field for B∥<100> is found using Eq. (15) of Ref.4.
The effective Bohr radius is now taken as $a^* = (a_t^2 a_l)^{1/3}$. The
chemical shift is proportional to the electron probability density
at the donor nucleus[4], i.e. $4\Delta \propto (a_t^2 a_l)^{-1}$. Fig. 1 shows the above
parameters as functions of B, relative to their zero field values.

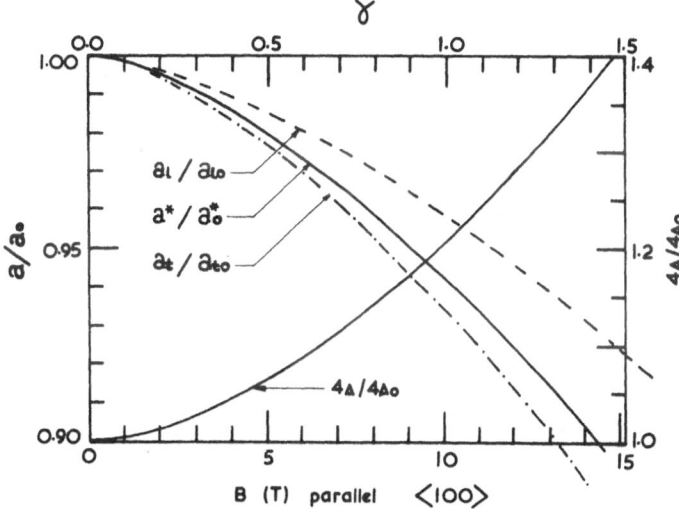

Fig. 1: The magnetic field dependence of the parameters a_t and a_l
in Eq. (1), the effective Bohr radius a^* and the chemical shift 4Δ.

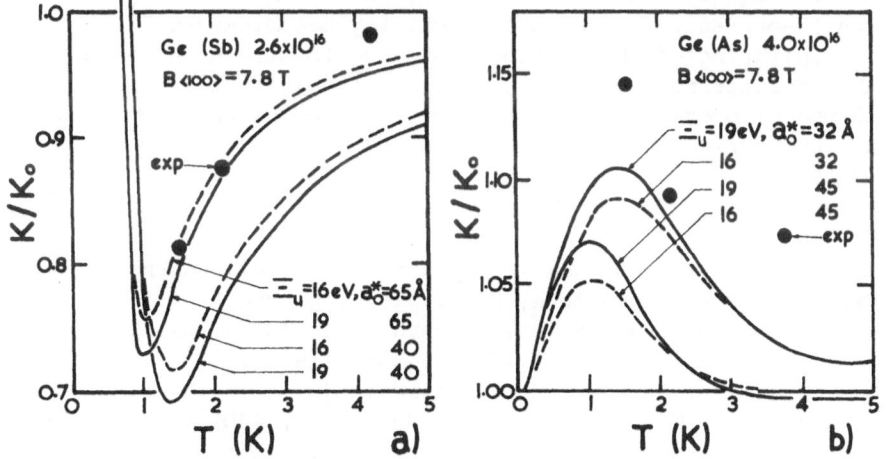

Fig. 2:   Calculated relative thermal conductivity at $B_{<100>} = 7.8$
Tesla, vs. T, for a) Ge(Sb), b) Ge(As). Points are experiment.[1]

The results of Fig. 1 were used to calculate the various scattering rates and the thermal conductivity by the expressions of SM. The values used for the shear deformation potential cons- tant Ξ_u and the zero field Bohr radius a_o^* are in the range usually found in the litterature, values of other parameters are as in Ref. 2. Some of the results for $K(B)/K_o$ are shown in Fig. 2.

The following conclusions may be drawn from the calculations: The effects of the magnetic field through a^* and through 4Δ are of comparable magnitude. The calculated $\Delta K/K_o$ with both effects included is negative for Ge(Sb) but positive for Ge(As), and the field dependence is approximately linear, in agreement with experiment. The temperature dependence agrees qualitatively with experiment, but the magnitude of $\Delta K/K_o$ calculated for Ge(As) is smaller than that measured (see Fig. 2).

Detailed computations for orientations of B other than <100> have not been made, since the use of a Hydrogen-like donor wave function[2] then is not acceptable, and since the degeneracy of the donor triplet state will be lifted, necessitating more modifications of the SM theory. However, by numerical diagonalization of the Hamiltonian matrix, where deviations from the effective mass approximation are included[3], we find that the splitting of the triplet becomes larger than $4\Delta_o$ for $B \gtrsim 4$ Tesla along <111> or <110> in the case of Ge(Sb), where $4\Delta_o = 0.32$ meV. A major decrease in the scattering may therefore be anticipated, as is observed. In Ge(As), with $4\Delta_o = 4.2$ meV, a very much larger field is necessary to give splitting comparable to $4\Delta_o$, so no large anisotropy is expected, - or observed.

The above agreement with experiment is considered satisfactory, in view of the approximations involved in the theory. Some possible reasons for the remaining discrepancy, and also the discrepancy between theory and zero field conductivity measurements [5],[6] are: a) splitting of the triplet state due to stress in the crystals[6], b) other scattering mechanisms, possibly involving two donor atoms[1],[6]
c) use of oversimplified wave function in the calculation of scattering matrix elements, d) inaccuracy in the variation calculation[4].

REFERENCES
1) L. Halbo and R.J. Sladek, Proc. 10th Int. Conf. on Physics of
    Semiconductors, Cambridge, Mass., 1970, p. 826.
2) K. Suzuki and N. Mikoshiba, J. Phys. Soc. Japan, 31, 186 (1971).
3) L. Halbo, Phys. Stat. Sol.(b), 59, 387 (1973).
4) N. Lee et al., J. Phys. Chem. Solids, 34, 1817 (1973).
5) B.L. Bird and N. Pearlman, Phys. Rev. B, 4, 4406 (1971).
6) K. Suzuki and N. Mikoshiba, Proc. 1st Int. Conf. on Phonon
    Scattering in Solids, Paris, 1972, p. 125.

THEORY OF ULTRASONIC ATTENUATION BY HOPPING PAIRS IN n-Ge[+]

Zbigniew W. Gortel

Institute of Theoretical Physics, Warsaw University

ul. Hoża 69, 00-681 Warsaw, Poland

It is known that the attenuation of ultrasound in n-Ge at microwave frequencies of the order of few GHz and in the temperature range from about 7 to 40 K is mainly due to the interaction of acoustic phonons with electrons localized on neutral donors [1,2] . Due to this mechanism the attenuation coefficient is strongly anisotropic and vanishes for longitudinal ultrasonic wave propagating along [100] and for transverse one with $\vec{q} \parallel$ [110] and polarisation along [1$\bar{1}$0] .

Our work is devoted to other possible mechanism of ultrasonic attenuation in lightly doped, partly compensated n-Ge at temperatures lower than 10 K. This attenuation is due to the absorption and emission of phonons interacting with electrons localized in the hopping-pair states. We used the eigenfunctions of an electron bound to the two donor ions in the presence of a compensating acceptor ion which were found by Miller and Abrahams' variational procedure [3] . Many-valley and nonspherical character of the bottom of the conduction band was taken into account. The interaction between phonons and electrons in the hopping pair states was treated in the deformation potential formalism. The rate of change of the number of phonons in the mode $(\vec{q},t)$, due to their interaction with an electron bound to a single pair, was found in the first order time-dependent perturbation theory. In order to sum the above quantity over all pairs present in the crystal, we assume that the temperature is low enough so that we can expect that one partner of each single-ionized donor pair is the donor nearest to the compensating acceptor ion. We assume also that the compensation ratio K is small ($K < 0.2$) so that the hop-

ping process take place on each pair independently. The single
-mode phonon reciprocal relaxation time $\tau_{\vec{q}t}^{-1}$ and the ultra-
sonic attenuation coefficient $\alpha_{\vec{q}t} = 1/2\,\tau_{\vec{q}t}v_{\vec{q}t}$ ($v_{\vec{q}t}$ - group
velocity of phonon) can be derived from the expression for
the rate of change of the number of phonons in the mode $(\vec{q},t)$
due to their interactions with all pairs. Two integrations
(from the six appearing in integration over all positions of
two donors) can be easily made and $\alpha_{\vec{q}t}$ has a form of a sum of
four terms: two of them (T_1 and T_2) being proportional to
fourfold and two (T_3 and T_4)proportional to sevenfold integrals
(in T_3 and T_4 we have two-center matrix elements of exp ($i\vec{q}\vec{r}$)).
The form of all integrals is such that so-called nearly homo-
polar pairs (i.e. pairs with the compensating acceptor ion
situated near the pair symmetry plane bisecting the distance R
between donors) give the main contribution to the attenuation.

For transverse phonons propagating along $[110]$ with polarisation parallel to $[1\bar{1}0]$ the coefficient $\alpha_{\vec{q}T} = 0$ (the same
phonons are also not scattered by neutral donors). For other
transverse phonons propagating along high symmetry directions
$[100]$, $[110]$ and $[111]$ (and also some other) only the term
T_3, proportional to twocentre matrix elements,does not vanish.
For all transverse phonons the attenuation coefficient is much
smaller than for longitudinal phonons with the same energy.
We have to state here that we are not able to calculate the
coefficient $\alpha_{\vec{q}T}$ for transverse waves with enough accuracy.

For longitudinal phonons the scattering due to the hopping
process is always present in contrary to the scattering by the
neutral donors. The integrals $T_1 \div T_4$ can be put into the form
of twofold integrals when the integrands depending on the direction of the vector $\vec{R}$ will be replaced by their averages over
all directions of $\vec{R}$. The very high anisotropy of these integrands were taken into account in order to calculate their
averages. In T_3 and T_4 the term exp ($i\vec{q}\vec{r}$) was approximated by
one. It leads to overestimation of $|T_3|$ and $|T_4|$, but even
for such values of T_3 and T_4 only the term T_1 is significant
and it gives

$$\alpha_{\vec{q}L}[dB/cm] = 4.323\left(16\pi^3 \epsilon\, K N_D^3/\varrho\, v_L^3\, e^2 \hbar\right)\left(X_L(\vec{q})/n\right)^2$$

$$\times \left[\left(\exp(\beta\hbar\omega_{\vec{q}L})-1\right)/\left(\exp(\beta\hbar\omega_{\vec{q}L})+1\right)\right]$$

$$\times \int_{R_{sw}}^{\infty} R\langle W_R^2\rangle\left(1-(\sin qR)/qR\right)E^{-1}\ell(R,E)\,dR,$$

where

$$\ell(R,E) = \int_{R_1}^{R_2} R_a^4 \left(1 - \epsilon R_a E/e^2\right)^{-3} \exp\left(-4\pi N_D R_a^3/3\right) dR_a ,$$

$$R_1 = e^2/\epsilon E + R/2 - \left[(e^2/\epsilon E) + (R/2)^2\right]^{1/2},$$

$$R_2 = -R/2 + \left[(R/2)^2 + Re^2/\epsilon E\right]^{1/2},$$

$$E = \left[\hbar^2 \omega_{\vec{q}L}^2 - 4(W_R^2)_{max}\right]^{1/2},$$

$$X_L(\vec{q}) = \sum_{i=1}^{n} \left(\Xi_d + (\hat{k}^{(i)} \hat{q})^2 \Xi_u\right)\left[1 + (qa/2)^2 - (a^2 - b^2)(\hat{k}^{(i)}\vec{q}/2)^2\right]^{-2} ,$$

$$\langle W_R^2 \rangle = (2n)^{-1}(\pi/\eta)^{1/2} (e^2/\epsilon a)^2 (a/R)^{5/2} \left[1 - 2(R/a)^2/3\right]^2 \exp(-2R/a),$$

$$(W_R^2)_{max} = (\sqrt{2} n^2)^{-1}(e^2/\epsilon a)^2 (a/R)^2 \left[1 - 2(R/a)^2/3\right] \exp(-2R/a),$$

$$\eta = (a^2 - b^2)/b^2.$$

In the above expressions $R_{s\omega}$ is the largest solution of equation $E = 0$, $\epsilon = 16$, N_D, $\rho = 5.4$ g/cm^3 , $n = 4$, $a = 70.8 \cdot 10^{-8}$ cm, $b = 15.91 \cdot 10^{-8}$ cm, $\Xi_d = -10.9$ eV and $\Xi_u = 16.2$ eV are the static dielectric constant, the donor concentration, the number of valleys, larger and smaller Bohr radii of the ground state of donor, and two deformation potential constants, respectively. $\langle W_R^2 \rangle$ and $(W_R^2)_{max}$ are the angular average of the resonance energy squared and a maximal value which W_R^2 reaches for fixed $|\vec{R}| = R$. $\hat{k}^{(i)}$ is the unit versor pointing from the Γ-point of the Brillouin zone to the i-th valley and $\hat{q} = \vec{q}/q$.

The phonon frequency ν considered in this theory is in the range 45 - 100 GHz. For $\nu < $ 45 GHz the expression for the transition probability derived in the adiabatic approximation is not valid; for $\nu > $ 100 GHz the Miller and Abrahams' procedure fails because overlap between donors becomes too large.

In Fig. 1a the temperature dependence of the ratio $\alpha_{\vec{q}L}/K$ for $\vec{q} \| [100]$ is plotted for different ν and N_D . The curves terminate at such temperatures above which the assumption, that the ionized donor is the nearest to the compensating acceptor, is not valid. In Fig. 1b the frequency dependence of $\alpha_{\vec{q}L}/K$ is plotted for different T and N_D . The curve for $N_D = 3.1 \cdot 10^{14}$ cm^{-3} indicates a broad maximum. The reason of this is that with the increase of phonon frequency, the most effective pair separation

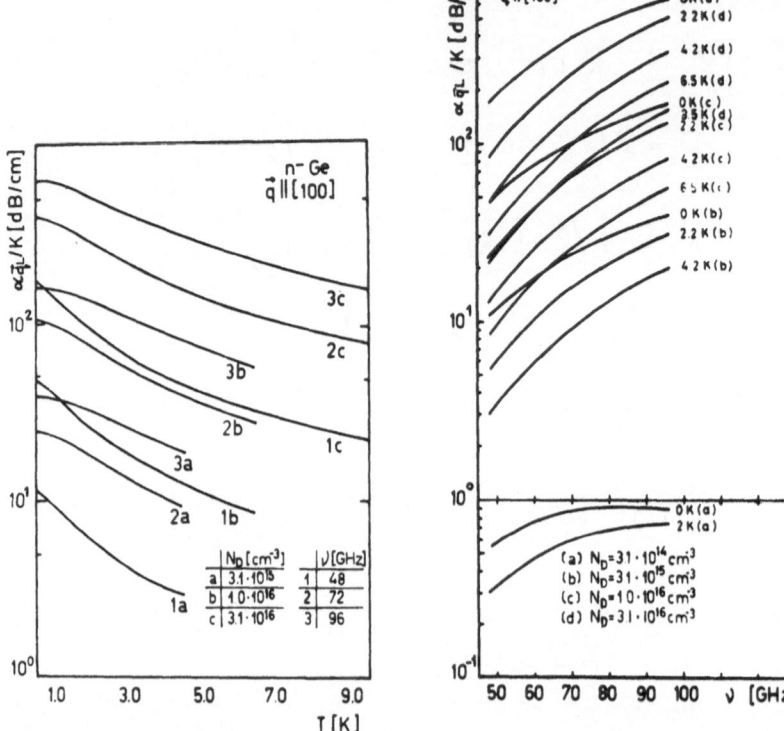

1a   1b

Fig.1. The acoustic attenuation coefficient per unit compensation for longitudinal 100 waves; 1a the temperature dependence
1b the frequency dependence.

$R_{S\omega}$ decreases being always smaller than the most probable distance between donors which is roughly $N_D^{-1/3}$. The maximum moves
toward higher frequencies with increasing N_D.

For other directions of propagation of longitudinal phonons,
for which the scattering by neutral donors is present, the hopping mechanism of scattering is more efficient than the scattering by neutral donors for the frequencies and temperatures
considered here.

+ Work supported by the Institute of Physics of the Polish
Academy of Sciences.
[1]. P.C. Kwok, Phys. Rev. 149 666 (1966),
[2]. M. Pomerantz, Phys. Rev. B1 4029 (1970),
[3]. A. Miller and E. Abrahams, Phys. Rev. 120 745 (1960).

ELECTRON-PHONON INTERACTION IN LIGHTLY AND HIGHLY DOPED n-TYPE GERMANIUM

Mahirdhwaj Singh and G S Verma

Department of Physics, B H U, Varanasi, India

The phonon conductivity of n-type Ge in the low temperature range has been studied by several workers. Bird and Pearlman[1] have carried out extensive measurements of n-Ge in the liquid helium range, where they observed resonance dips in the phonon conductivity vs temperature curves at about 0.7°K. Resonance scattering of phonons in the p-type Ge has been observed in several p-type materials and it is believed that the theory of Suzuki and Mikoshiba[2] accounts for the observed thermal conductivity as well as its magnetic field dependence. But this is not true for n-type Ge. The theory for n-type Ge has been given by Griffin and Carruthers[3], Keyes[4], and Kwok[5]. Suzuki and Mikoshiba[6] have given complete expressions for the elastic and inelastic relaxation rates for the phonon scattering by neutral donors in the case of light doped semiconductors. These expressions include inelastic scatterings of phonons from both singlet and triplet states, as well as phonon-assisted absorption processes. These expressions are in such a form that they can be easily applied to the analysis of the phonon conductivity results. Suzuki and Mikoshiba[6] also attempted to explain the results at temperatures higher than T_r, the temperature which corresponds to the resonance dips in the K vs T curve.

In the present paper an attempt has been made to investigate the phonon scattering by neutral donors in the resonance scattering range 0.3°K - 2°K.

The different phonon scattering relaxation rates are given by

$$\tau_{el}^{-1} (q,t) = A_t f^2 \left(\frac{\omega}{v_t}\right) F(x) \; Q(x)$$

$$\tau_{1,inel}^{-1}(q,t) = \frac{A_t}{2} f^2 \left(\frac{\omega}{v_t}\right) F(x + x_1) \, Q_1(x)$$
Triplet

$$\tau_{2,inel\ singlet,\hbar\omega>4\Delta}^{-1} = \frac{At}{2} f^2 \left(\frac{\omega}{v_t}\right) F(x - x_1) \, x \, Q_2(x)$$
phonon assisted $\hbar\omega<4\Delta$

where

$$A_1 = \frac{N_{ex}\ 48\ (4\Delta)^2 E_u^4}{81\hbar^4 \pi \rho^2 v_1^2\ 225}$$

$$A_2 = \frac{N_{ex}\ 32\ (4\Delta)^2 E_u^4}{81\hbar^4 \pi \rho^2 v_2^2\ 225}$$

$$A_3 = \frac{N_{ex}\ 40\ (4\Delta)^2 E_u^4}{81\hbar^4 \pi \rho^2 v_3^2\ 225}$$

$$f^2\left(\frac{\omega}{v_1}\right) = \left[1 + \left(\frac{a^* k_B}{2v_1 \hbar}\right)^2 x^2 T^2\right]^{-4}$$

$$F(x) = \frac{f^2\left(\frac{x}{v_1}\right)}{v_1^{-5}} + \frac{3}{2}\frac{f^2\left(\frac{x}{v_2}\right)}{v_2^{-5}}$$

$$Q(x) = \frac{x^4\left[1+e^{-x_1}\left\{2+\left(\frac{x_1}{x}\right)^2\right\}\right]}{\left[1+3e^{-x_1}\right]\left[(x_1^2-x^2)^2\right]}$$

$$Q_1(x) = \frac{(x+x_1)(1-e^{-x})}{x(3+e^{x_1})(1-e^{-(x+x_1)})}$$

$$Q_2(x) = \frac{(1-e^{-x})(x_1-x)}{x\left[e^{x_1-x}-1\right]\left[1+3e^{-x_1}\right]}$$

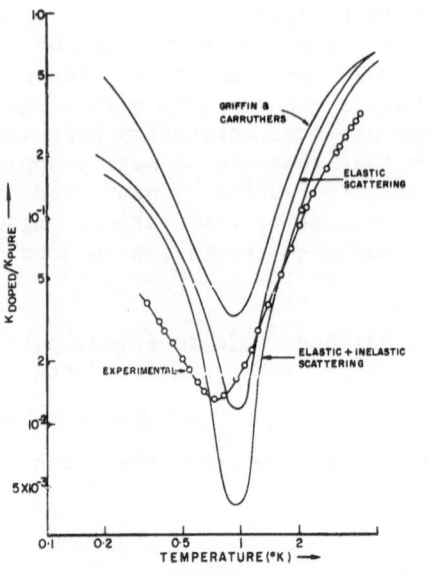

RESONANCE SCATTERING OF PHONONS BY NEUTRAL DONORS
IN DOPED GERMANIUM (Sb-306)

Fig 1

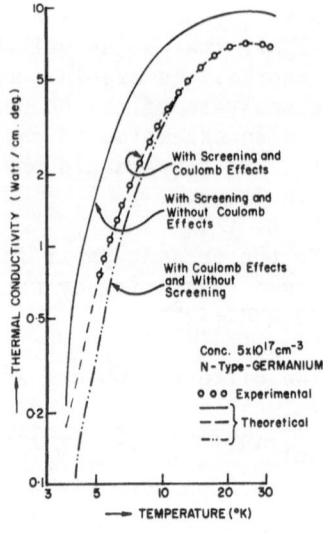

Fig 2

Here $x = \hbar\omega/k.T$, $\omega \equiv \omega_{qt}$, t = pol index, N_{ex} is the donor electron concentration, t is the polarization index, 4Δ is the energy difference between the singlet ground state and the next higher energy triplet state, E_u is the shear deformation potential, ρ is the density of the material, v_1 and v_2 are the average phonon velocities for the longitudinal and transverse modes, a^* is the donor electron radius, $x_1 = 4\Delta/k.T$.

The values of the various parameters used in the present calculations for the Sb-doped Ge (sample 306) are $\rho = 5.35$ gm/cm^3, $v_1 = 5.37 \times 10^5$ cm/sec, $v_2 = 3.28 \times 10^5$ cm/sec, $E_u = 19$ eV, $a^* = 40A^\circ$, $4 = 0.32$ meV, $N_{ex} = 3.0 \times 10^{16}$ cm^{-3}, $L = 0.455$ cm.

We calculate $\tau_{elastic}^{-1}$, τ^{-1}(inelastic, triplet), τ_2^{-1} (inelastic for $\hbar\omega > 4\Delta$, phonon assisted for $\hbar\omega < 4\Delta$) and $\tau_{2,r}^{-1}$ ($\hbar\omega \sim 4\Delta$) for different values of x at 1°K. These relaxation rates are used to evaluate the phonon conductivity at different temperatures. The agreement between theory and the experimental results in the resonance scattering region is not good. However, if one plots K_D/K_P vs Temperature, the experimental points near the resonance temperature indicate a well defined sharp dip in the ratio. The results of calculation are shown in Fig 1. It may be seen from this curve that the calculations based on the elastic scattering of phonons lead to reasonable agreement between theory and experiment. However, it is obvious from Fig 1 that none of the calculated curves explain the width of the dip.

Phonon conductivity of highly doped n-Ge has also been calculated in the temperature range 4 to 30°K. The results of calculation are shown in Fig 2. It may be seen from the curve that the donor ion electric field effects[7] are important in the high temperature range, whereas screening effects[8,9] are relevant at the low temperature end. The present calculations show that both these effects must be taken in consideration in the calculations of phonon conductivity of highly doped Ge.

REFERENCES

1. B L Bird and N Pearlman, Phys Rev 4B, 4406 (1971).
2. K Suzuki and N Mikoshiba, Phys Rev 3B, 2550 (1971).
3. A Griffin and P Carruthers, Phys Rev 131, 1975 (1963).
4. R W Keyes, Phys Rev 122, 1171 (1961).
5. P C Kwok, Phys Rev 149, 666 (1966).
6. K Suzuki and N Mikoshiba, J Phys Soc, Japan 31, 185 (1971).
7. V Kosarev, Sov Phys JETP 33, 793 (1971).
8. C R Crosby and C G Granier, Phys Rev B4, 1258 (1971).
9. M Singh and G S Verma, J Phys (Lond) C7, 3743 (1974).

DISCUSSION ON LOCALISED HOLES AND ELECTRONS

Phonon Scattering and Ultrasonic Absorption by Acceptors in Si and
Ge   T Ishiguro and H Tokumoto   Page 323

L Halbo: You suggest that the magnetothermal conductivity
observed in Ge(Ga) for B $\lesssim$ 6T can be explained by a second order
Zeeman effect. Is this a qualitative suggestion, or have quantitative calculations been made?

T Ishiguro: We have tried to obtain a fit by using the formulae
for the scattering rates in the presence of the magnetic field and
we found semiquantitative agreement. That is to say the absolute
values have the right order of magnitude but the calculation gives
a somewhat exaggerated variation of $\Delta K/K$ I think this can be
improved by a better choice of parameters.

K Lassmann: Two questions concerned with the temperature dependence of the acoustic attenuation in Ga doped Ge. 1. Is the under-
lying theoretical relaxation curve plotted beyond the validity range
$\omega\tau > 1$ of your formula? 2. Did you try to see saturation in the
low temperature part of your attenuation?

T Ishiguro: 1. For the temperature range lower than the relaxation
peak, $\omega\tau > 1$ is approximately satisfied. For $\omega\tau > 1$ one should
consider the problem of local equilibrium. (This problem is
discussed in the text). 2. In earlier stages of the study we
looked for but found no appreciable acoustic power dependence.
For the data shown here we have not checked the power dependence.

L J Challis: I understand that your view is that the $\sim$ 1K splittings of the acceptor ground state are due to dislocations and to
overlap of the acceptor wave functions of neighbouring acceptors.
I just wished to comment that the dislocation densities in our
samples were $\sim$ 1000 times too small and that the anomalous effects
we see are present at concentrations down to $10^{14} cm^{-3}$.

****

356

The Effect of Magnetic Field and Uniaxial Stress on the Thermal
Conductivity of p-Germanium   L J Challis, S C Haseler,
M W S Parsons and J Rivallin   Page 328

T Ishiguro: I wish to make a comment partly related to the comment
made by Professor Challis in the previous talk. In thermal conductivity measurement, samples are soldered to heat sinks. Though
stress caused by the difference in the thermal expansion constants
of adjacent materials will be greatest near the interface, I think
we should be careful about the remaining stress further down the
sample since the scattering of low frequency phonons at low temperature is very sensitive to stress.

S C Haseler: Hopefully this was minimised by use of indium solder
which is fairly soft.

****

Hole - Phonon Interactions in p-Type InSb by Heat Pulses
Experiments   F R Ladan and J P Maneval   Page 331

E Mosekilde: What processes do you think are causing the scattering
when $q > 2k_F$?

J-P Maneval: There is apparently no scattering for transverse
waves but there is indeed scattering for longitudinal waves and
at present we have no explanation for this.

L J Challis: Did you see any dispersion when the concentration is
such that the holes are localised?

F R Ladan: No.

****

Ultrasonic Attenuation in p-Type Germanium   E Ortlieb, Hp Schad
and K Lassmann   Page 334

T Ishiguro: I would like to ask you if the ω^2/T dependence is
observed in all of the samples, since the samples with higher
concentration used by you seem to be close to the impurity banding
regime.

K Lassmann: Yes, and apparently there is no correlation of δ_0 to
the acceptor concentration.

****

Ultrasonic Attenuation due to the Neutral Acceptor Indium in
Silicon   Hp Schad and K Lassmann   Page 337

J Jackle: Why don't you see relaxation by the direct process as
well?

Hp Schad: There probably is a small contribution but it is
negligible except in the wings of the relaxation peak.

****

Phonon Scattering Due to Deep Acceptors in Semiconductors
A de Combarieu and K Lassmann   Page 340

L J Challis: Can you account for the deep minimum in K(T) at
8T by a temperature <u>independent</u> scattering rate?

K Lassmann: An analysis taking into account all possible magnetic
field effects has not been tried. A simple elastic second order
resonant scattering does not give a minimum.

L Halbo: Is there any evidence of the excited 'non-hydrogenic'
states you discuss here from optical measurements or from electrical
resistivity and Hall measurements related to impurity conduction?

K Lassmann: In our preliminary far infra-red measurements and also
in one experiment with quasimonochromatic phonons from superconducting tunnelling junctions, such a level was not seen in Si(In).
But the experimental conditions were not optimised, and further
experiments have to be done. In a luminescence measurement a
satellite line to a line from an exciton to the In acceptor has
been seen at a distance of 4.2meV. Hall measurements have not
been made.

****

Scattering of Thermal Phonons in P Doped Si   D Fortier and
K Suzuki   Page 343

L Halbo: Is the concentration where your theory can no longer
explain the data the same as the concentration where electrical
impurity banding sets in? Furthermore, have you made attempts
to use other theories, e.g. the two-centre mechanism suggested
by Suzuki (Paris Conference 1972) for these high concentrations?

D Fortier: The works of other workers do not show the activation
energy ε_2 for such concentrations (4.7×10^{17} cm^{-3}). But spin
resonance experiments show the existence of donor pairings and
clusters at this concentration. For the second part of the
question the answer is no.

Discussion on the Low Temperature Magnetothermal Conductivity in
Lightly Doped Ge(Sb) and Ge(As)   L Halbo   Page 346

L J Challis: Which is more important, the field dependence of 4Δ
or that of a^*?

L Halbo: The effects in the thermal conductivity caused by the
change in a^* and in 4Δ are always comparable in magnitude. The
reduction in a^* always gives an increased scattering. The increase
in 4Δ also gives an increased scattering when $4\Delta_0 \lesssim 4kT$, since then
the scattering resonance will shift toward the peak in the phonon
distribution, due to the field. This is the case for Ge(Sb) in
the present temperature-range. However, the increase in 4Δ gives
a decrease in scattering when $4\Delta_0 \lesssim 4kT$, which is the case for
Ge(As) and also for Ge(Sb) below $\sim$ 1K. Since the phonon distribution decreases very fast with increasing frequency above the
peak, the effect of increasing 4Δ is the more important in the
latter case, giving positive $\Delta K/K_0$.

****

Theory of Ultrasonic Attenuation by Hopping Pairs in n-Ge
Z W Gortel   Page 349

L Halbo: It seems like the mechanism you discuss may also be
important in explaining thermal conductivity data at low temperatures, which have not been successfully explained by Suzuki-
Mikoshiba's theory. Do you know how the scattering will be at
frequencies higher than 100GHz?

Z W Gortel: Our theory does not apply when the donor overlap is
sufficient to give energy differences > 100GHz. However, we expect
that at these frequencies, the phonon relaxation rate for hopping
processes should start to decrease with frequency as can be seen in
fig. 1b for the lowest concentration. Since the scattering by
neutral donors is increasing as the resonance frequency is
approached, the hopping mechanism can be neglected. The problem
of scattering by transverse waves needs further study but it is
probably weak in comparison with that by neutral donors at these
frequencies.

L J Challis: Would this mechanism work for acceptors?

Z W Gortel: In principle yes, but at present it is difficult to
give even semiquantitative results because of the facts that
phonons are scattered mainly by the homopolar pairs, and also that
homopolar acceptor pairs have completely different energy structures
from the homopolar donor pairs. This problem was studied by Dr
Kaczmarek and myself in connection with the hopping infrared
absorption in p-type Ge and Si. In this process homopolar pairs
play a dominant role also and our results which were in reasonable

agreement with experiment were different from those for n-type
material. The second point is that maybe that one has to take
into account the possibility of splitting of the four-fold
degenerate acceptor ground state when two centre acceptor states
are constructed.

****

Electron-Phonon Interaction in Lightly and Highly Doped n-Type
Germanium   M Singh and G S Verma   Page 353

L Halbo: If I understand you correctly, you calculate the phonon
scattering in 5 x 10^{17}cm^{-3} Ge(Sb) as if the electrons are in an
ordinary conduction band whereas this concentration really
corresponds to impurity banding with different and not well understood properties. How for example do you know the effective mass
and the Fermi velocity to put into your calculation?

G S Verma: This particular donor electron concentration is enough
to allow overlap of the electronic wave functions and the merger of
the impurity states with the conduction band. This is also evident
from the electrical conductivity measurements. In the present
calculations we have shown that if one includes screening as well
as donor ion electric field effects, one can interpret the results
with just one value of the temperature independent density of
states effective mass, say m* = 0.2m. This agrees with the experimental results. The information regarding the reduced Fermi
potential as well as its temperature dependence can be obtained
from the thermoelectric data available at such low temperatures.

A C Anderson: In interpreting low temperature phonon transport
measurements, it should be kept in mind that the thermalization of
phonons at the sample boundaries can alter the apparent resonance
as happens at high temperatures when N and U processes are present.

G S Verma: As long as the boundary effects are frequency independent, the resonance scattering of phonons is not going to be affected by it.

L J Challis: Did your analysis on the scattering by localised
carriers indicate that inelastic scattering was weaker than theory
suggests?

G S Verma: As far as the magnitude of the resonance dip is conerned
one obtains good agreement with elastic scattering alone. However,
if one includes inelastic scattering also, the conductivity is
reduced further and the agreement is lost. The contributions by
elastic and inelastic scattering are comparable in the resonance
region.

MULTIPHONON ABSORPTION IN ALKALIHALIDES

H. Beck[*] and D.W. Pohl

IBM Zurich Research Laboratory

8803 Rüschlikon, Switzerland

The exponentially decaying absorption wing above the reststrahl frequency ω_R has recently attracted considerable attention.[1] Theoretically, the absorption is determined by the dipole-dipole response function. In alkalihalides the contribution linear in the ionic displacements seems to be the dominant one.. In order to calculate the displacement correlation functions and thus the absorption, we start from an anharmonic lattice Hamiltonian. The absorption β at frequency $\Omega \gg \omega_R$ is simply related to the imaginary part $\Gamma(\Omega)$ of the self-energy of the $q = 0$ transverse optic phonon:[2]

$$\beta = (4\pi e^{*2}/c\, m\, n_r\, V_c)\, \Gamma(\Omega)/\Omega^3 \quad . \tag{1}$$

Here e^{*} is the Born effective charge, c the speed of light, m the reduced mass, n_r the refractive index, and V_c the volume of the unit cell.

Γ is evaluated by means of anharmonic perturbation theory. The most important contributions describe the virtual decay into N other phonons:

$$\Gamma_N(\Omega) = \frac{\pi}{N!\,2^N} \sum_{\substack{\vec{q}_1 \cdots \vec{q}_N \\ \lambda_1 \cdots \lambda_N}} |V_{N+1}(\vec{q}_1,\lambda_1, \ldots \vec{q}_N,\lambda_N)|^2\, \Delta\left(\sum_{j=1}^{N}\vec{q}_j\right)$$

$$\times \delta\left[\Omega - \sum_{j=1}^{N}\omega(\vec{q}_j,\lambda_j)\right]\frac{1}{n(\Omega)}\prod_{j=1}^{N}R(q_j,\lambda_j) \quad . \tag{2}$$

Here V_N is the anharmonic coupling parameter of N-th order; $\omega(\vec{q},\lambda)$ the frequency of a phonon with wavevector $\vec{q}$ and polarization λ. The delta functions ensure momentum and energy conservation, respectively. Thermal occupation enters through the Bose distribution function n and

$$R(\vec{q},\lambda) \;=\; n[\omega(q,\lambda)] \;\; / \; \omega(q,\lambda) \qquad\qquad . \qquad\qquad (3)$$

The extensive experimental data on NaF[3,4] make this substance particularly suited for a quantitative comparison. We have evaluated β_2 and β_3 according to the above equations and higher orders by

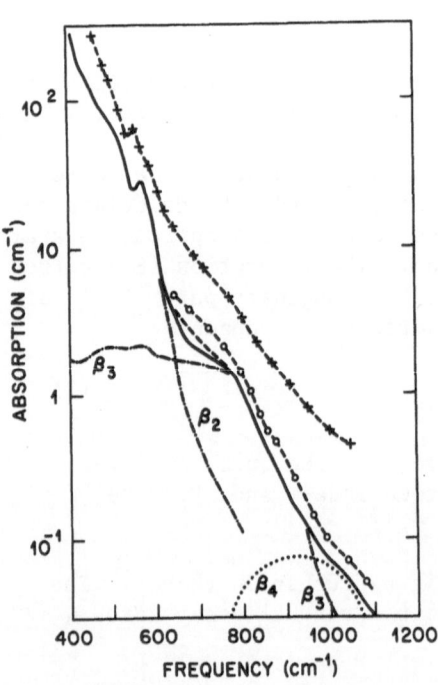

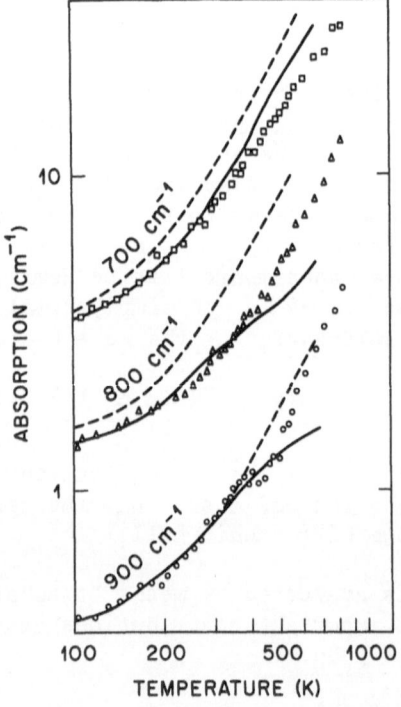

Fig. 1. Full curve: calculated total absorption at T = 0; dash-dotted curves: two-phonon (β_2) and three-phonon (β_3) contributions; dotted curve: estimated four-phonon contribution β_4; dashed curve: $\beta(\Omega)$ for smaller lifetime of intermediate phonons. Experiments: □ McNelly and Pohl (T = 100 K), + Hohls (300 K).

Fig. 2. $\beta(T)$ at Ω = 700, 800 and 900 cm^{-1}. Full curves include two- and three-phonon processes, evaluated with temperature-dependent phonon energies. Dashed curves: same, using the T = 0 phonon energies. The theoretical curves are shifted in such a way that they agree with the T = 0 exptrapolation of the experimental data.

means of a simplified formalism.[4] The harmonic frequencies $\omega(\vec{q},\lambda)$
were determined by a breathing shell model. Figure 1 presents β_2,
β_3 and β_4 as well as the resulting total absorption at $T = 0$ together with experimental data at $T = 100$ K and at room temperature.
Except for a constant factor, the experimental curve is well reproduced by the calculated spectrum. The shoulder between two- and
three-phonon processes, however, is less pronounced in the experiment indicating the effect of finite lifetimes of the intermediate
phonons. (This feature has been taken into account in the dashed
curve.)

The temperature dependence for selected frequencies is shown
in Fig. 2. It is seen that the curves calculated with a phenomeno-
logical temperature dependence of the phonon frequencies (full
lines) give a slightly better fit to the experimental data[4] than
those where the Bose factor is the only temperature-dependent
quantity.

References

[*] Postdoctoral Fellow from the Institute of Theoretical Physics of
the University of Zurich, 8001 Zurich, Switzerland.

[1] See, for example, Proc. Intern. Conf. on Highly Transparent Solids
(Waterville Valley, New Hampshire, Feb. 3-5, 1975), Plenum Publish-
ing Comp., New York.

[2] M. Sparks and L.J. Sham, Phys. Rev. B 8, 3037 (1973).

[3] H.W. Hohls, Ann. Physik 29, 433 (1937).

[4] T.F. McNelly and D.W. Pohl, Phys. Rev. Lett. 32, 1305 (1974).

RAMAN SCATTERING STUDIES OF ANHARMONICITIES IN

POTASSIUM HALIDES

M.S. Haque*, J.A. Taylor, J.E. Potts,
J.B. Page, Jr., and C.T. Walker

Physics Department, Arizona State University
Tempe, Arizona 85281 U.S.A.

In recent years, Raman scattering has become a widely accepted technique for investigating phonons in crystals. In this paper we demonstrate its use for studying anharmonicities in potassium halides. It has been shown[1] that, for dilute concentrations of a particular dopant, namely Tl+, the observed defect-induced first-order spectra of these crystals at low temperatures are best understood by assuming that essentially zero force-constant perturbations are induced by Tl+. Since the even parity type (A1g, Eg, and T2g) Raman active modes involve no defect motion, the observed spectra reflects the projected densities of states of these host crystals.

We have investigated anharmonic effects in the defect-induced spectra for KBr:Tl+, KCl:Tl+, and KI:Tl+, by observing the spectra as a function of temperature at atmospheric pressure, and as a function of hydrostatic pressure at constant temperatures, i.e. at 30, 85, and 300 K. We have also calculated the spectra theoretically as a function of crystal volume within the framework of a quasiharmonic model. Here, we have calculated the quasiharmonic eigenmodes of the host crystals at various crystal volumes, which resulted from the application of external pressures or thermal expansion. These new eigenmodes were then used to calculate the defect-induced spectra at these volumes by applying the formalism of Ref. 1. Only the host crystal anharmonicities were included, and these

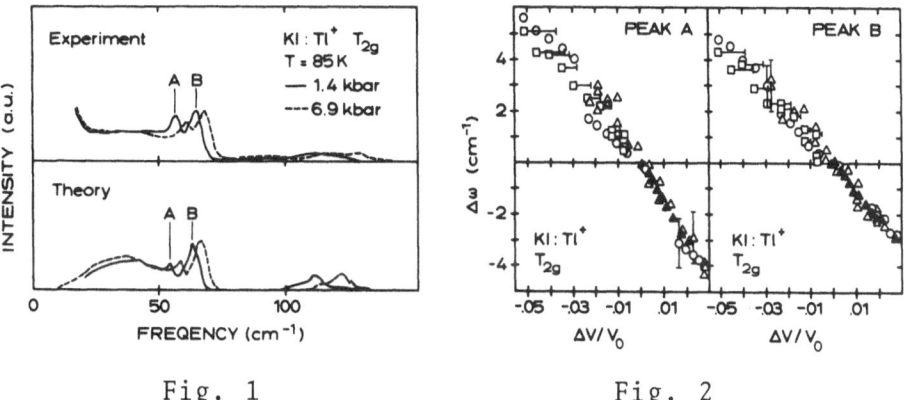

Fig. 1   Fig. 2

were restricted to third-order anharmonicities, which
were derived from a central potential between nearest
neighbors. Details of experimental[2] and the theoretical[2,3] techniques have been published elsewhere.

In Fig. 1 we show the experimental and calculated
T2g spectra for KI:T1+ at 85 K for two different pressures. We have followed the frequencies of the peaks
A and B as functions of fractional volume change of the
crystal, and the results are shown in Fig. 2. Here the
shifts in frequency and volume are with respect to
quantities at 0 K and zero pressure. The darkened
triangles correspond to the temperature dependence
measurements at zero pressure, while the open triangles
and squares represent the pressure dependence measurements at 300 K and 85 K respectively. The circles are
the calculated results. Temperatures and pressures
were converted to crystal volume by using published
values for the coefficient of linear expansion, P-V data
at 300 K, and the bulk modulus at 85 K. Details of
these conversions along with the accompanying uncertainties are to be found in Ref. 2.

Note that the results of the temperature dependence
measurements coincide with those of the pressure dependence measurements. Hence the frequency shifts of
these peaks result primarily from volume changes of
the crystal with negligible contributions from phonon-
phonon interactions. Note also the very good agreement
between theory and all three sets of experimental
results. Since the calculated spectra employed only
pure crystal phonons and anharmonicities, we conclude

that the observed temperature and pressure dependence
of the spectra reflects the anharmonicities of pure KI.
Similar results have been observed for both the Eg and
T2g spectra of KBr:T1+ and KC1:T1+.

The anharmonic model that was used in these calculations, has been successfully extended to calculate the
effect of phonon-phonon interactions that are observed
in the dielectric properties of pure potassium halides[3].
We view this as a further confirmation of the validity of
our theoretical model of anharmonicity.

We concluded from results, such as we have shown in
this paper, that first-order Raman scattering from T1+-
doped potassium halides can be used as a probe for pure
crystal anharmonicities in these host crystals.

*Present address: Fachbereich Physik, Universität
Regensburg, D-84 Regensburg, W. Germany

1.   R.T. Harley, J.B. Page, Jr., and C.T. Walker, Phys.
     Rev. B3, 1365 (1971).

2.   J.A. Taylor, M.S. Haque, J.E. Potts, J.B. Page, Jr.,
     and C.T. Walker, Solid State Commun., 16, 1179 (1975).

3.   M.S. Haque, Phys. Rev. (in press).

THERMAL RESISTANCE OF DIELECTRIC CRYSTALS AT HIGH TEMPERATURES*

P. G. Klemens and D. J. Ecsedy**

Dept. of Physics and Inst. of Materials Science

Univ. of Connecticut, Storrs, Conn. 06268, U.S.A.

At $T > \theta$, the intrinsic thermal resistivity should vary as T, but is often of the form $W = aT + bT^2$, with the quadratic term up to 30% of W around 1000 K. This has been attributed to four-phonon processes.[1]

The relaxation rate for three-phonon U-processes is given approximately by[2]

$$1/\tau_{3U} = 2\gamma^2 (T/T_0)\, \omega^2/\omega_D \qquad (1)$$

where ω is the (angular) phonon frequency, ω_D the Debye frequency, γ the Grueneisen constant and $T_0 = Mv^2/k_B$ is a characteristic temperature typically between 50,000 and 100,000 K; T/T_0 is the mean square thermal shear. Based on the same approximation one can show[3] that the relaxation rate for four-phonon processes is

$$1/\tau_4 = 2\gamma^2 A (T/T_0)^2\, \omega^2/\omega_D \qquad (2)$$

where A is somewhat smaller than unity. This follows because to every three-phonon interaction there corresponds a group of four-phonon processes, with the extra phonon ranging over the entire zone, and the square of the interaction Hamiltonian containing the square of the strain as an additional factor. Since (2) is less than (1) by at least the factor T/T_0, typically 0.02 at 1000 K, the term bT^2 in W should be much smaller than what is observed.

Note, however, that $W = aT$ predicted for three-phonon processes holds only if the crystal volume is constant. Thermal expansion changes the volume, perhaps only by 2% to 1000 K, but the effect is amplified by the sensitivity of the conductivity to dilatation. If $\chi_0(T) = 1/W_0(T)$ is the thermal conductivity at constant volume, and if $\Delta(T)$ is the thermal dilatation,

$$ 1/W(T) = \chi(T) = \chi_0(T)\left[1 - p\Delta(T)\right] \qquad (3) $$

where $p = (1/\chi)d\chi/d\Delta$, the fractional variation of χ with dilatation, is related to the pressure dependence. There is very little experimental data on the pressure dependence of χ, but from the theory of Mooney and Steg[4]

$$ p = 3\gamma + (2/3) + 2\gamma'/\gamma \qquad (4) $$

where γ' is the derivative of γ with Δ. With $\gamma' \approx \gamma \approx 2$ - see Thomsen[5] - p is typically about 9. If $\Delta(T) \propto T$ and $W_0 = aT$, W has an added term bT^2 typically around 15% at 1000 K. One can use measured values of $\Delta(T)$, together with p given by (4), to estimate W_0, the thermal resistivity corrected to constant volume. In many dielectric crystals there is uncertainty about $\chi(T)$ at high temperatures due to internal radiation. This is reduced but not eliminated in the sintered oxides discussed in reference.[1] The data is not inconsistent with W_0 being linear in temperature.[3]

In single crystal germanium and silicon,[6] where radiation is no problem, thermal expansion accounts only for a part of the bT^2 term in W. To explain the remainder, we invoke an old suggestion of Pomeranchuk.[7] Momentum and frequency conservation prevent direct three-phonon interaction between low-frequency longitudinal modes and modes of thermal frequency. These longitudinal modes interact only with modes of comparable frequency, with a rate[2]

$$ 1/T_{3N} = 60\pi^2\gamma^2 (T/T_0)\,\omega^4/\omega_D^3 \qquad (5) $$

Let ω_0 be the frequency for which $1/T_4 = 1/T_{3N}$, so that $\omega_0 = (T/T_0)\,\omega_D/17$. For $\omega < \omega_0$, T is limited by four-phonon processes and long, so that χ_L, the added conductivity due to these modes, compared to χ_{th}, that due to the major part of the phonon spectrum, is

$$ \chi_L/\chi_{th} = \tfrac{1}{3}(\omega_0/\omega_D)(T_0/T) = (1/50)(T_0/T)^{1/2} \qquad (6) $$

With $T_0 = 50{,}000$ K, ω_0/ω_D is less than 10^{-2} at 1000 K, and χ_L/χ_{th} is about 15% at 1000 K and 20% at 500 K. This would lead to a conductivity (at constant volume) of the form

$$ \chi_0(T) = \alpha/T + \beta/T^{3/2} \qquad (7) $$

Over a limited temperature range near 500 K this would appear as $\chi_0 \propto T^{-1.1}$. At lower temperatures the second term would be limited by boundary scattering. From (2), $l(\omega_0) = (a/2)(T_0/T)^3$, where a is the interatomic distance. In our example, $l(\omega_0) = 1.5 \times 10^{-2}$ cm at 500 K, and 0.25 cm at 200 K. Below 200 K, $\varkappa_L$ is size-dependent even in single crystals. In poly-crystals, $\varkappa_L$ is much smaller even at 500 K, which may explain why the thermal expansion correction suffices for sintered oxides but not for single crystals.

In germanium[6] between 500 and 700 K, $\varkappa \propto T^{-1.3}$. Corrected to constant volume $\varkappa_0 \propto T^{-1.2}$. Thus $\varkappa_L/\varkappa_{th} = 0.4$, compared to our expected ratio of 0.2. In silicon[6] the discrepancy is even larger. However, all these estimates are rough, since we took $A = 1$ in (2).

Since $\varkappa_L$ is negligible for polycrystalline aggregates, one may use the temperature dependence of $\varkappa(T)$ to estimate p, and thus deduce indirectly the pressure dependence of the thermal conductivity.

*Supported by the U.S. Army Research Office, Durham, North Carolina.
**Now at Planning Systems Inc., McLean, Virginia 22101.
(1) D. Billard and F. Cabannes, High Temps.- High Pressures 3, 201 (1971).
(2) P. G. Klemens, in Thermal Conductivity, vol. 1, Acad. Press, London, 1969.
(3) D. J. Ecsedy, Doctoral Thesis, Univ. of Connecticut, 1975.
(4) D. L. Mooney and R. G. Steg, High Temps.- High Pressures 1, 237 (1969).
(5) L. Thomsen, J. Phys. Chem. Solids 31, 2003 (1970).
(6) C. J. Glassbrenner and G. A. Slack, Phys. Rev. 134, A1058 (1964).
(7) I. Pomeranchuk, J. Phys. U.S.S.R. 4, 259 (1941).

THREE PHONON RELAXATION TIMES

R. A. H. Hamilton and G. P. Srivastava

Department of Applied Physics and Electronics, UWIST

Cardiff, CF1 3NU

We have computed three phonon relaxation times for acoustic phonons assuming a dispersion law of the form

$$\omega_s(q) = \frac{2c_s q_D}{\pi} \sin\left(\frac{\pi q}{2q_D}\right) \qquad (1)$$

where q is the wavenumber, q_D the Debye radius and C_s the sound velocity for the mode s. One longitudinal mode with velocity C_L and two transverse with velocities $C_T = 0.6C_L$ are assumed.

The transition rate for a process in which a phonon (q,s) is annihilated and phonons (q,'s') and (q,"s") created is taken to be of the form

$$B\omega\omega'\omega''nn'n''e^{A\omega}\delta(\omega-\omega'-\omega'')\delta_{q+g,q'+q''} \qquad (2)$$

where $n \equiv n(w)$ is the Plank function, $A = \hbar k^{-1} T^{-1}$ and q is a vector of the reciprocal lattice. Both (1) and (2) are of the right form for high symmetry directions for a simple cubic lattice with nearest neighbour interactions only and represent plausible interpolations between these directions. Liebfried and Schlomann have justified their use in some detail.

Our intention is to analyse the computed results so as to represent them sufficiently accurately by formulae depending on $x = q/q_D$ and $C = \hbar c_L q_D/kT$.

Reciprocal relaxation times for U processes were calculated assuming a reciprocal lattice vector of length $1.612q_D$, being the length of the shortest r.l.v. Our calculations show that the

effect of U processes involving longer r.l.v.'s is negligible. The
U process inverse relaxation times are averaged over r.l.v.
direction. We find inverse relaxation times for both N and U
processes have the form

$$\tau^{-1} = a\left(\frac{1}{e^{A\xi}-1} + \frac{1}{2}\right) + b\left(1+yA^3\right)\left(\frac{1}{e^{A\eta}-1} - \frac{1}{e^{A(\eta+\omega)}-1}\right)$$

where a, b, ξ, η and y depend on x only. This is accurate to
within a few percent down to about 30°K and gives the right order
of magnitude down to about 15°K for Ge, for example. The first
term is absent for transverse phonons. We have produced forms
for a,b, etc., but work on them is not yet complete.

It is intended to test the formulae by using them to calculate
the lattice thermal conductivity of a range of electrical insulators.
To have a standard for comparison we have first of all used the
computed results to do this for Ge. We calculated the Debye term
$\sum_{qs} n(n + 1)\tau\omega^2 (\partial\omega/\partial q)^2$ where τ is the total relaxation time. Included in τ are boundary scattering (estimated by use of a
boundary scattering length) and isotope scattering determined so as
to be consistent with the dispersion law used (This gives a $q^2\omega^2$
rather than ω^4 behaviour). We also calculated corrections due to
the effect of drift using formulae based on Callaway's theory and
its extension to include U processes (Hamilton). The results are
given in Table 1 (HS). Results of a variational calculation using
a non-dispersive model by Hamilton and Parrott (HP) are shown for
comparison. The experimental results are taken from Geballe and
Hull, and Slack and Glassbrenner.

Table 1. Thermal Conductivity of Ge(w/cm.deg).

Temp.	HS	HP	Expt.
4.4	3.3	3.2	3.3
8.7	9.3	8.7	8.7
17.4	13.3	10.9	11
34.8	6.9	7.1	8.2
69.7	3.0	3.2	3.4
139	1.58	1.6	1.5
279	.91	.93	.6
557	.52	.6	.25

The HS results fit less well near the conductivity maximum.
We put this down to use of a much simpler angular dependence for
the 3rd order coupling constant than HP and do not regard it as
very significant. Despite the use of different models and methods
HS and HP give practically identical results apart from this and
in particular show the same deviation from experiment at high
temperatures. This had been thought to be due to neglect of
dispersion by HP, but it seems now that another scattering process

is needed.   We hope to try acoustic-optical interactions as a possible mechanism.

The corrections for off-diagonal term behaviour were negligible near the conductivity maximum and rose to 18% for N-processes (Callaway correction) and 31% for N and U.

References

1.   G.Liebfried and E.Schlomann
         Nachr.Akad.Wiss.Göttingen, Math.Phys.K14,71(1954)
2.   J.Callaway, Phys.Rev.113,1046 (1959).
3.   R.A.H.Hamilton,J.Phys.C.,6,2653 (1973)
4.   R.A.H.Hamilton and J.E.Parrott,
         Phys.Rev.,178,1284 (1969)
5.   T.H.Geballe and G.W.Hull,Phys.Rev.110,773 (1958)
6.   G.A.Slack and C.J.Glassbrenner,Phys.Rev.120,782 (1960)

OPTICAL-ACOUSTICAL PHONON SCATTERINGS IN THE LATTICE THERMAL

CONDUCTIVITY OF ROCK SALT CRYSTALS

M D Tiwari

Department of Physics, Birla Postgraduate College

Garhwal University, Srinagar, Garhwal (U P), India

An exact knowledge of the lattice thermal conductivity of solids is very scant, most being qualitative estimates only. At high temperatures, one should consider the dispersion of acoustical branches, contribution of optical phonons, electron-phonon scattering, four-phonon processes, etc. We have seen in cases of Si[1] and Ge[2] that electron-phonon scattering does not affect the high temperature thermal conductivity to any appreciable extent and four-phonon scattering plays an important role above 350°K. At temperatures compared with the Debye temperature the optical phonons play a significant role in heat transfer and scattering, although such a mechanism is usually not considered[2,3]. As the thermal conductivity is sensitive to the shape of the vibrational spectrum, one requires first a dynamic model giving a spectrum close to the experiment. We have studied here the lattice thermal conductivity of NaCl in the temperature range 50-250°K and have used the lattice dynamics of NaCl calculated by Raunio and Rolandson[4].

The relaxation time $\tau_s(\vec{k})$ of a phonon $(\vec{k}, s)$ where $\vec{k}$ is the wave vector of s - the vibrational branch, is determined by the collision diagonal matrix element as given in Ref. 5.

In our case the thermal conductivity K is given by

$$K = \frac{1}{24\pi^3} \sum_s \int \tau_s \, v_s^2(\vec{k}) \, C_s(\vec{k}) \, d\vec{k}, \qquad (1)$$

where $C_s(\vec{k})$ is the heat capacity of $(\vec{k}, s)$ mode.

CALCULATIONS AND RESULTS

The calculation of relaxation time involves the knowledge of the frequency spectrum and the polarization vector, the eigen values and eigen vectors of the dynamic matrix and the tensor of anharmonic coefficients. We have used the frequency spectrum and polarization vector calculated by Raunio and Rolandson[4] in an extended shell model. The group velocity of optical branches within the brillouin zone is comparable to the slope of acoustic branches, hence we can say that the effect of optical phonons on heat transfer is not negligibly small.

We have assumed the central ionic interactions in the determination of anharmonicity constants, and the non-diagonal elements of the tensor, have been taken as zero[6].

In analyzing the lattice thermal conductivity of NaCl one must remember that the calculation was performed in the τ approximation and the transport relaxation time $\tau_{\ell\gamma}$ is kept to $\tau_{N,U}^{-1} = \tau_N^{-1} + \tau_U^{-1}$ just as done by Guthrie[7]. Our calculated results are compared with the experiment in Table 1.

TABLE1: Experimental and calculated values of the lattice thermal conductivity of NaCl, K_L, K_T and K_O represent the contributions of acoustic longitudinal and transverse branches and optical branch, respectively.

TEMPERATURE (0°K)	K_{exp}	K_T	K_L	K_O
		(Watts/cm degree)		
50	0.460	0.213	0.108	0.139
60	0.265	0.125	0.058	0.082
100	0.200	0.092	0.040	0.068
150	0.125	0.059	0.028	0.038
200	0.090	0.040	0.016	0.034
250	0.070	0.033	0.012	0.025

It is almost the accepted point of view that for practically all directions an acoustic phonon τ increases with increasing $|\vec{k}|$. This can be explained by the significant role of optical phonons in the scattering. From the spectral shape it can be seen that short wave acoustic phonons cannot generally participate in processes of the type A (acoustic) + 0 (optical) $\to$ 0 (optical) and at the same time these processes are possible with an increase in the acoustic phonon wavelength only. Small $\vec{k}$ phonons are scattered in such processes by large $\vec{k}$ phonons with a high density of states. For processes $A + A \to 0$ we will consider first that the δ-function integral in the $\vec{k}$ space reduces to a surface integral with conservation laws on the surface, i.e.

$$\int \Delta(\omega) \; \vec{k}' \to \int \frac{ds'}{v'_n}$$

where $v'_n = \partial/\partial \vec{k}'_n \{\omega(\vec{k}) \pm \omega(\vec{k}') - \omega(\vec{k}'')\}$. We consider now the dependence of (ds'/v'_n) on ω_A. Suppose $\omega' = B \sin(|k'|a/2)$, where B is the limiting acoustic branch frequency and a is the lattice parameter. It can be seen here that

$$\int \frac{ds'}{v'_n} \sim \frac{\left| \text{arc sin} \frac{\omega_o - \omega_A}{B} \right|}{\cos(\text{arc sin} \frac{\omega_o - \omega_A}{B})}$$

From this we can see that for ω_A decreasing towards $\omega_o - B$, when $\vec{k}'$ approaches the zone boundary, the scattering probability of the phonon increases with decreasing v_n. Since in NaCl ω_o and B are close and in several directions it may even happen that $\omega_o < B$; this situation may occur even for the same long-wave phonons. Our calculated results agree with these qualitative considerations, showing that optical phonons are more important in long-wave acoustic phonon scattering than in short-wave scattering. Thus, the presence of an optic part in the spectrum makes the frequency dependence of relaxation time $\tau(\omega)$ at high temperatures quite different than is usually assumed[8,9].

In conclusion, we can say that our results confirm that optical phonons play an essential role in acoustic phonon scattering. While k_o contributes only about one third of the total thermal conductivity by heat transfer, processes with participating optical phonons become dominant in several phonon group scatterings. In other words, if neglecting the contribution of k_o to K leads to a quantitative error, neglecting the optical branches in scattering processes may qualitatively affect the frequency and temperature dependence of τ. Moreover, if the approximation $K_o \sim 0$ can be justified by the smallness of the group velocity v_o, precisely owing to $v_o \sim 0$ the processes $A + 0 \to 0$ may become more important[5,6]. This, in turn, may cause a stronger T^{-1} dependence of $\tau(T)$ and $K(T)$ at temperatures $T < \hbar\omega_o/k_B$, since scattering processes with participating optical phonons start freezing much earlier than processes $A + A \to A$.

ACKNOWLEDGMENT

The author is grateful to Dr Bal K Agrawal and Professor G S Verma for their interest in the present work. The University Grants Commission, New Delhi, is thanked for providing a computer grant.

REFERENCES

1. Y P Joshi, M D Tiwari and G S Verma, Phys Rev B 1, 642 (1970).
2. M D Tiwari and Bal K Agrawal, Phys Rev B 4, 3527 (1971).
3. C J Glassbrenner and G A Slack, Phys Rev 134, A1050 (1964).
4. G Raunio and S Rolandson, Phys Rev B 2, 2098 (1970).
5. Handbuch der Physik (S Flugge ed) Vol 7/1, Springer, Berlin
   (1955), pp 104-324.
6. P B Gate, Phys Rev 139, A1666 (1965).
7. G L Guthrie, Phys Rev 152, 801 (1966).
8. R F Caldwell and M V Klein, Phys Rev 158, 851 (1967).
9. J M Ziman, Electrons and Phonons, Clarendon Press, Oxford (1960).

THE THERMAL CONDUCTIVITY OF SOLID KRYPTON BETWEEN 2.4 AND 39K DETERMINED FROM MEASUREMENTS OF THE THERMAL DIFFUSIVITY WITH A HEAT WAVE METHOD

P Korpiun, J Moser, F J Pieringer and E Lüscher

Physik Department, Technische Universität München

8046 Garching, Germany

Solid rare gases behave at low pressures as ideal dielectric crystals. They should be appropriate to study scattering mechanisms of the crystal lattice. Because of the simple type of forces acting between the atoms it should be possible to describe these scattering phenomena relatively correct. Nevertheless, there exist only few investigations on thermal transport properties of solid rare gases /1-5/. One reason for this situation is that it is very difficult to produce free standing crystals generally needed for such experiments and to couple heater and thermometers to them thermally. We have used a method which made us independent from the quality of the thermal contact between the crystal and the surrounding. We have measured the velocity of propagation of sinusoidal heat waves in a quasi semi-infinite medium. Sinusoidal heat waves travelling through a solid are damped and phase shifted but their characteristic shape will not be altered.

An arrangement of a flat electrical heater with an area of 2.5 x 2.5 cm^2, two carbon resistors and a germanium resistor has been mounted in a cylindrical chamber with an inner diameter of 5.5 cm. The crystal which has been grown afterwards in this chamber enclosed tightly heater and thermometers. A sinusoidal voltage of frequency ω applied to the heater generated a temperature wave with the frequency 2ω in the crystal. The variation of the temperature with the time has been determined with two carbon resistors at distances of $x_1 = (0.56 \pm 0.01)$ cm and $x_2 = (1.29 \pm 0.01)$ cm from the heater. The resistors have been measured with an Anderson-bridge and the technique of Lock-in-amplifying. Between the two thermometers the phase of a plane temperature wave is shifted by the amount $\Delta\rho = (x_2 - x_1)(\omega/2D)^{\frac{1}{2}}$; D is the thermal diffusivity. We have measured the phase shift in solid krypton in the temperature region from

377

2.39 to 39.2K. To proof the consistence it has been measured at
fixed temperatures for several frequencies between 0.07 and 0.3 Hz.
We have investigated two samples. The results for the thermal diffusivity D are plotted in Fig 1. All measurements have begun at

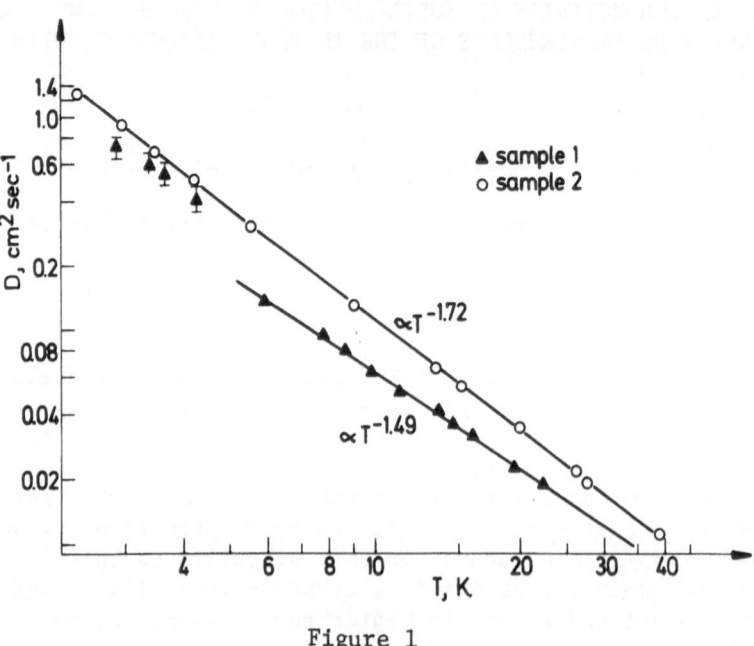

Figure 1

low temperatures. Sample 1 has been of a bad quality being completely opaque; at temperatures above 4.2K, even cracks could be
observed. Log D seems to depend linearly on log T. The straight
lines fitted to the experimental data of both samples satisfy the
relation $D \propto T^{-1.49}$ and $T^{-1.72}$ respectively.

The thermal conductivity λ is connected with the diffusivity D
by the relation $\lambda = DC\rho$. Since the data for the specific heat C
/6,7/ and the density ρ /8,9/ are very accurately known λ can be
obtained from D, Fig 2. Measurements of other authors are added
/1,4/. The maximum in λ (T) is near 12K. In the temperature range
above 25K, our values agree within one percent with the results of
measurements from Krupskii and Manzhelii /4/.

At temperatures below the maximum of the conductivity, λ (T)
rises less than proportional to T^3. Between 2.4 and 4K, λ is in a
good approximation proportional to $T^{1.6}$. This temperature behaviour may be explained by scattering processes on impurities and
different types of dislocations /10,11/. Even for an ideal crystal
one should expect the T^3 - behaviour of λ only for temperatures
below 2K. Also the specific heat is proportional to T^3 only in the

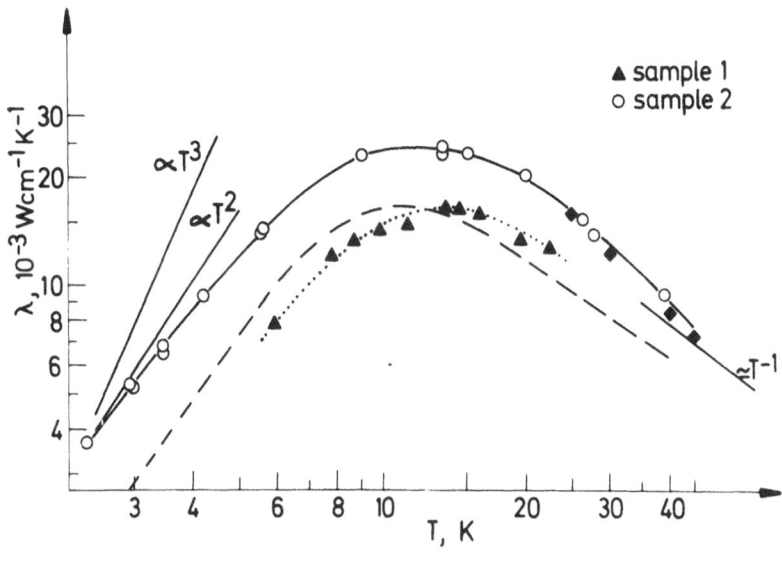

Figure 2

temperature region below 2K as Finegold and Phillips /7/ have found.
At temperatures above $\theta/2 \sim 36K$ when the phonons are scattered by
Umklapp-processes the relation $\lambda \propto \exp(\theta/2T)$ describes fairly well
the experimental results. The Debye-temperature $\theta = 71.9K$ has been
obtained from measurements of the specific heat /7/.

To get more information on the scattering mechanisms of phonon
spectra in solid rare gases one would need crystals of better quality. But with all methods of crystal growth that we know it is not
even possible to produce real single crystals /12/.

/1/ G K White and S B Woods, Phil Mag 3, 785 (1958).
/2/ D J Lawrence, H T Stewart and G W Guptill, Can J Phys 37, 1069
    (1959).
/3/ A Berne, G Boato and M de Paz, Il Nuovo Cimento 46B, 182 (1966).
/4/ I N Krupskii and V G Manzhelii, Soviet Physics JETP 28, 1097
    (1969).
/5/ F Clayton and D N Batchelder, J Phys C6, 1213 (1973).
/6/ R H Beaumont, H Chiara and J A Morrison, Proc Phys Soc (London)
    A78, 1462 (1961).
/7/ L Finegold and N E Phillips, Phys Rev 177, 1383 (1969).
/8/ D L Losee and R O Simmons, Phys Rev 172, 944 (1968).
/9/ P Korpiun and H-J Coufal, Phys Stat Sol (a) 6, 187 (1971).
/10/ P G Klemens in Solid State Physics, Vol 7, ed by F Seitz and
    D Turnbull, Acad Press Inc, New York (1958).
/11/ P Gruner and H Bross, Phys Rev 172, 583 (1968).
/12/ P Korpiun, H Meixner, H Peter and E Lüscher (to be published).

DISCUSSION ON ANHARMONIC EFFECTS AND PHONON-PHONON SCATTERING

Multiphonon Absorption in Alkali Halides   H Beck and D W Pohl
Page 361

R O Pohl: Do you understand the discrepancy between your theoretical curves for $\beta(T)$ and experimental data for high T values?

H Beck: In principle the experimental data are fairly well represented by the more simple-minded dashed curves of Fig 2, which are calculated by using T-independent phonon frequencies. However at low temperatures, frequencies with a T-dependence which is an extrapolation of the known behaviour at the Γ-point to the full zone, give a better fit. For T $\gtrsim$ 500K these full curves fail since 4-phonon processes were completely neglected in Fig.2. They would contribute considerably at high T and would yield the stronger rise of β with T shown by the data.

G A Gehring: Could you comment on the expected effect from the direct coupling of the radiation to higher powers of the displacement operators which you neglected.

H Beck: Unfortunately the corresponding coupling parameters are not known. Estimates of these effects have shown that for reasonable couplings the effect of the non-linear dipole moment will be small in alkali-halides. The temperature of the main quadratic dipole contribution is very similar to that of the 2-phonon contribution Γ_2. There are however also interference terms, which could change the Ω- and T-dependence of β.

****

Raman Scattering Studies of Anharmonicities in Potassium Halides
M S Haque, J A Taylor, J E Potts, J B Page Jr and C T Walker
Page 364

M D Tiwari: You have shown that there is 10% stiffening in the force constant change in the case of $KCl:Tl^+$; have you ever seen that this result is reflected by other properties of $KCl:Tl^+$ such as specific heat, thermal conductivity etc?

M S Haque: The 10% change is in the longitudinal overlap force-constant between n.n within the framework of a formal force-constant model for a particular type of displacement pattern. It may not be appropriate to generalise such a change to some other symmetry type.

J G Collins: What are the values of the Gruneisen parameters resulting from your measurements of $\Delta\omega$ as a function of $\Delta V/V_0$ for these spectral peaks?

M S Haque: I shall send you the results later. Bear in mind however that the Gruneisen parameters do not refer to any parti-cular mode in the crystal and that it is an average property of several modes that contribute to the peak in the spectrum.

****

Thermal Resistance of Dielectric Crystals at High Temperatures
P G Klemens and D J Ecsedy   Page 367

J E Parrott: What is effect of anisotropy as discussed by Herring, on your use of the Pomeranchuk suggestion?

P G Klemens: The effect should be small. The Herring processes are additional to the three-phonon processes used here, but are dominant only at lowest frequencies. At those low frequencies the mean free path would be limited by four-phonon processes. At lower temperatures the Herring processes would be important.

P Kocevar: For microwave attenuation at low temperatures the broad-ening of the thermal phonon states can weaken the restrictions due to energy-momentum conservation so strongly as to gain an order of magnitude for the longitudinal absorption rate. Do you have any guess of this effect in your cases?

P G Klemens: The same higher-order processes (two successive three-phonon processes) also occur here but they are effectively four-phonon processes, so that the picture will not change materially. The effect you mention is the special case when one of the parti-cipating phonons has a very low frequency.

H Weinstock: What would the effect (on high temperature thermal conductivity) be of a high density of lattice defects?

P G Klemens: The conductivity is reduced, because phonons are scattered. Scattering affects mainly the mean free path of the phonons of highest frequency. The overall effect is to reduce the temperature dependence; in the limit of strong point defect scattering $K(T) \propto 1/\sqrt{T}$.

G P Srivastava: 1. What do you expect the role of optical phonons compared to the role of four-phonon processes is in lattice thermal resistivity at elevated temperatures? 2. Do you expect that the phonon conductivity is sensitive to the sound velocity ratios? 3. Can you give a simple qualitative picture how one gets $\omega\tau^{4}$ dependence for τ^{-1} for high frequency longitudinal phonons?

P G Klemens: 1. Optical phonons reduce the mean free path of acoustic phonons, but since the interaction involves three phonons (A+A$\leftrightarrow$O or A+O$\leftrightarrow$O) the resulting resistivity varies as T at sufficiently high temperatures. 2. Yes, because this ratio controls the relaxation times. 3. Energy and momentum conservation requires that all participating modes have a frequency comparable to that of the scattered mode. Thus all phonon occupation numbers are proportional to T and $1/\tau$ varies as T. From Herring's scaling law $1/\tau \propto T^{n}\omega^{m}$ where $n + m = 5$ it then follows that $m = 4$.

****

The Thermal Conductivity of Solid Krypton between 2.4 and 39 K Determined from Measurements of the Thermal Diffusivity with a Heatwave Method   P Korpiun, J Moser, F J Pieringer and E Luscher

Page 377

D W Pohl: 1. What are the amplitudes of the thermal wave? 2. What was the damping? 3. What was the total energy input?

P Korpiun: 1. The temperature amplitudes have been estimated to have been between about 10^{-2} K at the heater and about 10^{-5} K at the thermometers. 2. The exponent in the damping term has the same form as the phase shift mentioned in my talk. Therefore it depends on the temperature and frequency. 3. The heat power input varied between about 1mW near 4.2K to about 100mW near 23K.

J E Parrott: What impurities are present in the krypton?

P Korpiun: The gas has had a purity of 99.997% krypton as it has been delivered by the 'Deutsche L'Air Liquide'. But the crystals have contained additional impurities. Different sorts of molecules initially absorbed at the walls and gaskets of the chamber have been condensed in the crystal during crystal growth.

THE ANISOTROPY OF PHONON PROPAGATION IN SOME CUBIC AND TRIGONAL CRYSTALS

J. DOULAT, M. LOCATELLI, J. RIVALLIN
Centre d'Etudes Nucléaires de Grenoble
Service des Basses Températures
BP 85 Centre de Tri - 38041 GRENOBLE-CEDEX (France)

It is now well known that the propagation of acoustic phonons in the ballistic regime is anisotropic in all crystals. This is because the propagation is entirely governed by the elastic modulus tensor, which is a fourth rank tensor and hence can have the same symmetry elements as, say, a cubic crystal, and yet not be spherical.

CALCULATION OF ACOUSTIC WAVE GROUP VELOCITY

If ρ is the specific mass of the crystal and C_{iklm} its elastic modulus tensor, the propagation properties of a wave of wave vector $\vec{q}$ can be derived from the second rank tensor [1] :

$$\Lambda_{im} = (1/\rho) \ C_{iklm} \ n_k \ n_l \quad (\text{where } \vec{n} = \vec{q}/|\vec{q}|)$$

Namely, its eigenvectors $\vec{n}$ are the displacement vectors, its eigenvalues are the squares of the corresponding phase velocities v. The group velocity is then given by

$$s_j = (1/\rho v) \ C_{ijlm} \ n_j \ u_j \ u_m \quad (\vec{u} = \text{unit vector})$$

A program has been written and run on a Telemecanique T1600 computer. Given the elastic modulus tensor and the specific mass, this program sets up a uniform distribution of wave vector directions on the unit sphere and computes the corresponding group velocities. Currently two versions are adapted to cubic and trigonal crystals respectively. The group velocity directions are plotted on a plane representation of the unit sphere which insures area conservation : if θ and ϕ are the Euler angles in the crystal axis system, then the plane circular coordinates are $\rho = 2 \sin \theta/2$ and ϕ. The plot is restricted to the useful sector, according to symmetry. The quasi-longitudinal and quasi-transverse modes are sorted and plotted separately. Such plots are given in figure 1. In every case, no drastic effect appears for the longitudinal mode (the plot of which is given only

383

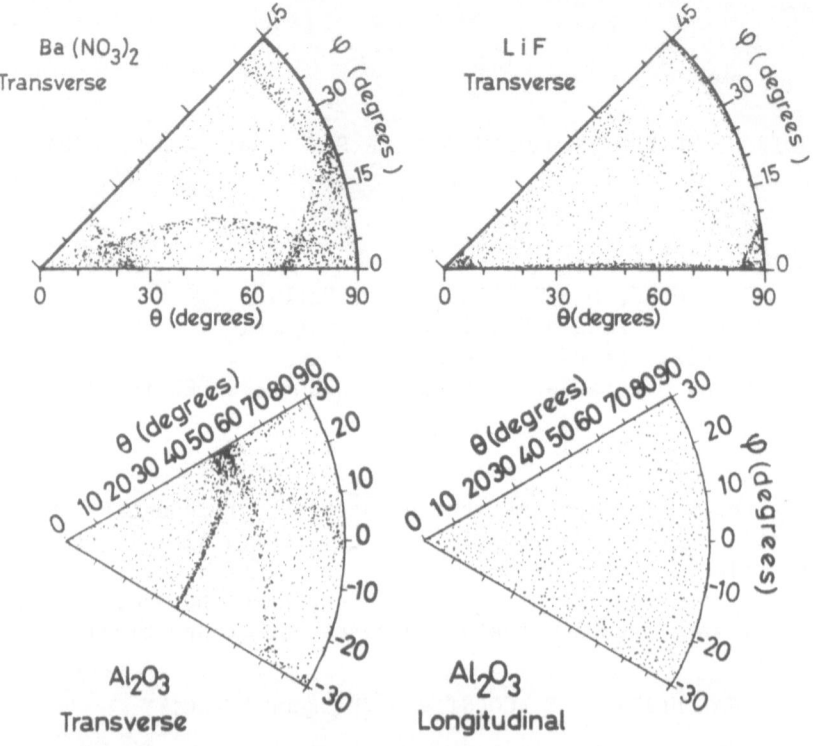

Fig. 1 – Distribution of group velocity directions resulting from
a uniform distribution of wave vector directions.

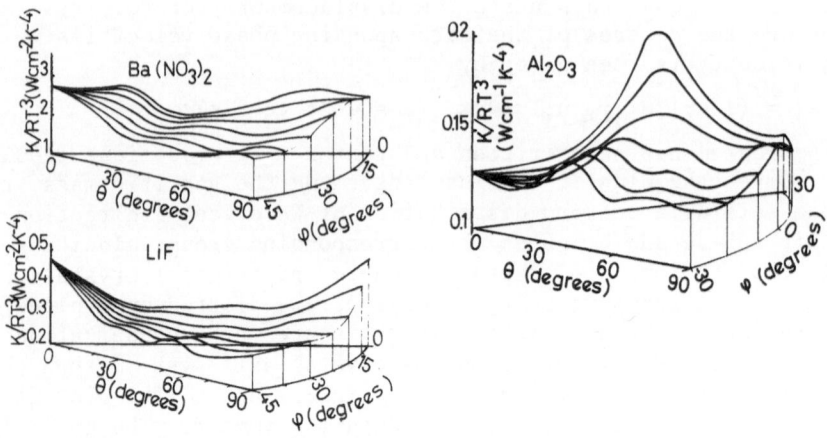

Fig. 2 – Calculated thermal conductivity as a function of heat flow
direction.

for Al_2O_3), but for the transverse modes one can see regions of intense concentration, which clearly illustrate the "focusing" effect.

BOUNDARY SCATTERING THERMAL CONDUCTIVITY

The thermal conductivity in circular rods at the low temperature limit (boundary scattering regime) is calculated, following roughly the method given by Mc CURDY et al.[2], with the same assumption that there is only diffuse scattering at the surface. Adding what can be left of Debye's approximation (no dispersion other than angular, summation on wave vectors within the Debye's sphere) results into the following formula which gives the thermal conductivity κ of an infinite length sample as a function of the temperature T :

$$\kappa = \frac{4}{45} \frac{k^4}{h^3} RT^3 \sum_p \int_\Omega \frac{s_z^2}{v^3 s_1} \, d\Omega$$

where R is the radius of the sample, p = 1, 2, 3 the index of polarisation, v the phase velocity, s_z the component of the group velocity along the sample axis, s_1 its perpendicular component. The integral is over a 4π solid angle. The correction for finite length has also been calculated, but the more complicated resulting formula is not given here.

A program computes κ by replacing the integral by a summation over 100 directions of $\vec{q}$ regularly spaced on the useful sector and as many on each of the other sectors, using symmetry considerations. Perspective computer-drawn plots (fig 2) show the variations of κ as a function of the heat flow direction (θ,ϕ). They follow with some smoothing the density variations of the group velocity vectors.

Experimental determinations of κ down to 0.05K have been done on samples of Al_2O_3 cut with four different orientations chosen in the maxima and minima regions (Table I). One finds a ratio $\kappa_{meas}/\kappa_{calc}$ very nearly constant. Its value, about 1.5, can be explained by specular reflection, which the calculation does not take into account.

TABLE 1

θ	ϕ	κ_{calc}/RT^3 (corrected for length)	κ_{meas}/RT^3	$\kappa_{meas}/\kappa_{calc}$
38	-30	0.146	0.208	1.43
58	-30	0.132	0.196	1.48
54	+30	0.191	0.276	1.45
10	-30	0.127	0.196	1.54

REFERENCES

[1] Fedor I. FEDOROV - Theory of Elastic waves in Crystals, Chap. 3 Plenum Press. N.Y. 1968
[2] A.K. Mc CURDY, H.J. MARIS, C. ELBAUM - Phys Rev. B, 2, 10, p 4077

INTERFACE EFFECTS ON PHONON FOCUSING IN RUTILE

R.A.M. van Lopik and H.W. de Wijn

Fysisch Laboratorium der Rijksuniversiteit

Utrecht, The Netherlands

The phenomenon of phonon focusing, $i.e.$, the influence of elastic anisotropy on the propagation of heat pulses through solids, was initially studied by Taylor $et\ al.$ (1). The framework of elastic continuum theory appears to be an appropriate starting point for calculations, but agreement is mostly qualitative. One way to improve upon the situation is to include the effects of reflection and refraction at the heater-crystal and the crystal-detector interfaces, again by using classical elastic theory. Another motivation for such calculations and their experimental verification is that these ideas are also applied in the more basic theories of the Kapitza resistance (see $e.g.$ Little (2)). We have chosen to study the angular dependence of the focusing of the longitudinal wave around the [110]-direction in tetragonal rutile (TiO_2), because in this case the focusing is of almost two-dimensional nature, making it more amenable to calculation.

When interfaces are not considered, a uniform distribution of injected $\vec{k}$-vectors is assumed within the crystal. For a given direction of $\vec{k}$ the phase velocities and displacements are calculated from the Christoffel equations (see Fedorov (3)). The relevant elastic constants are taken from Lange (4). Then the energy velocities $\vec{s}$, with polar angles θ and ϕ, are calculated, and sorted into 'boxes', where each box represents a certain range of θ, ϕ, and $1/|\vec{s}|$. It is essential to include $1/s$ in calculating the actual pulse shape at the detector in cases where s varies substantially with θ or ϕ. The dotted 'focusing function' in Fig. 1a is the result of a three-dimensional (3D) calculation of the longitudinal pulse height versus the detector angle ϕ relative to the [110]-direction in the (001)-plane. The boxes had dimensions $\Delta\theta = 12°$,

$\Delta\phi = 6.5°$, and $\Delta(1/s) = 25$ ns, while allowance was made for the
actual pulse duration of 100 ns. The $\vec{k}$-vectors searched had
$60° \leqslant \theta \leqslant 90°$ and $0° \leqslant \phi \leqslant 90°$. The dashed curve is the result of
a 2D calculation, $i.e.$, $\theta = 90°$ and $\Delta\theta = 0°$. It is seen that
focusing contributions from outside the (001)-plane are small for
the longitudinal mode.

Under the more realistic assumption of a uniform $\vec{k}$-distribu-
tion within the heater, rather than the crystal, it is necessary
to consider refraction at the heater-crystal interface. We only
consider wave vectors lying in the (001)-plane, the plane of incidence. The problem thus becomes 2D, as (001) is a symmetry plane
which must contain the resulting energy velocities. The interface
problem is solved by a suitable modification of the well-known
Snell's construction in optics. The calculations were checked for
having the sum of all reflection and refraction coefficients equal
to unity. Both incoming longitudinal and transverse waves must be
considered, since mode conversion may occur at the interface.
Their contributions are weighted with their relative abundances
within the (isotropic) heater material and subjected to the box-
sorting procedure described above. (See drawn curve in Fig. 1.)
At the crystal-detector interface refraction is towards the normal
so that, apart from the mismatch of acoustic impedances, there will
be little angular dependence. This has been confirmed by further
calculations. In our case the effect of the interfaces is to lift
the wings of the focusing function. The details of the calculations
show mode conversion to be of minor importance. The 2D calculation

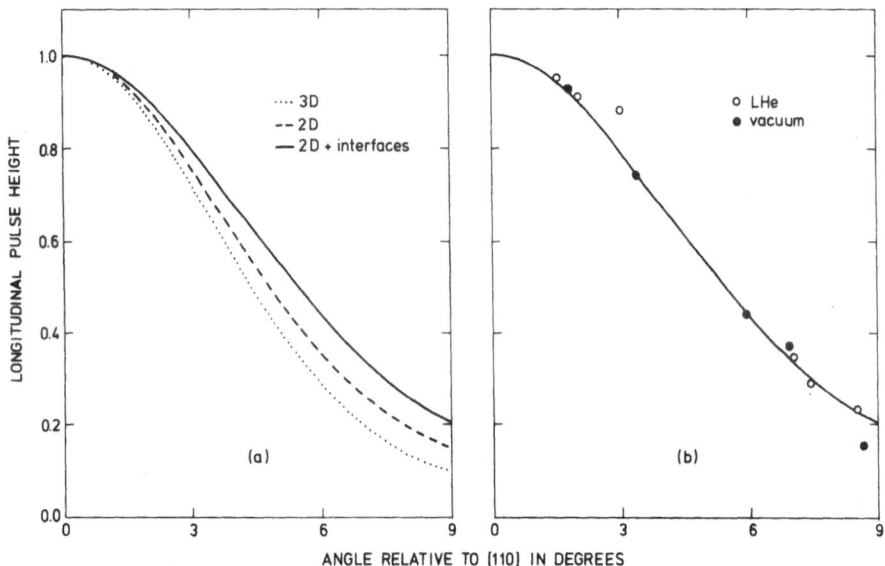

Fig. 1. Focusing function of longitudinal phonons.

is not suitable to predict the ratio of, say, the longitudinal
pulse height over the transverse one, since a 3D calculation without
interfaces shows the transverse wave with polarisation along [001]
to defocus by a factor of two.

The experiments were performed on a 10.2 mm long rutile single
crystal (obtained from Nakazumi Crystal Co., Japan) with the two
end faces normal to [110]. Five meander-shaped indium bolometers,
positioned at increments of 5° in the (001)-plane, were simultaneously vacuum deposited. The remaining slight deviations of the
residual resistances of the detectors at 4.2 K were used to correct
their individual sensitivities. This procedure has been checked
by observing the non-focusing transverse wave. With allowance for
the heater dimensions, each of the detectors subtended an area of
6.5° × 13°, with the shorter angle in (001). For vanishing $\Delta\phi$,
the calculated half-value width of the focusing function is 8°.
In the curves of Fig. 1 the 6.5° resolution of the detectors has
been included, which appeared to cause a broadening of only 1°.
Measurements were taken with the sample immersed in liquid helium
as well as with the sample in vacuum. Typical input pulse powers
were 2 W/mm² in the LHe case and 0.06 W/mm² in the vacuum case.
The bolometer signals, typically 25 μV, were wide-band amplified
and averaged with a boxcar system. The results are shown in
Fig. 1b. A normalisation was derived from the pulse height of the
transverse wave nearest to [110] of each run; there the experimental
L/T ratio was 2.2. For $\phi > 7°$, when the longitudinal pulse partly
coincides with a stronger oblique transverse wave, the interpretation had to be guided by the calculated pulse shape. It will be
seen that inclusion of the interface effects results in a better
agreement between continuum theory and experiment. Finally we note
that the serious heat loss to the liquid helium (at 3.4 K) of about
97% does not affect the focusing function or the L/T ratio. The
latter means that either the phonon transmission from heater to
liquid helium is equal for all modes, or that thermalisation is
complete within the 300 Å thick constantan heater.

(1) B. Taylor, H.J. Maris, and C. Elbaum, Phys. Rev. Lett. 23,
    416 (1969); Phys. Rev. B3, 1462 (1971).
(2) W.A. Little, Can. J. Phys. 37, 334 (1959).
(3) F.I. Fedorov, *Theory of Elastic Waves in Crystals*
    (Plenum Press, New York, 1968).
(4) J.N. Lange, Phys. Rev. 176, 1030 (1968).

CONICAL REFRACTION AND ACOUSTIC ACTIVITY IN QUARTZ

P Levinson and J Jouffroy

Laboratoire de Physique de la Matiere Condensee

College de France, 75005 Paris, France

Acoustic phonons of transverse polarisation which propagate in quartz with a wave vector parallel to the optic axis Oz can be considered to be degenerate when their wavelength is sufficiently large that the symmetry seen by the excitations should be that of the point lattice. Under these conditions the wave surface for this mode shows a conical point on the axis Oz. Using the method of heat pulses with an emitter of small dimensions energy can be detected at 17° from the axis. This is internal conical refraction.

On the other hand in the direction of the axis Oz, energy propagation can be detected which corresponds to the points on the wave surface where the tangential plane is perpendicular to Oz. These points form a line of contact with a particular tangential plane. The corresponding wave vectors form a cone the opening of which slowly varies with azimuth and is in the vicinity of 20°. This is external conical refraction or focussing (Fig 1). It is this which produces the more intense signal seen in the direction Oz.

These two phenomena are described by a calculation of the 'elastic' type where the waves propagate in the middle of a continuous medium having the symmetry of the point lattice and where the phase

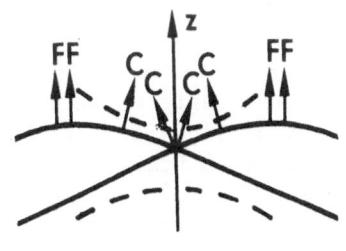

Fig 1: Wave surface for quartz : section of the plane containing the ternary axis and a binary axis. The normals C correspond to conical propagation, the normals F to focussed propagation. The dashed line is a 'surface of dispersion' near to Oz.

velocity, like the group velocity, is only dependent on the direction
of the wave vector.

When the wave vector increases (that is to say when the temperature of the emitter is raised to generate phonons of higher frequency) the symmetry seen by the excitations is the true symmetry of
the crystal. The phase velocity and the energy velocity depend on
the size of the wave vector. The description of the propagation is
made with the help of a dynamical matrix where the 'elastic' matrix
is the term of order zero in q (the length of the wave vector). The
shape of the wave surface then depends on q but for sufficiently
small wave vectors compared to the size of the Brillouin zone this
distortion is hardly perceptible except in the region near the Oz
axis. Indeed the conical point of the wave surface then makes two
separate sheets each having a tangential plane perpendicular to the
axis. Thus for these frequencies conical refraction does not exist
any more but is energy propagation parallel to the axis with double
valued velocity, this is acoustic activity.

The detection of this phenomenon by means of heat pulses is made
awkward by the presence of a focussed peak at a velocity near to
those of the double valued velocities. We have succeeded in ident-
ifying the various observed signals by simulating 'elastic' propag-
ations for all directions of the wave vector. The peaks arising
from the acoustic activity are then recognised as those which are
not reproduced by this simulation.

The simulation calculation takes account of the following hypo-
theses: the phonons in the emitter have a Boltzmann energy distribution, acoustic mismatch at the interfaces, and the high frequency

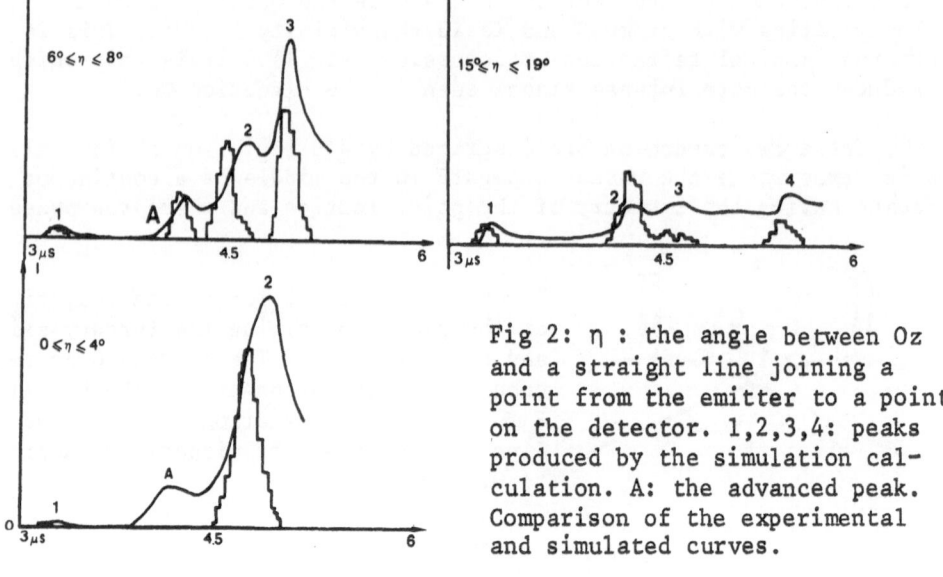

Fig 2: η : the angle between Oz
and a straight line joining a
point from the emitter to a point
on the detector. 1,2,3,4: peaks
produced by the simulation calculation. A: the advanced peak.
Comparison of the experimental
and simulated curves.

limitation comes from the temperature and not from particular cut offs for each mode. It takes account of all the observed signals except for one or two. The signal corresponding to the increased speed is present each time the arrangement has the appropriate geometry; the retarded signal is most of the time obscured by focussing or by diffusion phenomena which lengthen the signal.

We have used two types of detectors for studying the position and intensity of these peaks: a superconducting bolometer and a superconducting tunnel junction sensitive only to frequencies higher than 2.8 10" Hz (2).

The intensity ratio R of a 'normal' signal like that from the focussed peak to that of a signal of acoustic activity like that from the advanced peak reflects the difference in their frequency composition arising from variations in emission temperature.

As the temperature increases at low temperatures, advanced peak detected by the bolometer grows more quickly than the normal peak since the first is composed only of the highest frequencies. It stops growing when the speed of the highest frequency phonons has reached the region where the dispersion curve makes a parallel tangent to the tangent at the origin. When one replaces the bolometer by a tunnel junction the advanced peak and the focussed peak begin by growing at the same rate because the contribution from the low frequencies of the focussed peak are not detected by the junction (cf table below).

TABLE

Emission voltage (volts)	2	3	4	5	8,4
Approximate temperature of the emitter	(3,8)	(4)	(4,5)	(5)	(6,5)
R measured by the junction	5,8	5,8	6	5,9	6,2
R measured by the bolometer	11,6	7,8	6,6	6	6,3

The speed of the advanced peak reaches its maximum (point of inflexion on the dispersion curve) at frequencies corresponding to an emission temperature of near 5K. In order to see this the measurements must be made at low heater voltages (2 to 5V) and in a helium bath whose temperature is lower than 1.5K. The relative shift in the peak is small (of the order of 2%) but under good conditions it can be seen by using as a reference the arrival time of longitudinal phonons since this is practically independent of temperature.

(1) J Jouffroy, P Levinson, J Physique 36, (1975) 709.
(2) P Levinson, J Jouffroy, C Acad Sc Paris 280 b (1975) 571.

ACCURACY OF THE BOUNDARY RELAXATION TIME APPROXIMATION

W. Bausch

Fachhochschule Darmstadt

D61 Darmstadt, Germany

Phonon transport in crystals of finite diameter d is usually described by the use of a boundary relaxation time $\tau_B = d/c$ [1] (c is the phonon velocity, no distinction will be made between longitudinal and transverse phonons). To test the validity of this approximation, we start from the Boltzmann equation for the disturbed distribution function N [2]:

$$\vec{c}\nabla N = c_x \frac{\partial N}{\partial x} + c_y \frac{\partial N}{\partial y} + c_z \frac{dT}{dz}\frac{dN_0}{dT} = -\frac{N - N_0}{\tau_R} - \frac{N - N(\vec{\lambda})}{\tau_N} \quad (1)$$

Here τ_R is the relaxation time for the resistive processes (without τ_B) and τ_N that for the normal (N-)processes; $\vec{\lambda}$ is the velocity of the drifting phonon distribution to which the N-processes relax. Usually the x,y-dependence of N is approximated by the use of τ_B, we take the radial dependence of N exactly into account but replace $\lambda(x,y)$ by its mean value $\bar{\lambda}$. $\bar{\lambda}$ is determined from the requirement that the change of phonon momentum $\hbar\vec{q}$ due to N-processes should be zero. We then obtain on averaging over the sample radius R

$$\int_0^R r\,dr \int \frac{N(\bar{\lambda}) - N}{\tau_N}\,\vec{q}\,d^3q = 0 \quad (2)$$

N being known from (1) and (2) the thermal conductivity can be calculated with the result

$$K = G\,T^3 \left[\langle \tau_C\,F(\Delta)\rangle + \frac{\langle \frac{\tau_C}{\tau_N}F(\Delta)\rangle^2}{\langle 1/\tau_N - \frac{\tau_C}{\tau_N^2}F(\Delta)\rangle}\right] \equiv K_1 + K_2 \quad (3)$$

where $\tau_C^{-1} = \tau_N^{-1} + \tau_R^{-1}$, $G = \frac{k}{2\pi^2}c\,(\frac{k}{\hbar})^3$, $\Delta = d/\tau_C c = \tau_B/\tau_C$

and $\langle A\rangle = \int_0^{\theta/T} A(x)\,\frac{x^4 e^x}{(e^x - 1)^2}\,dx.$

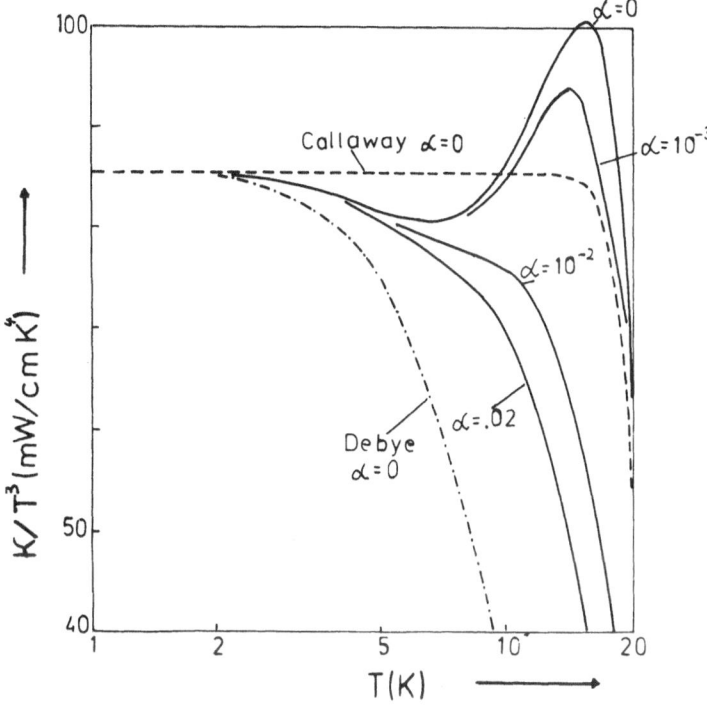

Fig 1. κ/T^3 vs. T of LiF for the expanded Callaway model. Results due to the original Callaway and the Debye model are shown for comparison.

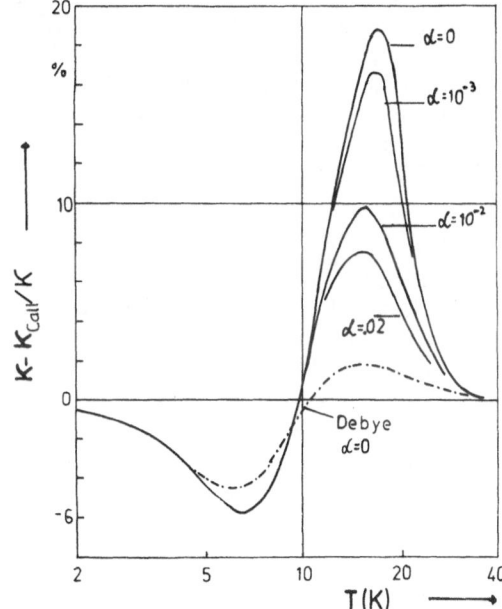

Fig 2. Relative difference of thermal conductivities of LiF according to the expanded and original Callaway model vs. T

The function $F(k)$ was introduced by Dingle [3] as the ratio of the electrical conductivities of a wire of thickness d and the bulk:

$$F(k) = 1 - \frac{12}{\pi} \int_0^1 (1-x^2)^{1/2} S_4(kx) \, dx \tag{4}$$

where

$$S_4(u) = \int_1^\infty e^{-ux}(x^2-1)^{1/2}x^{-4} \, dx \tag{5}$$

Dingle gives approximate expressions for $F(k)$ for $k < 0.1$ and $k > 5$. In order to calculate $F(k)$ for all values of k we have fitted $S_4(u)$ by a Chebyshev polynomial and have then integrated (4).

In (3) the thermal conductivity is written as the sum of two terms, the Debye term K_1 and the drifting phonon term K_2. For the usual boundary relaxation time approximation the Debye term is written as

$$K_1^B = G \, T^3 \langle \tau_{CB} \rangle \tag{6}$$

with $\tau_{CB}^{-1} = \tau_C^{-1} + \tau_B^{-1}$.

To compare the results of our model with those from the Callaway model, we have used the experimental relaxation time for LiF [4] :

$$\tau_C^{-1} = \tau_N^{-1} + \tau_{Umklapp}^{-1} + \tau_{imp}^{-1} = 35xT^4 + 600x^2T^4e^{-170/T}$$
$$+ \alpha x^4T^4 \quad s^{-1} \tag{7}$$

with $\alpha \approx 5 \times 10^{-6} \times$ isotope concentration in ppm and $\tau_B = 10^{-6}$ s. The comparison of the Debye terms K_1 (from our expanded model, first term in (3)) and K_1^B (6) shows that the results always agree within 4.8% for different α (fig.2). The total conductivities K according to (3) and to the Callaway model are shown in fig.1 and compared in fig.2. For small values of the impurity concentration, there is an indication of Poiseuille flow in this model, whereas the Callaway model cannot explain Poiseuille flow even for $\alpha = 0$.

References

[1] H.B.G. CASIMIR, Physica 5, 495 (1938).
[2] J. CALLAWAY, Phys.Rev. 113, 1046 (1959).
[3] R.B. DINGLE, Proc. Roy. Soc. A201, 545 (1950).
[4] R. BERMAN and J.C.F. BROCK, Proc. Roy. Soc. A289, 46 (1965).

SCATTERING OF PHONONS BY DISLOCATIONS IN CRYSTALS

Yoshiaki Kogure and Yosio Hiki

Tokyo Institute of Technology

Oh-okayama, Meguro-ku, Tokyo 152, Japan

THEORY

On the basis of nonlinear elasticity theory, the scattering of lattice waves by a static strain field has been treated with special consideration of the anisotropy of wave-propagating medium [1]. The equation of a small-amplitude elastic wave propagating in a statically strained crystal is

$$\rho_0 \ddot{u}_i = (c_{ismr} + D_{ismr})(\partial^2 u_m / \partial X_s \partial X_r). \tag{1}$$

Here ρ_0 is the density of undeformed material, u_i is the displacement due to the wave, and D_{ismr} is a combination of the second- and the third-order elastic constants, c_{ismr} and C_{ismrtu}:

$$
\begin{aligned}
D_{ismr} = {}& c_{ksmr} U_{i,k} + c_{ikmr} U_{s,k} + c_{iskr} U_{m,k} + c_{ismk} U_{r,k} \\
& + c_{srtu} \delta_{im} U_{t,u} + C_{ismrtu} U_{t,u},
\end{aligned} \tag{2}
$$

where U_i is the displacement due to the static strain field and $U_{i,k} = \partial U_i / \partial X_k$ is the displacement gradient. The amplitude of scattered waves can be determined by regarding the second term in eq.(1) as a perturbation and by adopting the Born approximation. The scattering cross section for phonons by the static strain field can thus be calculated. The merit of our treatment is two-fold, namely, the anisotropic scattering of phonons is fully taken into account and the contributions of phonons with different modes can be evaluated separately. The general method was applied to the case of scattering by an edge and a screw dislocations in a cubic crystal [2]. The results of numerical calculation showed that the scattering was very anisotropic with regard to the

incident direction and the scattering angle of the phonons. It
was found that, contrary to the dominance of purely forward scattering in isotropic medium, obliquely forward scattering occurred
intensively in crystals. The total amount of the effective scattering becomes large when the anisotropy is taken into account,
because purely forward scattering has no contribution to the effective scattering of incident phonons. It was also found that
the scattering cross widths of both an edge and a screw dislocations for transverse mode phonons were larger than those for
longitudinal phonons.

By using the calculated scattering cross widths and by adopting the relaxation-time approximation, the thermal resistivity of
the dislocation was calculated. It was argued that the slow
transverse phonons had dominant contribution to the decrease of
thermal conductivity due to dislocations. In the case of metals,
the total thermal conductivity can be expressed as

$$\kappa = \kappa_e(\text{electronic}) + \kappa_g(\text{lattice}) = AT + [DE/(D+E)]T^2, \quad (3)$$

where D and E are the coefficients of the lattice thermal conductivity when the scattering of phonons by dislocations and by
conduction electrons are considered separately. Here D is inversely proportional to the dislocation density N. Our numerical
calculation for copper resulted that $DN \simeq 5 \times 10^7$ W/cm$^3 \cdot$deg^3,
which was noticeably smaller than the values previously obtained
by other authors [3]. Namely, our treatment concluded much larger
scattering by dislocations.

EXPERIMENTS

Experiments have been done for copper-aluminum polycrystals
with 2, 5, 8, and 12 at.% aluminum by using the modified tempera-
ture-wave (Ångström) method [4] in the temperature range 1.4 -
7.5 K. A sinusoidal variation of temperature propagating in a
specimen is produced by an AC current in a heater attached to the
specimen. The thermal diffusivity is obtained from the phase difference of the temperature wave between two thermometers on the
specimen, and the thermal conductivity is determined from the
build-up static temperature gradient between the thermometers.
Specific heat can be obtained from the thermal diffusivity and
conductivity. Dislocations were introduced in the annealed specimens by tensile deformation, and the densities of dislocations
were determined by measuring the tensile flow stresses. Dislocation densities in the deformed specimens were about 10^{10} cm^{-2}.

As was expected from eq.(3), a reasonably linear relationship
was observed in the κ/T -vs- T plots for both the annealed and
the deformed specimens. The experimental values of DN determined

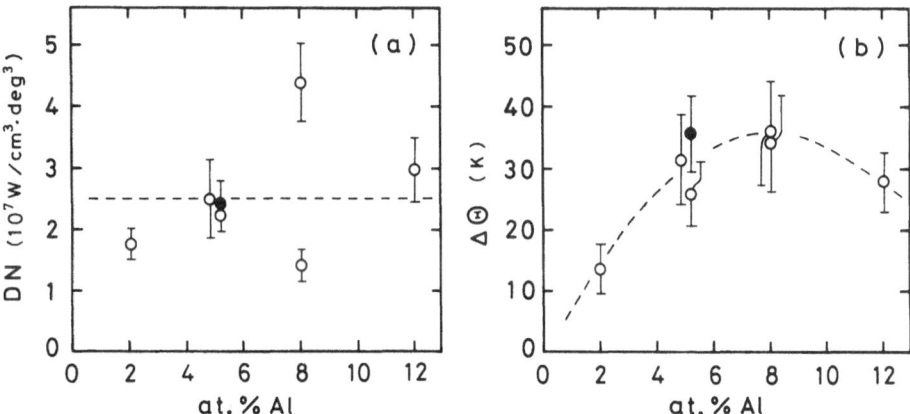

Fig. 1.   The coefficient of the lattice thermal conductivity due
          to dislocation scattering DN—(a), the change of Debye
          temperature due to dislocations $\Delta\Theta$—(b).

from these plots are shown in Fig. 1(a). No apparent dependence
of DN on the aluminum content can be seen, and the mean value of
DN is 2.5×10^7 W/cm$^3 \cdot$deg^3. The solid circle represents the result for a specimen deformed and then irradiated with Co^{60} γ-rays
to a total dose of 10^7 R. No meaningful change of DN is produced
by the γ-irradiation. It was further found, from the specific
heat data, that the Debye temperatures of alloy crystals were
strongly increased by the introduction of dislocations. The differences of the Debye temperatures in deformed and annealed specimens, $\Delta\Theta$, are shown in Fig. 1(b). The effect is considered to be
characteristic in alloy crystals [5]. Since the phonon conductivity κ_g is proportional to Θ^{-2}, the experimental values of DN
should be corrected when they are compared with the theoretical
value, which is derived after assuming that the Debye temperature
is not changed by the introduction of dislocations. The corrected
value averaged over all specimens was $DN = 4.8 \times 10^7$ W/cm$^3 \cdot$deg^3,
being very close to our theoretical value.

                              REFERENCES

[1] Y. Kogure and Y. Hiki, J. Phys. Soc. Japan, 36, 1597 (1974).
[2] Y. Kogure and Y. Hiki, J. Phys. Soc. Japan, 38, 471 (1975).
[3] for example, P. G. Klemens, Proc. Phys. Soc. A68, 1113
     (1955).
[4] Y. Kogure and Y. Hiki, Japan. J. Appl. Phys. 12, 814 (1973).
[5] Y. Kogure and Y. Hiki, J. Phys. Soc. Japan, 39, No.3 (1975);
     also see Y. Hiki, T. Maruyama and Y. Kogure, J. Phys. Soc.
     Japan, 34, 725 (1973).

DISCUSSION ON DEFECT SCATTERING AND PHONON FOCUSSING

The Anisotropy of Phonon Propagation in some Cubic and Trigonal Crystals   J Doulat, M Locatelli and J Rivallin   Page 383

R A M van Lopik: As interface effects are important when considering focussing, where did you fix your heaters on to the crystal?

J Doulat: Heater and sink are fixed laterally at the ends of the sample, but we hope the interface effect is not important because the thermometers are pretty far from the ends.

R Berman: Do you expect the conductivity enhancement due to specular reflection to be independent of angle, or is it too complicated to calculate?

J Doulat: The comparison of experimental and calculated values (table 1) suggests that the relative enhancement due to specular reflection does not depend on the sample orientation. It is probably complicated to calculate but we intend to try.

****

Accuracy of the Boundary Relaxation Time Approximation   W Bausch
                                                           Page 392

P G Klemens: I always believed that a Poiseuille effect arises from $\vec{\lambda}$ being a function of position. In your calculations you took an average value $\langle \vec{\lambda} \rangle$. I do not see how you could get a Poiseuille effect with this assumption.

W Bausch: You are right. I was also surprised that there is a rise in the Poiseuille flow region. We have now preliminary results for a position dependent $\vec{\lambda}$, which show that the above results give a lower limit for Poiseuille flow.

H Maris: Another complication in the T^3 region comes from phonon focussing which can enhance the conductivity by as much as a factor of 2. It is interesting that the phonons responsible for this enhancement have velocities very near to the direction of the rod

398

axis and therefore they travel a long way between successive
scatterings at the wall. Thus, the enhancement expected is
reduced by even a small amount of phonon scattering in the bulk
and so one obtains deviations from T^3 behaviour at temperatures
substantially below the temperature for which $\Lambda \sim d_0$.

W Bausch: I did not take phonon focussing into account. The
effect you mention is important at low temperatures when $\Delta < 1$.
But as shown here the boundary approximation is not quite exact
in the region $\Delta \sim 1$; this also remains true for transport without
N-process.

****

Scattering of Phonons by Dislocations in Crystals   Y Kogure and
Y Hiki   Page 395

R Berman: In your abstract you say that your values of DN are
much smaller than previous calculated values. Can you say by how
much they are smaller?

Y Kogure: We show the DN values (in unit of $10^7 W/cm^3 deg^3$) for Cu
calculated by other authors together with our calculated value and
our experimental value

Klemens	220
Ohashi	1.59
Bross et al	14.3
Present (Cal)	5.1
Present (Expt)	4.8

HIGH FREQUENCY PHONON EMISSION FROM SUPERCONDUCTING Al-TUNNELLING

JUNCTIONS

W. Eisenmenger

Universität Stuttgart, Phys. Inst., Teilinst. 1

7 Stuttgart 80, Germany

In single particle tunneling between identical superconductors phonons are generated by relaxation – and recombination transitions
/1/ of injected quasiparticles. Phonon detection /1/ is mediated by
the breaking of Cooper-pairs by phonons with energy exceeding the
superconducting energy gap $2\Delta_D$. This leads to an increase of the
quasiparticle population and a corresponding contribution to the
tunneling current in the thermal tunneling regime $0 < eV < 2\Delta_D$
(V = battery voltage).

Voltage tunable phonon absorption spectroscopy /2/ makes use
of the sharp upper limit of the continuous relaxation phonon or
phonon Bremsstrahlung spectrum /2/ at the energy $\Omega = eV - 2\Delta_G$. In
strong coupling superconductors as Sn and Pb the accessable energy
range extends from zero energy to the gap energy $2\Delta_G$ since phonon
reabsorption by Cooper-pairbreaking /3/ prevents the escape of relaxation phonons with higher energy than $2\Delta_G$. In weak coupling superconductors as Al, phonon reabsorption is reduced and in addition
injected quasiparticles have relatively long relaxation time constants leading to quasiparticle diffusion from the barrier to the
substrate boundary before phonon emission by relaxation takes place
/4/. This results in a finite phonon escape probability also at energies exceeding the limit of $2\Delta_G$ as has been given evidence in experiments /4/ /5/ using Al junctions as phonon generator and Sn
junctions as detector. At these high phonon energies the shape of
the spectrum at the upper edge $eV - 2\Delta_G$ is changed as compared to
the simpler steplike behavior /3/ for maximum energies below $2\Delta_G$.

Theoretical and experimental information on the high energy
phonon spectrum at finite temperature or quasiparticle occupation
is important for spectroscopic applications but also gives valuable

insight into the different quasiparticle transitions.

The relevant quasiparticle transition processes leading to different contributions to the phonon spectrum are indicated in Fig. 1 as follows:

A: recombination of quasiparticles at the gap edge leading to phonon emission at $2\Delta_G$;

B: relaxation of quasiparticles injected from completely occupied states leading to a continuous spectrum with cutoff at $\Omega = eV - 2\Delta_G$;

C: direct recombination of the same quasiparticles with e.g. thermally excited quasiparticles resulting in a continuous spectral background with a peak at $\Omega = eV$;

D: relaxation of quasiparticles injected from thermally occupied states also resulting in a peak $\Omega = eV$;

E: direct recombination of quasiparticles injected from thermally occupied states. The peaklike phonon contribution slightly above $\Omega = eV + 2\Delta_G$ has a sharp <u>low</u> frequency cutoff at $\Omega = 2\Delta_G + eV$ /6/ /7/.

Within the limits of the BCS approximation the complete spectrum can be obtained quantitatively by numerical convolution procedures /4/ /5/, as shown in Fig. 2, where the contributions of the transitions B, C, D are clearly depicted. The contribution of process E is too small as to appear on this scale, also the major contribution to transitions resulting from recombination of already relaxed quasiparticles, i.e. secondary transitions, is not shown. The calculation has been also extended to the normal conductor limiting case for T = 0 introduced in Fig. 2 by dotted lines. In this case the sharp upper frequency edge is changed to a linear intensity decay cutting the zero level at $\Omega = eV$.

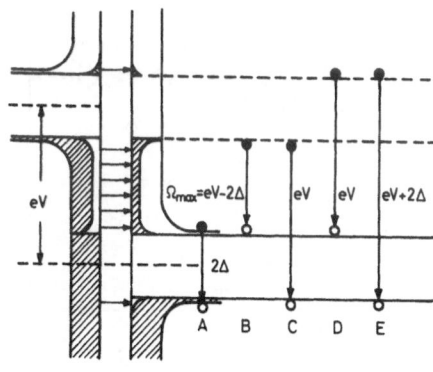

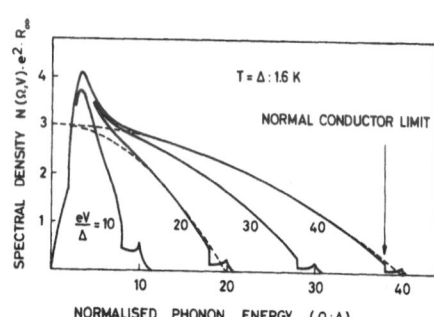

Fig. 1 Quasiparticle Transitions and Phonon Energies, c.f. text. Ref. /6/

Fig. 2 Calculated Generator Phonon Spectrum. Ref. /5/

For experimental comparison an Al generator and a Sn detector
with Si-substrate have been used /6/ /7/. The signal derivative with
respect to the generator - current (or voltage) as function of the
generator voltage, as shown in Fig. 3 /7/, has the same shape as
the calculated spectrum but with inverted energy scale. This can be
interpreted in terms of the sharp detector sensitivity threshold at
$\Omega = 2\Delta_D$ and the common modulation technique by superimposing a
small pulse or AC signal on the generator DC voltage. The detector
modulation signal then corresponds to the first signal derivative
with respect to the generator voltage. As can be seen from Fig. 2 a
small change in the generator voltage produces an almost parallel
shift of the spectrum to higher energies. The resulting change in
the detector signal under this condition is determined by the spectral intensity at the threshold frequency $\Omega = 2\Delta_D$. The spectral
structures of Fig. 2 are found in Fig. 3: The steep signal rise at
$eV = 2\Delta_{Sn} + 2\Delta_{Al}$ corresponds to the onset of the main relaxation
process B at $\Omega = eV - 2\Delta_G = 2\Delta_D$; the maximum at $eV = 2\Delta_{Sn}$ corresponds
to processes C and D with $\Omega = eV$; the smaller maximum at $eV =$
$2\Delta_{Sn} - 2\Delta_{Al}$ is clearly revealed at higher amplification and reflects
process E with $\Omega = eV + 2\Delta_G$ /6/ /7/. The weak oszillatory structure
at $eV = 2\Delta_{Sn} + 2\Delta_{Al}$ can be attributed to longitudinal phonon interferences in the generator film of 1125 Å total film thickness. Time
of flight pulse separation indicates that only longitudinal phonon
interferences are observed. With larger film thickness values the
interference structure disappears.

For spectroscopic applications the modulation method for taking
the first signal derivative $di_s: dV$ /2/ is appropriate for a gen-

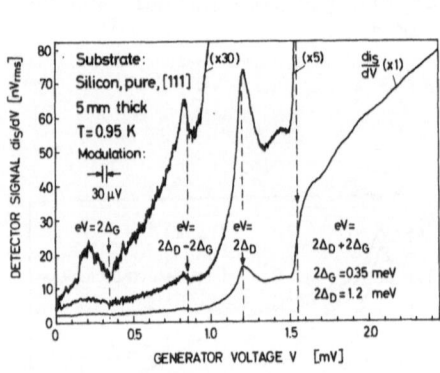

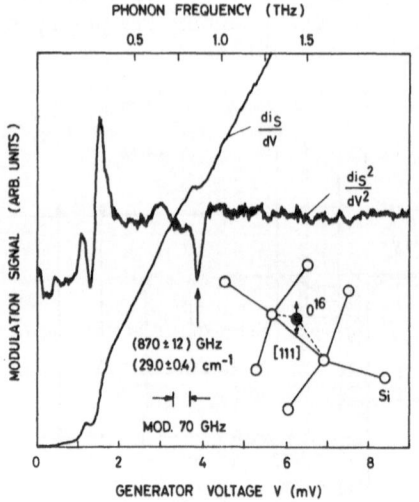

Fig. 3 Differential Signal for
an Al-Generator and Sn-
Detector on Pure Si Sub-
strate. $2\Delta_{Al}$ = 0.35 meV,
$2\Delta_{Sn}$ = 1,2 meV. Ref. /7/

Fig. 4 Phonon Absorption in Si: 0,
Al-Generator and Sn-De-
tector. Ref. /5/

erator spectrum with a large step at the high frequency cutoff at $\Omega = eV - 2\Delta_G$. Under this condition the modulation results in a narrow effective phonon band at $\Omega = eV - 2\Delta_G$. In the limit of very high generator voltages compared to Δ_G the cutoff step at $eV - 2\Delta_G$ is reduced in height and becomes comparable to the thermal struc- tures. This situation can be approximated by the normal conductor spectrum. It can be demonstrated /5/ /6/ that this spectrum also re- sults in a narrow effective phonon band if the A.C. modulation tech- nique is extended to the second derivative of the detector signal $d^2i_s : dV^2$ by recording the second harmonic of the detector modu- lation.

This method has been used in measurements of the phonon reso- nance scattering of oxygen impurities in silicon /5/, as shown in Fig. 4. Again an Al generator and a Sn detector were evaporated on the doped Si crystal. The structures of the first signal derivative at low generator voltages correspond to the results in Fig. 3. In addition, a pronounced plateau appears at 3.9 meV as expected for the oxygen resonance scattering. In the limit of high generator voltages the signal derivative approaches a linear dependence on voltage in accord with the detector sensitivity increasing in pro- portion to phonon energy. Phonons of energy $\Omega > 4\Delta_D$ are reabsorbed by pairbreaking and the succeeding relaxation of quasiparticles gives rise to additional phonons of $2\Delta_D$.

The second signal derivative $d^2i_s : dV^2$, introduced in Fig. 4, reveals the absorption maximum in the proper shape indicating that the generator spectrum at these energies $eV \approx 2\mathbf{4}\Delta_G$ can be approxi-

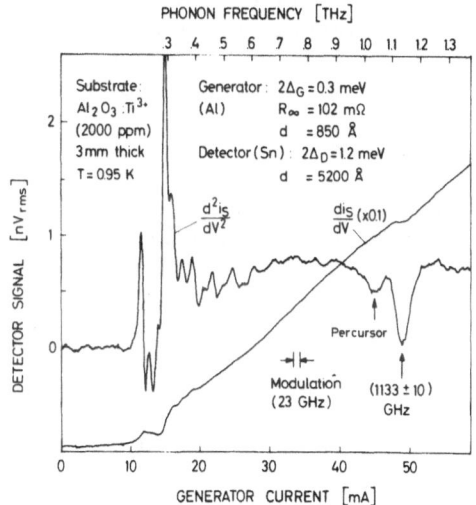

Fig. 5 Phonon Absorption in $Al_2O_3:Ti^{3+}$, Al-Generator and Sn-Detec- tor. Ref. /7/

mated by the normal conductor limiting case. The resonant frequency of the oxygen atom results with 870 ± 12 GHz in good agreement with far infrared data /5/.

Evidence for phonon emission from Al-junctions at still higher frequencies has been found by W. Forkel very recently in using resonant scattering by electronic transitions between ground state levels of Ti^3 in Al_2O_3 /7/. The result of the experiment being similar to the one discussed before, is presented in Fig. 5. The first and second signal derivative show strong absorption at the phonon energy $\Omega = eV - 2\Delta_G$ corresponding to a frequency of 1133 ± 10 GHz in close agreement with the far infrared value /8/ of $37.\overline{8}$ cm^{-1}. The precursor of the absorption line in the second derivative results from a smaller phonon contribution at $\Omega = eV$, being detected as a consequence of the high experimental resolution. The observed linewidth of 55 GHz exceeds the experimental resolution of 23 GHz and also the linewidth of far infrared absorption. This may be due to the high Ti concentration of nominally 0.2 %. The second derivative also shows strong oscillations in the range of the detector energy gap structure resulting from longitudinal phonon interferences in the generator film of 850 Å total thickness. The sound velocity results with $6.3 \cdot 10^5$ cm/s in agreement with the bulk value.

Besides its application to phonon spectroscopy the observed high energy phonon escape from Al-tunneling junctions is expected to provide more information on the absolute strength of the phonon - electron interaction in Al.

I wish to thank W. Forkel, K. Laßmann and M. Welte for valuable discussions. Recent experimental results /7/ were kindly supplied by W. Forkel prior to publication. Financial support by the Deutsche Forschungsgemeinschaft is gratefully acknowledged.

References:

/1/ W. Eisenmenger and A.H. Dayem, Phys. Rev. Lett. <u>18</u>, 125, 1967
/2/ H. Kinder, Phys. Rev. Lett. <u>28</u>, 1564, 1972
/3/ H. Kinder, K. Laßmann and W. Eisenmenger, Phys. Lett. <u>31 A</u>, 475, 1970
/4/ M. Welte, K. Laßmann and W. Eisenmenger, J. de Physique, <u>33</u>, C4-25, 1972
/5/ W. Forkel, M. Welte and W. Eisenmenger, Phys. Rev. Lett. <u>31</u>, 215, 1973
/6/ W. Forkel, in "Microwave Acoustics" (E.R. Dobbs and J.K. Wigmore ed.), Proc. Eighth Int. Congr. on Acoustics, Inst. of Phys. London 1974.
/7/ W. Forkel, supplied prior to publication, 1975
/8/ R.R. Joyce and P.L. Richards, Phys. Rev. <u>179</u>, 375, 1969.

TIME REVERSED ULTRASONIC ECHOES AS AN EXPERIMENTAL TOOL

N. S. Shiren and R. L. Melcher

IBM Thomas J. Watson Research Center
P. O. Box 218
Yorktown Heights, New York   10598

I.   INTRODUCTION

Ultrasonic parametric backward wave interactions were first observed by Damon and Van de Vaart[1] who utilized the "parallel pumping" magnetoelastic interaction in YIG. In this method the backward propagating wave, at frequency ω, is generated through the parametric interaction of a forward propagating wave at ω, with a non-propagating pump field at 2ω. Subsequently, a similar interaction, second order piezoelectricity, was investigated by Thompson and Quate[2] in $LiNbO_3$. This effect exists in all piezoelectric crystals, and has been extensively utilized in nonlinear surface wave devices. More recently, two other ultrasonic backward wave interactions, ultrasonic spin echoes[3], and polarization echoes[4], have been discovered. For both these phenomena the pump frequency is the same as that of the forward and backward waves, ω[5]. The source of the nonlinearity in the case of polarization echoes occurs through electric field induced ionization of trapped electrons[6], resulting in a space and time modulated free electron distribution, and/or a space modulated trapped charge distribution. We define two types of polarization echoes, parametric and holographic, according as the backward wave is generated by interaction of the forward wave with either the free or trapped distributions, respectively.

All four phenomena are generally termed echoes because of their observational similarity to spin echoes; i.e. when a pump pulse is applied at time τ following the forward wave launching pulse at time 0, the backward wave pulse is detected at time 2τ. In addition, for ultrasonic spin echoes and holographic polarization echoes, application of a subsequent pulse at time T results in a stimulated echo

at time $T + \tau$. Furthermore, all nonlinear echo phenomena have in
common the property of time reversal. In spin echoes, as is well
known, the motion of the spin isochromats is reversed in time following application of the pump pulse. Similarly (see Sec. II), for the
backward wave echoes forward wave vectors are reversed in the echoes,
which implies time reversal. Elastic scattering processes in the
forward propagation path then proceed backward in time for the backward wave. (Note that this is not the case for a simple back reflection.) We therefore use the terminology, Time Reversed Ultrasonic
Echoes, or TRUE.

The theories, experimental characteristics, and material properties for the various echo phenomena have been extensively reported
in the literature and will not be reviewed here. We confine ourselves in Section II, to a discussion of some of the properties of
TRUE that make it attractive as an ultrasonic experimental technique.
In Sec. III we present two experimental demonstrations of the useful-
ness of the technique and some speculation on other possibilities
which have been under investigation at our laboratory.

II. USEFUL PROPERTIES OF TRUE

A. Time Reversal

Although the original observations of Damon and Van de Vaart
were reported in 1965, the presence of time reversal, and its importance to ultrasonic physics experiments seems to have been recognized
only recently[7,8,9]. A major problem in ultrasonic measurements at
high microwave frequencies is the necessity for flat and parallel
transducing and reflecting surfaces. This has made it almost impossible to work with bonded samples at frequencies higher than 10GHZ.
However, the problem is obviated by utilizing TRUE since, to the
extent that the interaction is isotropic, the time reversal property
implies that regardless of wave front distortion in the propagation
paths the original wave front is reconstructed and the echo arrives
back at the transducer in phase over the whole surface. In practice
the interaction is never completely isotropic; nevertheless, as will
be seen in the examples of Sec. III, the improvement obtained in
ultrasonic attenuation experiments can be considerable.

B. Correlation

For the holographic polarization echo, interaction of the forward wave with the pump creates a stored charge pattern. If the pump
pulse is short (ideally a δ function), the charge pattern is a
replica of the forward wave (first pulse). Phase, frequency, and

amplitude modulation are stored. The stimulated echo is then the
cross correlation of the first and third pulses. Stated another way,
the response of the stored distribution to the third pulse is that
of a matched filter for a replica of the first pulse, and thus provides a matched transducer for conversion of ultrasonic input to
electrical output, and vice versa. The efficiency of this charge
grating transducer can be considerably larger than that of a piezoelectric transducer, and we have measured conversion efficiencies as
high as 10%. Furthermore, the grating is electronically variable
and so can be matched to any input.

C. Signal Averaging Through Integration

In holographic echoes the spatial phase of the stored charge
pattern is equal to the phase difference between the carriers of the
first and second pulses. If this phase difference is maintained constant from one pulse pair to the next, then the second grating will
add to the first. In this way large gratings may be integrated up
from small signals to provide a signal averaging function.

III. APPLICATIONS

For all the applications discussed below the relevant experimental geometry is shown in Fig. 1. The part labeled crystal may
be a single piezoelectric crystal which also is the active nonlinear
material, a piezoelectric transducer bonded to the nonlinear material, or a transducer - test sample - nonlinear crystal combination,
depending upon the specific application. The forward ultrasonic
wave is launched from the transducer face in cavity 1. The pump
pulse is applied to cavity 2, and the 2-pulse echo is received in
cavity 1. The third pulse may be applied to either cavity 1 or
cavity 2 and the echo detected in cavities 2 or 1, respectively.

A. Use of Unpolished or Non-planar Transducer

According to Sec. II-A the received echo is in phase over the
transducer surface, therefore it is not necessary to use a flat polished transducer. Fig. 2 shows an ultrasonic spin echo spectrum
obtained in an MgO crystal. The transducer consisted of ZnO powder
suspended in G. E. 7031 varnish and painted onto the surface of the
MgO[10]. Comparison of this spectrum with those shown in Ref. 3, for
which a polished transducer was used, shows that while there is some
reduction in signal to noise, there is still a vast improvement over
the usual ultrasonic attenuation technique.

B. Attenuation Measurements

The time reversal of elastic scattering is demonstrated in
Fig. 3, which shows the ultrasonic attenuation spin resonance line
shape of the Fe^{2+} $\Delta M=2$ transition in MgO, as seen by the usual pulse
reflection method and by the TRUE technique. Fe^{2+} ions tend to
cluster in MgO, and this causes the ultrasonic attenuation and dispersion to differ along different propagation paths[11]. The resulting variation of phase velocity causes distortion of the total wave
front at the receiving surface in a reflection experiment. As magnetic field is scanned through resonance, the net alignment of the
wave front with the receiving surface varies, causing large amplitude
variations in the received signal and gross distortion of the line
shape. However, since dispersion is an elastic scattering process
it does not affect the signal observed by the TRUE technique; the
measured line shape agrees with the theoretically predicted one for
Fe^{2+} in MgO.

C. Large Attenuations

One problem that the parametric technique has in common with
direct transmission or reflection measurements is poor signal to noise
in the case of very strong attenuation. However, this may be over-
come by utilizing the integration property of holographic polarization echoes. The integrated charge distribution can be strong enough
so that conversion of electrical power to acoustic power presents a
significant electrical loss to cavity 2. Then it is not necessary
to detect the echo (which will be strongly attenuated); the change
in cavity coupling may be measured directly. This is demonstrated
in Fig. 4 which shows a frequency sweep and cavity dip, with and
without a stored charge grating present.

D. Further Speculations

Using TRUE, it should be possible to make attenuation measurements at high microwave frequencies in the millimeter wave region.
We have had positive results at 18GHZ and are currently exploring
35GHZ and 70GHZ applications.

The possibility of ultrasonic Fourier transform spectroscopy
is also under investigation. This arises from the cross-correlation
property of holographic echoes. It is well known[12] that the cross-
correlation of a chirp with an amplitude modulated, but otherwise
identical chirp, is the Fourier transform of the modulation. In
Fig. 5 we show a Fourier transform obtained by first storing a
chirped signal and then probing it with a chirp modulated by the wave
form in the figure.

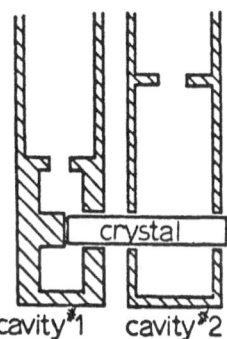

Fig. 1.   Two cavity experimental geometry discussed in the text.

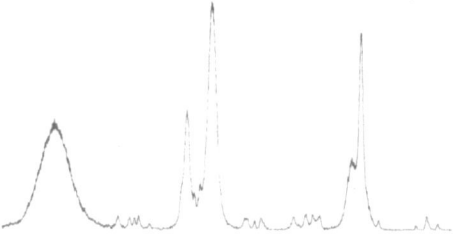

Fig. 2.   Ultrasonic spin echo spectrum of Mn^{2+}, Fe^{3+} and Ni^{2+} in MgO.
          A ZnO paint ultrasonic transducer was utilized.

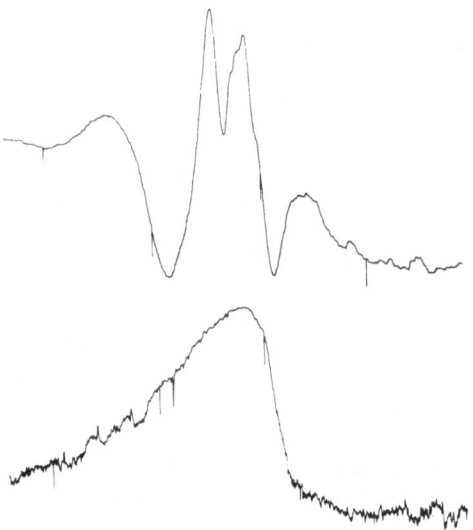

Fig. 3. Fe^{2+} ($\Delta M=2$) ASR attenuation line shape, as observed by pulse
          reflection (above) and TRUE (below).

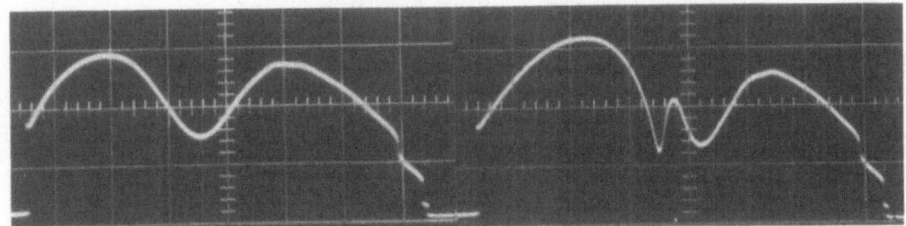

Fig. 4.   Effect of stored charge grating on cavity coupling. Left
          Trace: Klystron mode sweep and cavity dip. Right Trace:
          Same in presence of stored grating.

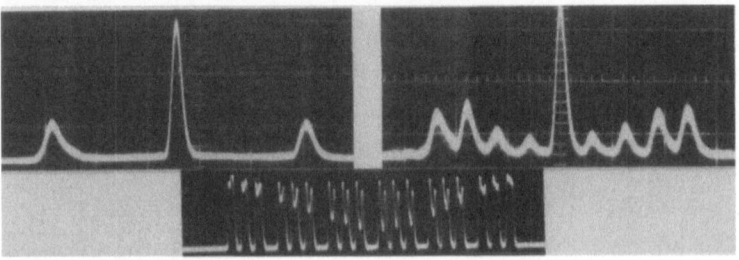

Fig. 5.   Fourier transforms obtained by holographic echo. Left: 8MHz
          square wave. Right: 8MHz square wave gated off at 2MHz rate
          as shown below. Only frequency components between ±8MHz are
          are shown in the transforms.

References

1. R. W. Damon and H. Van de Vaart, Appl. Phys. Lett. 6, 194 (1965).
2. R. B. Thomson and C. F. Quate, J. Appl. Phys. 42, 907 (1971).
3. N. S. Shiren, Phys. Rev. Lett. 28, 1304 (1972).
4. References to this work are listed in the papers cited in refer-
    ences 6 and 8.
5. Subharmonic pumping may also be utilized: R. L. Melcher and
    N. S. Shiren, Phys. Rev. Lett. 34, 731 (1975).
6. N. S. Shiren and R. L. Melcher, Proceedings of the 1974 IEEE
    Symposium in Sonics and Ultrasonics, (IEEE, New York, 1974),
    p. 572.
7. N. S. Shiren, R. L. Melcher, and D. K. Garrod, Bull. Amer. Phys.
    Soc. 19, 297 (1974).
8. J. Joffrin, Proceedings of the Symposium on Microwave Acoustics
    (Institute of Physics, London, 1974) E. R. Dobbs and J. W.
    Wigmore, Ed., p. 23.
9. Envelope time reversal was pointed out in Ref. 2.
10. The authors are grateful to Dr. C. H. Anderson for informing
    them of the use of ZnO paint as a transducer.
11. J. K. Wigmore and H. M. Rosenberg, Phil. Mag. 15, 701 (1967).
12. L. Mertz, Transformations in Optics (Wiley, New York 1965).

X-RAY DIFFRACTION STUDY OF NON-EQUILIBRIUM PHONON

DISTRIBUTIONS IN GaAs

E. Mosekilde, A. Segmuller, and D.G. Carlson
Physics Laboratory III
The Technical University of Denmark
and IBM Watson Research Center
Yorktown Heights, New York, U.S.A.

The application of X-ray diffraction techniques[1-3] to study the spectral distribution of piezoelectrically amplified sound flux has permitted a detailed investigation of linear and nonlinear acoustoelectric effects[4] in the quantum regime. In particular, these experiments have revealed the high frequency cut-off of the acoustoelectric coupling in a very direct manner[2]. This cut-off occurs when the acoustic wavenumber exceeds twice the characteristic electron deBroglie wavenumber, so that the phonon momentum becomes too large to be given up or absorbed by an electron in an energy conserving process.

In this note, an anomalous variation of the acoustic parametric interaction will be discussed. This anomaly, which is a replica of the cut-off of the linear acoustoelectric coupling, gives rise to an enhancement of the acoustic sum frequency generation when the acoustic wavenumber becomes comparable with the characteristic electron deBroglie wavenumber. Experimental results on acoustic sum frequency generation and on the angular distribution of the amplified sound flux are reported.

By virtue of the nonlinear response of the free electron gas, the current density associated with one acoustic wave will contain nonlinear terms arising from the presence of other waves. For the case of three interacting waves this can be expressed as

$$j_3 = \sigma_3 F_3 + \gamma F_1 F_2 \qquad (1)$$

where F_1, F_2 and F_3 are the selfconsistent electric fields accompanying the two pump waves and the sum frequency signal, respectively. $\sigma_3 = \sigma(q_3, \omega_3)$ is the linear conductivity evaluated at the wavenumber and angular frequency of the sum frequency signal. γ is the nonlinear conductivity coefficient to which the efficiency of acoustic sum frequency generation is proportional.

A quantum mechanical calculation of the nonlinear conductivity[3] shows that γ can be expressed in terms of the linear conductivity. For the case of acoustic second harmonic generation

$$\gamma = i(me/\hbar^2 q^3)\left[\sigma(q,\omega) - 2\sigma(2q,2\omega)\right] \tag{2}$$

where m is the effective electron mass and e the elementary charge. If we were to use semiclassical expressions, the difference $\sigma(q,\omega) - 2\sigma(2q,2\omega)$ would be very little, since for a collisionless electron gas the linear conductivity[5]

$$\sigma(q,\omega) \cong -i\omega\varepsilon_o(q_D^2/q^2)(1 - i\sqrt{\pi}yv_s/v_T) \tag{3}$$

varies as q^{-1}. ε_o is the static dielectric constant, q_D the Debye screening wavenumber, v_s the velocity of sound, v_T the thermal electron velocity and $y=(v_d-v_s)/v_s$ the drift parameter, v_d being the electron drift velocity. The electron gas has been assumed to be non-degenerate, the essentials of the problem being independent of the degree of degeneracy.

In the quantum theory, the screening wavenumber g_D is a function of the acoustic wavenumber[6]

$$g_D^2(q) \cong q_D^2(2k_T/q)W\{q/2k_T\} \tag{4}$$

where k_T is the thermal electron deBroglie wavenumber, and $W(x)$ is Dawsons integral. This removes the near cancellation between $\sigma(q,\omega)$ and $2\sigma(2q,2\omega)$ and hereby causes a quantum anomaly in the sum frequency generation. For small acoustic wavenumbers $g_D^2(q) \cong q_D^2$. As q becomes comparable with k_T, $g_D^2(q)$ starts to decrease and approaches zero in the limit of very high acoustic wavenumbers. A transition region then exists in which $g_D^2(q)$ is rather close to its low wavenumber value while $g_D^2(2q)$ has already decreased significantly. In this region, the numerical value of $q(\sigma(q,\omega)-2\sigma(2q,2\omega))$ attains a maximum at $2q \cong 3.2k_T$, i.e., beyond the cut-off of the linear acoustoelectric coupling for the sum frequency signal.

Our experiments on acoustic sum frequency generation were performed at 4o K on a 2o μm thick epitaxial single crystal of GaAs grown onto the (1oo) face of a semiinsulating GaAs substrate. The free carrier density of the active layer was $4.2 \cdot 10^{16}$ cm^{-3}. A 12 mm long sample was cut along the [o11]-direction in which there is a strong piezoelectric coupling to acoustic shear waves polarized along the [1oo]-direction. 5 μsec voltage pulses were applied to the sample with an amplitude ajusted such that a slight acoustoelectric current saturation could be observed. This indicates the amplification of a certain spectrum of thermal acoustic lattice vibrations by 4-6 orders of magnitude.

A beam of CuKα$_1$ X-rays was incident on the sample, and the scattered X-ray intensity in the vicinity of the (4oo) Bragg peak was recorded. By rotating the crystal through a small angle Ω of the order of o.5° while keeping the directions of the incident and the scattered X-ray beams constant, a region of reciprocal space corresponding to acoustic frequencies up to 16o GHz was scanned with a frequency resolution of about 5 GHz.

In Fig. 1, curve A shows the form of the normal Bragg peak whereas curve B shows the variation of the scattered X-ray intensity with Ω as measured during the voltage pulses. The difference between the two curves represents the excess scattered X-ray intensity due to the presence of the sound flux. Two shoulders are visible on curve B corresponding to two distinct peaks in the acoustic spectrum. The low frequency peak (Ω = o.17°) approximately occurs at the maximum for the linear acoustoelectric coupling. The second peak (Ω = o.32°) is ascribed to sum frequency generation from the linearly amplified sound flux and explained in terms of the above discussed enhancement of the acoustic parametric interaction in the quantum regime. The nonlinear character of this peak was established from the fact that the intensity of the peak varies almost proportional to the square of the intensity of the fundamental peak.

A detailed comparison[3] shows a very good agreement between the observed acoustic spectrum and that expected from the linear theory of acoustoelectric coupling supplemented by the quantum theory of acoustic parametric interaction. In contrast to the results obtained with light scattering for the classical collision dominated regime[4], a pronounced downshift of the acoustic spectrum at high flux levels has never been observed in the X-ray experiments. It seems that the frequency downshift in

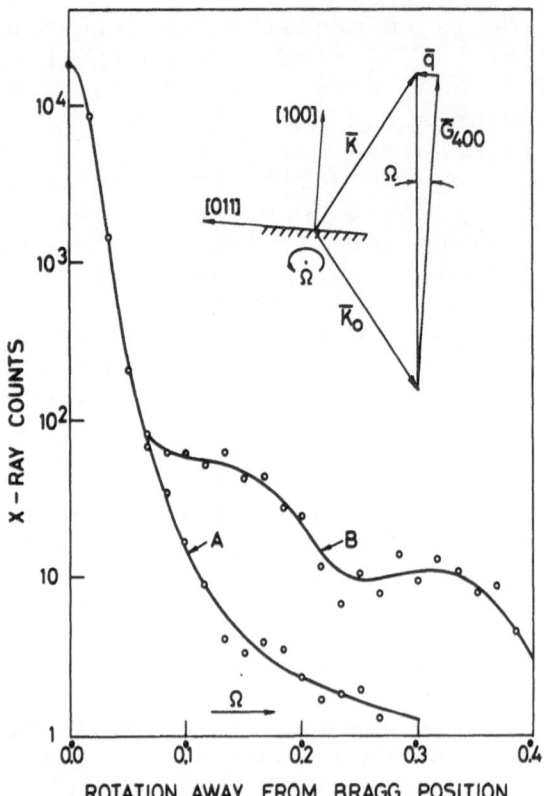

Fig. 1. The form of the GaAs 4oo Bragg peak (curve A) is distorted (curve B) in the presence of the strongly amplified sound flux. The insert shows the momentum matching condition for a one-phonon scattering process.

the collision dominated regime requires not only parametric interactions to efficiently convert the energy between the various acoustic modes but also nonlinear bunching effects[4] to direct the flow of energy towards lower frequencies. For the collisionless electron gas in our experiments, nonlinear bunching effects are expected to be less significant.

Fig. 2 shows the angular distribution of the piezoelectrically amplified sound flux for a GaAs sample with a free carrier density of $2.0 \cdot 10^{17}$ cm^{-3} at 3o K. These results were obtained by performing a two-dimensional scan in reciprocal space around the 511 reciprocal lattice point. The scattering angle was preset to a number of values around that corresponding to the Bragg peak, and for each of these values a scan perpen-

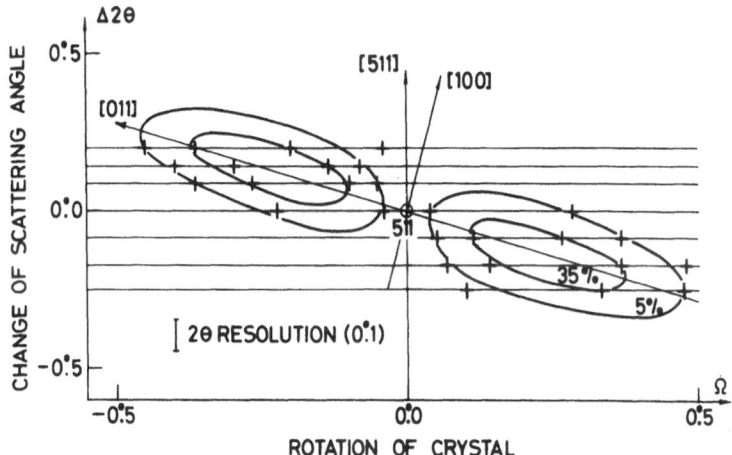

Fig. 2. Contour curves showing the angular distribution of the piezoelectrically amplified sound flux. The curves correspond to 5 and 35 per cent of the peak acoustic intensity, respectively.

dicular to the $\left[511\right]$-direction was performed by rotating the crystal about the $\left[01\bar{1}\right]$-axis. As expected, the intense sound flux is confined to directions close to the $\left[011\right]$-direction in which the drift field is applied.

1. CARLSON D.G., SEGMÜLLER A., MOSEKILDE E., COLE H. and ARMSTRONG J.A., Appl. Phys. Lett. 18, 33o (1971)

2. CARLSON D.G. and SEGMÜLLER A., Phys. Rev. Lett. 27, 195 (1971)

3. MOSEKILDE E., CARLSON D.G. and SEGMÜLLER A., J. Phys. C Solid State Phys. 7, 4281 (1974)

4. For extensive references on linear and nonlinear acoustoelectric effects see MEYER N.I. and JØRGENSEN M.H., Adv. Solid State Phys. 1o, 21 (197o)

5. SPECTOR H.N., Phys. Rev. 165, 562 (1968)

6. GUREVICH V.L. and KAGAN V.D., Sov. Phys.- Solid State 4, 1788 (1963)

TERAHERTZ-PHONON PULSE GENERATION BY LASER

O.Weis

Institute für Angewandte Physik, Universität Heidelberg

69 Heidelberg, Albert-Überle Str. 3-5, Germany

In this paper a survey is given on recent experiments[1,2] and calculations[3] concerning the excitation of short phonon pulses in the
THz-frequency range by means of far infrared laser pulses. As we
will see, this new technique allows to generate (spatial) coherent
THz-phonons in X-quartz and to observe lattice dispersion directly
by observing the frequency dependence of the group velocities of
these phonon pulses. Since we will be concerned with coherent as well
as with incoherent phonon radiators the difference is explained in
Fig. 1 for an excitation at the X-face of a quartz crystal. If the
phonon excitation takes place with equal phase over the radiator
area we have a special kind of a coherent radiator which emits a
highly collimated beam of phonons (in analogy to laser pulse emission in optics), whereas a statistically independent emission of
phonons from adjacent radiator elements leads to an incoherent phonon
emission, i.e. to an emission into all directions (in analogy to
photon emission from a flash lamp). For incoherent phonon radiation
a calculation of the phonon irradiance of the detector demands the
detailed knowledge of the whole anisotropic phonon propagation in
the crystal and the knowledge of the special q-space source distribution in the crystal near the radiator interface. One advantage of
coherent phonon pulses consists in the fact that the whole phonon
pulse energy remains concentrated in a small volume and can be transmitted over large distances (assuming no interactions), for example
to the detector.
Mostly, coherent phonons ('hypersound') are generated in the GHz range
by piezoelectric surface excitation and are detected coherently too
using the wellknown pulse-echo method. Such experiments were performed
up to 114 GHz [5]. The use of far infrared lasers as electromagnetic
source and of incoherent phonon detectors like superconducting tunnel
junctions or bolometers was already proposed[6] in 1966.

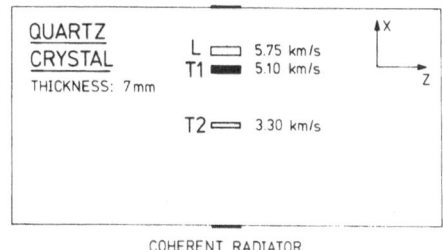

COHERENT RADIATOR

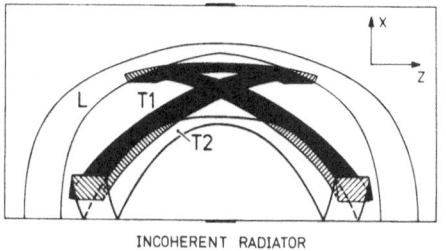

INCOHERENT RADIATOR

Fig.1: Phonon-pulse emission from a spatial coherent radiator and from an incoherent radiator into an X-cut quartz. The positions of the emitted phonons 1 μs after emission are indicated assuming free ballistic propagation. The X-axis in quartz is a pure mode axis for all three phonon polarizations, i.e. group velocity and phase velocity are the same in this direction.
Quartz thickness: 7 mm, radiator diameter: 1 mm, phonon pulse length: 50 ns.

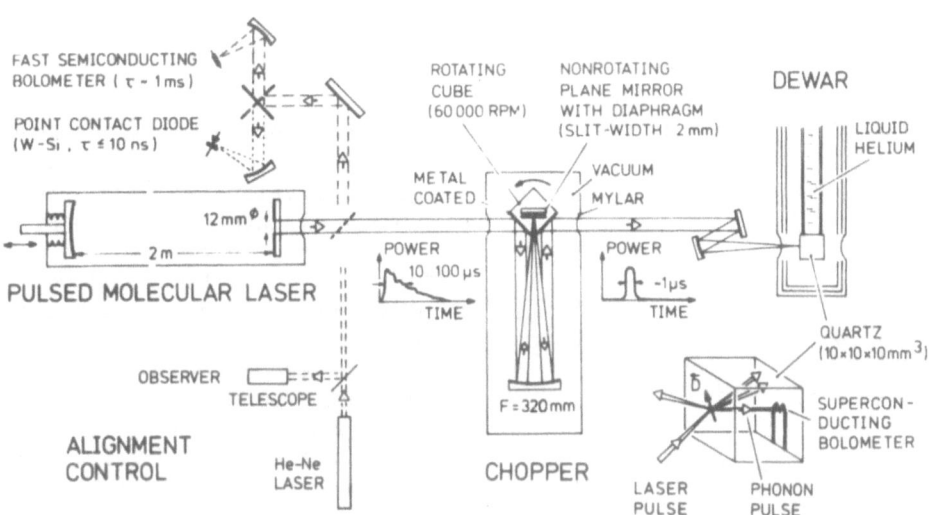

Fig. 2: Experimental apparatus for the generation of 1 μs phonon pulses by means of a chopped far infrared laser radiation, focussed onto the X-cut face of a cooled quartz cube. Phonon detection is done at the rear side using an evaporated superconducting tin bolometer which covers 1 mm^2 and has the shape of a meander (metal-stripe width and free space between: 0.1 mm). Typical peak-power outputs of the laser are 10 W at 0.891 THz (HCN-gas), 10 W at 2.53 THz (H_2O) and 50 W at 10.7 THz (H_2O). Spectral purity and wavelengths are controlled by a compact Michelson interferometer (not shown).

In our first experiments[1,2] on piezoelectric surface excitation of
coherent THz-phonons we used the experimental apparatus shown in
Fig.2. Short far infrared laser pulses are focussed onto the X-face
of a 10 x 10 x 10 mm^3 quartz cube by means of a spherical mirror. Since
no electromagnetic resonator is used for rising the field strength
the efficiency for transforming the electromagnetic power into hypersound via piezoeffect is quite small (-60...-70 dB), but this arrangement has the advantage that it is possible to move the spot of phonon
excitation over the crystal face. This allows new experiments exploi-
ting the strong collimation of the generated coherent phonons.
In order to get an information about the expected power conversion
and about the polarization of the excited phonons for different crystal faces, angles of incidence and electromagnetic polarizations
the complicated boundary problem was investigated and numerically
solved for special cases[3]. Results for the phonon excitation at the
X-face are presented in Fig.3. At normal incidence only the excitation
of fast and slow transverse phonons is to be expected. Contrary, at
a Y-face mainly longitudinal phonons are generated with the same
efficiency.
The bolometer signals observed in our experiments are given for a
laser frequency of 0.891 THz in Fig.4, for 2.53 THz in Fig.6 and for
10.7 THz in Fig.8. Fig.5 represents for comparison the bolometer
signal of an incoherent thermal radiator using the same crystal geometry. Besides the direct laser pulse of 1 μs width a broader phonon
pulse is observed at 0.891 THz. The high collimation of these phonons

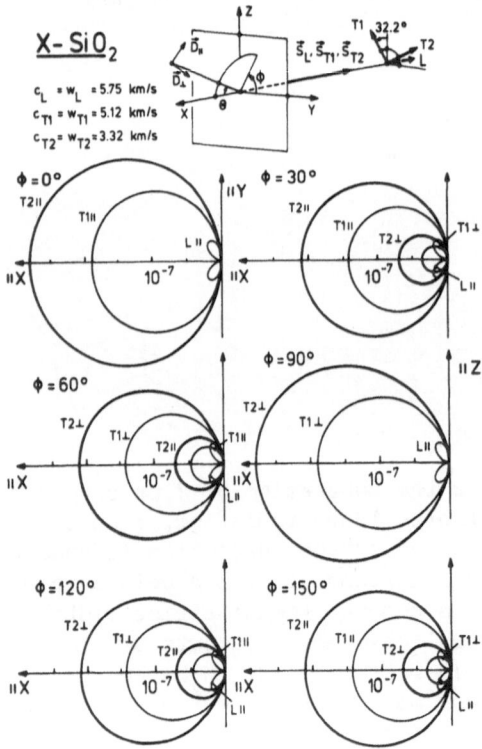

Fig. 3: Power conversion from an
incident electromagnetic wave of
given polarization and angle of
incidence into hypersound at the
X-face of a quartz crystal within
validity of continuum acoustics.
Phase velocity $c_6(\vec{e}_1)$ and group
velocity $w_6(\vec{e}_1)$ coincide in this
cut for each of these three
generated sound waves. As a
consequence of the large difference between sound velocities
and the velocity of light only
sound waves can be generated with
a wave vector normal to the crystal
face independent of the angle of
incidence of the exciting electromagnetic plane wave.

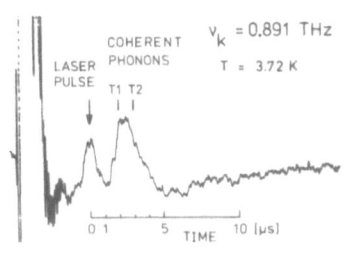

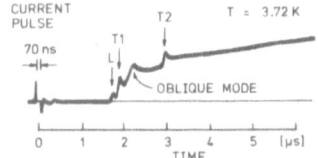

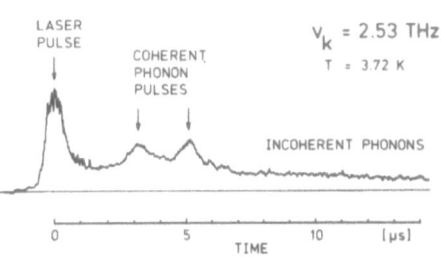

Fig.4: Bolometer signal of surface excitation at 0.891 THz showing the laser pulse and the coherent phonons. The trace is recorded by a boxcar integrator. The expected positions of fast and slow transverse phonons are merked assuming validity of continuum acoustics.

Fig.5: Bolometer signal from an incoherent radiator (thermal phonon radiator) at an X-cut quartz taken from the oscilloscope. The position of the oblique mode is predicted by the lower diagram of Fig.1 .

Fig.6: Bolometer signal of phonon generation at 2.53 THz. Note that the time of flight of the coherent phonon pulses is much longer than those of the 0.891 THz phonons of Fig.4 resulting in a clear resolution of both pulses. The shapes correspond to that of the exciting laser pulse.

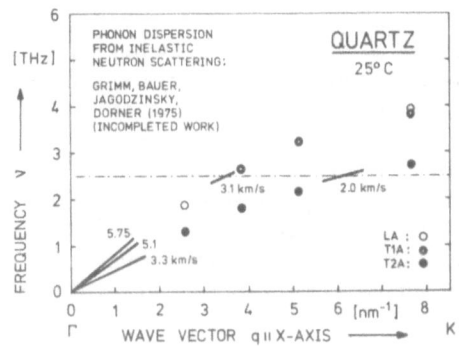

Fig.7: Comparison of phonon dispersion along the X-axis in quartz with the slopes due to the observed group velocities of 3.1 and 2.0 km/s at 2.53 THz. The q-positions of the measured tangents are chosen for best fit to the points measured by inelastic neutron scattering. The indicated low frequency slopes are taken from continuum acoustics.

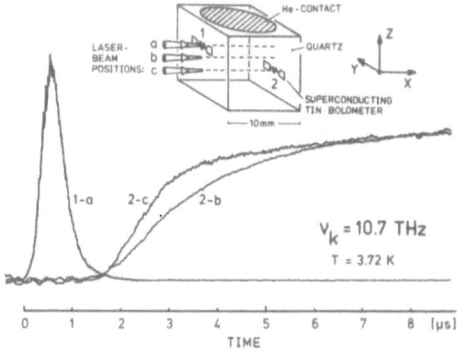

Fig.8: Bolometer signals of laser generated phonons at 10.7 THz. The three signal traces are taken with different positions of laser focus and of bolometer as shown in the inset. Phonons are also generated in laser position a) by direct infrared absorption in the free space between the bolometer stripes. Note that these phonons contribute to the short tail in trace 1-a, but give practically no signal at later times.

can be proved by moving the focal spot of the laser beam over the
crystal face and observing at the same time the bolometer signal.
An enhancement of directional phonon intensity can of course also
occur with incoherent phonons as a consequence of 'phonon focussing',
but this possibility can be excluded in this case as experiments and
detailed numerical studies reveal. Since the exciting laser pulse is
still too long and the strong laser-power fluctuations are too much
disturbing, the time-of-flight of the different phonon groups can
not be determined from Fig.4. Note the difference to the incoherent
phonon signal of Fig.5 where a strong ramp occurs which may be due
to scattered incoherent phonons and eventually to some extent also
to phonon dispersion since these measurements were made with a broad-
band Planck spectrum centred at 1.2 THz. The observed bolometer trace
using 2.53 THz laser excitation gives two well resolved coherent phonon
pulses which have a much longer time-of-flight as those at 0.891 THz.
As Fig.7 shows, these group velocities are within experimental error
in accordance with the results of inelastic neutron scattering[7]. Due
to this diagram the first phonon pulse can be identified by phonons
of the T1A-mode with a wavelength of 1.8 nm and the second coherent
pulse is due to T2A-phonons with a wavelength of 1.0 nm! From these
experiments it is evident that the method of piezoelectric surface
excitation allows to study in future in detail the propagation and
interaction properties of phonons having a wave vector far in the
Brillouin zone.
On the other hand, direct absorption of infrared photons in the bulk
produces incoherent phonons which can be detected by the tin bolometer too. At 2.53 THz the weak but finite bulk absorption of photons
gives rise to the generation of acoustic phonons of half the frequency
and opposite wave vector in a two-phonon process. The incoherent phonon background in Fig.6 is due to these phonons. Fig.8 shows the results at 10.7 THz where many optical branches exist in quartz and
photon absorption becomes so strong that practically all photons are
converted into phonons within a very thin layer near the surface. The
recorded bolometer traces show that for a relative long time no homogeneous diffuse phonon field appears in the crystal after photon absorption. This may be explained by a relative long life time of the
generated phonons in combination with an extremely small group velocity.

References:
1 W.Grill, O.Weis, Satellite Symposium of the 8th International Congress on Acoustics on 'Microwave Acoustics', Lancaster 1974, E.R.
   Dobbs and J.K.Wigmore (Editors), Institute of Physics, p. 179,
2 W.Grill, O.Weis, Phys. Rev. Letters 35 (1975), 588,
3 O.Weis, Z. Physik B 21 (1975), 1,
4 H.Bömmel, K.Dransfeld, Phys. Rev. Letters 1 (1958), 234,
5 J.Ilukor, E.H.Jacobsen, Science 153 (1966), 1113,
6 J.Ilukor, E.H.Jacobsen, in 'Physical Acoustics' (W.P.Mason ed.)
   Vol. V , p. 221, Academic Press, New York, 1968,
7 H.Grimm, K.H.W.Bauer, H.Jagodzinsky, B.Dorner (incompleted work).

DISCUSSION ON NEW TECHNIQUES

High Frequency Phonon Emission from Superconducting Al Tunnelling
Junctions   W Eisenmenger   Page 400

A F G Wyatt: As the generator is essentially normal then this
system could be used in a high magnetic field.

W Eisenmenger: That is correct, however in very thin films superconductivity exists even in the kilogauss range. Experimentally
the superconducting properties are important for establishing a
well defined voltage in the films.

W Bron: I do not want to make a race out of it but we have phonons
at 1.34 THz.

W Eisenmenger: Very good.

****

The Reversed Ultrasonic Echoes as an Experimental Tool
N S Shiren and R L Melcher   Page 405

W C Overton Jr: In holography your third pulse samples the grating
produced by the second pulse. The backward travelling echo then
comes off as the first order diffraction from the grating. There
is also a second order diffracted signal from the grating that
contains information. Can you comment on this signal?

N S Shiren: I have not considered the problem of detection of the
second order wave and, offhand, I can think of no way to observe it.

D Pohl: What is the upper frequency limit for this technique?

N S Shiren: In the case of the holographic storage there are two
limiting factors (1) The wavelength must be longer than the separation of the trapping centres. (2) The local electric field
produced is proportional to the wave vector so at very high frequencies the field is large enough to cause tunnelling of charge
out of the traps.

X-Ray Diffraction Study of Non-Equilibrium Phonon Distributions
in GaAs   E Mosekilde, A Segmuller and D G Carlson   Page 411

T Ishiguro: I wish to make a comment on the derivation of acoustic
intensity from the observed X-ray scattering intensity. As I
presented in this conference, there is a problem with the assumption in the formula used and we found that some down shift in
frequency appears in the acoustic intensity spectra.

E Mosekilde: Even if one considers your proposal for interpreting
the acoustic spectrum from the excess scattered X-ray intensity,
a pronounced downshift will not result from it. However, I think
that the relatively wide angular spread of the flux in the plane of
the epitaxy means that your proposed corrections are exaggerated.

****

Terahertz-Phonon Pulse Generation by Laser   O Weis   Page 416

F Persico: Could you tell me the results which lead you to believe
that the pulses are really coherent?

O Weis: There are several reasons. 1. There is no time broadening
i.e. we get a 1μs pulse of phonons from a 1μs laser pulse. 2.
There is only defocussing along the x direction in the quartz but
the phonons are highly collimated; we see this when we move the
excitation spot around on the crystal face and the detector signal
varies accordingly. 3. The velocity of the phonons is given by
the group velocity found from the dispersion (ω–k) curve. I only
claim that the phonons are spatially coherent i.e. we excite wave
vectors in only one direction. I have not said whether they are
monochromatic or not.

K Dransfeld: You observe two sound waves, one you ascribe to
piezoelectric generation and the other to parametric decay of
photons in the crystal. The former signal should be proportional
to the laser power but the latter to the square of the power.
Have you observed such dependences?

O Weis: We use a discharge laser and it has instabilities which
change the output power so we have not measured the power dependence.

J K Wigmore: With the bolometer on the exciting surface do you
see any evidence for surface wave generation?

O Weis: Up to now we have not observed experimentally any effect
which can be attributed to the generation of surface waves.

H Maris: The phonons you are generating are of very short wavelength i.e. 10 or 20Å. It seems to me that it may be possible that

at these wavelengths, materials which are not microscopically piezoelectric may exhibit a "surface piezoelectricity". One might expect
that the generation efficiency would be down by a factor of a/λ or
$(a/\lambda)^2$ compared to a microscopic piezoelectric. (a = lattice
parameter).

O Weis: All our calculations of piezoelectric surface excitation
were done within continuum acoustics. At higher frequencies a
lattice dynamical treatment is necessary and I also believe that
a surface excitation might be possible even in nonpiezoelectric
material. In practice one has to take into account that the first
atom layers differ in structure from that of the bulk material.
This will also strongly influence the surface excitation of very-
high frequency phonons.

D Pohl: Would it be possible to drive the acoustic waves contin-
ously using laser radiation of lower power?

O Weis: This would be possible. But mostly there will exist an
uncertainty whether an observed signal in a CW-experiment is due
to the generated phonons or to the exciting laser radiation or
even comes from decaying phonons. Moreover, the directions of wave
vector and of polarisation of the phonons are in this case not well
defined.

J R Fletcher: How important is the quality of the surface polish?
Is it necessary for the surface to be flat and smooth on the scale
of 10 Å?

O Weis: The samples of quartz came from Brazil and are optically
polished. It is important to avoid surface contamination from the
residual gas in the cryostat.

LIST OF DELEGATES

Dr M Abou-Ghantous | Physics Department, Faculte des Sciences, Universite Libanaise, Al-Hadet, Beirut, Lebanon.

Professor H J Albany | CEN Saclay, 91190 Gif-sur-Yvette, France.

Dr C Alquie | Laboratoire d'Electricite Generale, Ecole Superieure de Physique et de Chimie, 10 Rue Vauquelin, 75005 Paris, France.

Professor A C Anderson | Physics Department, University of Illinois, Urbana, Illinois 61801, U S A.

Dr C H Anderson | RCA Laboratories, Princeton, New Jersey 08540, U S A.

Dr W Arnold | Max-Planck Institut fur Festkorperforschung, BP 166, Avenue des Martyrs, F-38 Grenoble, France.

Dr B Barnett | Soreq Nuclear Research Centre, Yavne, Israel.

Dr H W T Barron | Natural Philosophy Department, Aberdeen University, Aberdeen, Scotland.

Dr C A Bates | Physics Department, Nottingham University, University Park, Nottingham NG7 2RD, U K.

Dr W Bausch | Fachhochschule, 61 Darmstadt, Schoeffer Str, West Germany.

Dr H Beck | Institut fur Theor. Physik der Universitat, Schonberggasse 9, 8001 Zurich, Switzerland.

Dr G Bellessa | Laboratoire de Physique des Solides, Batiment 510, Faculte des Sciences, 91405 Orsay, France.

Dr R Berman	Clarendon Laboratory, Parks Road, Oxford, OX1 3PU, U K.
Mr L Bjerkan	Fysikkseksjonen Nth, 7034 Trondheim, Norway.
Professor W E Bron	Physics Department, Indiana University, Bloomington, Indiana 47401, U S A.
Dr M A Brown	Physics Department, Loughborough University of Technology, Loughborough, Leics, U K.
Dr R Brown	Macquarie University, North Ryde, NSW 2113, Australia.
Professor R Buisson	Laboratoire de Spectrometre Physique, Domaine Universitaire, BP 53, 38041 Grenoble Cedex, France.
Dr F Canal	15 Rue Alsace-Lorraine, 31 Toulouse, France.
Dr C M Care	Physics Department, Sheffield University, Sheffield, S3 7RH, U K.
Professor P Carrara	Universite Paul Sabatier, 118 Route de Narbonne, 31077 Toulouse, France.
Professor L J Challis	Physics Department, Nottingham University, University Park, Nottingham NG7 2RD, U K.
Dr J M Chamberlain	Physics Department, Nottingham University, University Park, Nottingham NG7 2RD, U K.
Professor J D N Cheeke	Department de Physique, Universite de Sherbrooke, Sherbrooke, J1K 2RI, Quebec, Canada.
Dr J G Collins	National Measurement Laboratory, CSIRO, Sydney 2008, Australia
Mr G N Crisp	Physics Department, Nottingham University, University Park, Nottingham NG7 2RD, U K.
Dr M T Cross	c/o Physics Department, Nottingham University, University Park, Nottingham NG7 2RD, U K.
Dr J Daubert	Institut fur Kernphysik der Universitat Frankfurt/Main, August Euler Str 6, West Germany.
Professor J G Daunt	Cryogenics Center, Stevens Institute of Technology, Hoboken, New Jersey 07030, U S A.

Mr J Y Desmons	Centre Universitaire de Valenciennes, 59326, France.
Ms N Devismes	DTCE - SBT, CENG, BP 85, Centre de Tri, 38041 Grenoble-Cedex, France.
Dr W Dietsche	Physik Department E10, der T U Munchen, D-8046 Garching, West Germany.
Dr G S Dixon	Physics Department, Oklahoma State University, Stillwater, Oklahoma 74074, U S A.
Mr J Doulat	CENG, BP 85, Centre de Tri, 38041 Grenoble Cedex, France.
Professor K Dransfeld	Laboratoire de Champ Intense, Avenue des Martyrs, Grenoble, France.
Mr J Dykhuis	Solid State Department, Fysisch Laboratory, Sorbonnelaan 4, Utrecht, Netherlands.
Professor W Eisenmenger	Physikalisches Institut, Teil 1, D-7000 Stuttgart 80, Pfaffenwaldring 57, West Germany.
Mr W Eisfeld	Universitat Regensburg, Fachbereich Physik, D-84 Regensburg, Universitatsstr 31, West Germany.
Dr J R Fletcher	Physics Department, University of Nottingham, University Park, Nottingham NG7 2RD, U K.
Mr W Forkel	Physikalisches Institut, Teil 1, D-7000 Stuttgart 80, Pfaffenwaldring 57, West Germany.
Dr D Fortier	Laboratoire de Physique des Materiaux, Service d'Electronique Physique, CEN Saclay, 91190 Gif-sur-Yvette, France.
Dr G A Gehring	Department of Theoretical Physics, Parks Road, Oxford, U K.
Professor A N Gerritsen	Physics Department, Purdue University, West Lafayette, Indiana 47906, U S A.
Dr S K Ghatak	Groupe des Transitions de Phases, CNRS, BP 166, Centre de Tri, 38042 Grenoble, France.
Mr A A Ghazi	Physics Department, University of Nottingham, University Park, Nottingham NG7 2RD, U K.

Dr M Givon	Solid State Physics Department, Soreq Nuclear Research Centre, Yavneh, Israel
Dr M Goda	Faculty of Engineering, Niigata University, Nagaoka, 940, Japan.
Dr A-M de Goer	Service des Basses Temperatures, CENG, BP 85, Centre de Tri, 38041 Grenoble-Cedex, France.
Dr J M Goodwin	California State College, Stanislaus, Turlock, California 95380, U S A.
Mr Z W Gortel	Institute of Theoretical Physics, Warsaw University, ul Hoza 69, 00-681 Warsaw, Poland.
Professor P Gosar	Institut J Stefan, University of Ljubljana, 39 Jamova, 61000 Lhubljana, Yugoslavia.
Dr J Graebner	Room 1A-132, Bell Laboratories, Murray Hill, New Jersey 07974, U S A.
Dr D Greig	Physics Department, University of Leeds Leeds LS2 9JT, U K.
Mr M Gross	c/o Segalovitz, 146 Boerhavelaan, Leiden, Netherlands.
Professor R Guermeur	Universite de Provence, Department d' Electronique, 13397 Marseille Cedex 4, France.
Dr P Haen	Centre de Recherches sur les Tres Basses Temperatures, CNRS, BP 166, Grenoble Cedex, France.
Dr L Halbo	Central Institute for Industrial Research, Blindern, Oslo 3, Norway.
Dr B Halperin	Soreq Nuclear Research Centre, Yavne, Israel.
Dr R A H Hamilton	Applied Physics & Electronics Department, UWIST, Cathays Park, Cardiff CF1 3NU, U K.
Dr M S Haque	Fachbereich Physik, Universitat Regensberg, 84 Regensburg, West Germany
Dr J P Harrison	Physics Department, Queens University, Kingston, Ontario, Canada
Mr S C Haseler	Physics Department, Nottingham University, University Park, Nottingham NG7 2RD, U K.

Mr B G M Helme	Physics Department, Lancaster University, Bailrigg, Lancaster LA1 4YB, U K.
Dr J A Herb	California Institute of Technology, California, U S A.
Professor Y Hiki	Tokyo Institute of Technology, Oh-Okayama, Meguro-ku, Tokyo 152, Japan.
Dr O H Hughes	Physics Department, Nottingham University, University Park, Nottingham NG7 2RD, U K.
Professor A Ikushima	Institute for Solid State Physics, University of Tokyo, Roppongi, Minato-ku, Tokyo 106, Japan.
Dr T Ishiguro	Electrotechnical Laboratory, Mukodai 5-4-1, Tanashi, Tokyo 188, Japan.
Professor J Jackle	Fachbereich Physik, Universitat Konstanz, D-7750 Konstanz, Postfach 7733, West Germany.
Mr G Jaracs	Villamosipari Kutato Intezet, 1601 Budapest, Cservenka Miklos ut 86, Hungary.
Professor J Joffrin	Laboratoire Ultrasons, Tour 13, 4 Place Jussieu, 75230 Paris Cedex 05, France.
Dr J Jouffroy	Laboratoire Physique de la Matiere Condensee, College de France, 11 Place Marcellin Berthelot, 75231 Paris Cedex 05, France.
Professor L Kaiser	Laboratoire de Physique Statistique, Centre Universitaire de Perpignan, 66000 Perpignan, France.
Dr K Kajimura	Electrotechnical Laboratory, Tanashi, Tokyo 188, Japan.
Mr S Kelham	Clarendon Laboratory, Parks Road, Oxford OX1 3PU, U K.
Professor H Kinder	Physik Department der T U Munchen, 8046 Garching, West Germany.
Dr P J King	Physics Department, University of Nottingham, University Park, Nottingham NG7 2RD, U K.
Dr T Kitchens	Office of Naval Research - London Branch, 223 Old Marylebone Road, London W1, U K.

Professor P G Klemens	Physics Department, University of Connecticut, Storrs, Connecticut 06268, U S A.
Mr A Knowles	Physics Department, University of Nottingham, University Park, Nottingham NG7 2RD, U K.
Dr P Kocevar	Institut fur Theoretische Physik, A8010 Graz, Universitatsplatz 5, Austria.
Mr Y Kogure	Tokyo Institute of Technology, Oh-Okayama, Meguro-ku, Tokyo 152, Japan.
Dr J Kopp	Physics Department, University of the Witwatersrand, Johannesburg, South Africa.
Mr Y Korczynskyj	Physics Department, University of Nottingham, University Park, Nottingham NG7 2RD, U K.
Dr P Korpiun	Technische Universitat Munchen, Physik Department E13, D-8046 Garching, West Germany.
Mr F R Ladan	GPS de l'ENS, 24 Rue Lhomond, 75231 Paris Cedex 05, France.
Dr C Laermans	Physics Department, Kath. Universiteit Leuven, Celestijnenlaan 200D, 3030 Heverlee, Belgium.
Professor J Lange	Physics Department, Oklahoma State University, Stillwater, Oklahoma 74074, U S A.
Dr J-C Lasjaunias	CRTBT-CNRS, Avenue de Martyrs, BP 166, Centre de Tri, 38042 Grenoble, France.
Dr K Lassmann	Physikalisches Institut, Teilinstitut 1, D-7 Stuttgart 1, Pfaffenwaldring 57, West Germany.
Mr P Levinson	Laboratoire Physique de la Matiere Condensee, College de France, 11 Place Marcellin Berthelot, 75231 Paris Cedex 05, France.
Professor J Lewiner	Laboratoire d'Electricite Generale, Ecole Superieure de Physique et de Chimie, 10 Rue Vauquelin, 75005 Paris, France.

Professor P Lindenfeld	Physics Department, Rutgers University, New Brunswick, New Jersey 08903, U S A.
Dr M Locatelli	SBT, CENG, BP 85, Centre de Tri, 38041 Grenoble, France.
Mr N Lockerbie	Physics Department, University of Nottingham, University Park, Nottingham NG7 2RD, U K.
Mr K V Loftus	Clarendon Laboratory, Parks Road, Oxford OX1 3PU, U K.
Dr A R Long	Department of Natural Philosophy, University of Glasgow, Glasgow G12 8QQ, U K.
Dr R A M van Lopik	Physical Laboratory, State University Utrecht, Sorbonnelaan 4, Utrecht, Netherlands.
Professor K Luszczynski	Physics Department, Washington University, St Louis, Missouri, U S A.
Professor I Malecki	IPPT PAN, Swietokrzyska 21, 00049 Warsaw, Poland.
Dr J P Maneval	Physique des Solides ENS, 24 Rue Lhomond, 75231 Paris 5, France.
Professor H J Maris	Physics Department, Brown University, Providence, Rhode Island 02912, U S A.
Professor J J Martin	Physics Department, Oklahoma State University, Stillwater, Oklahoma 74074, U S A.
Mr M Martin	Centre Universitaire de Valenciennes, 59326, France.
Mr M Martinez	Clarendon Laboratory, Parks Road, Oxford, OX1 3PU, U K.
Dr K J Maxwell	Physics Department, University of Nottingham, University Park, Nottingham NG7 2RD, U K.
Dr R Maynard	Service SBT, BP 85, Centre de Tri, 38041 Grenoble Cedex, France.
Dr D J Meredith	Physics Department, University of Lancaster, Lancaster, U K.
Dr F Michard	Department de Recherches Physiques, Tour 22, Universite Pierre et Marie Curie, 4 Place Jussieu, 75230 Paris Cedex 05, France.

Mr I F I Mikhail	Applied Mathematics Department, Faculty of Science, Ein Shams University, Cairo, Egypt.
Dr W S Moore	Physics Department, University of Nottingham, University Park, Nottingham NG7 2RD, U K.
Dr I Morton	Physics Department, University of Southampton, Southampton SO9 5NH, U K.
Dr E Mosekilde	Physics Laboratory III, Technical University of Denmark, 2800 Lyngby, Denmark.
Professor F Moss	Physics Department, University of Missouri, St Louis, Missouri 63121, U S A.
Dr T Nakajima	Research Institute for Iron, Steel & Other Metals, Tohoku University, Sendai 980, Japan.
Dr T Nakayama	Department of Engineering Science, Faculty of Engineering, Hokkaido University, Sapporo, Japan.
Dr H Namaizawa	Institute of Physics, College of General Education, University of Tokyo, Komaba 3-8-1, Meguro-ku, Tokyo, Japan.
Dr V Narayanamurti	Room ID466, Bell Laboratories, Murray Hill, New Jersey 07974, U S A.
Mr S G Oates	Physics Department, University of Nottingham, University Park, Nottingham NG7 2RD, U K.
Dr W C Overton Jr	Los Alamos Scientific Laboratory, P O Box 1663, Los Alamos, New Mexico 87544, U S A.
Mr G J Page	Physics Department, University of Nottingham, University Park, Nottingham NG7 2RD, U K.
Mr J H Page	Clarendon Laboratory, Parks Road, Oxford, OX1 3PU, U K.
Mr B Pannetier	Physique des Solides ENS, 24 Rue Lhomond, 75231 Paris 5, France.
Professor W H Parker	Physics Department, University of California, Irvine, California 92664, U S A.

Dr J E Parrott	Department of Applied Physics & Electronics, UWIST, Cathays Park, Cardiff, CF1 3NU, U K.
Dr M W S Parsons	Physics Department, University of Nottingham, University Park, Nottingham NG7 2RD, U K.
Dr J L Patel	Physics Department, The University, Lancaster, LA1 4YB, U K.
Mr G Pauli	Universitat Regensburg, Fachbereich Physik, D-84 Regensburg, Universitatsstr 31, West Germany.
Dr F Persico	Istituto di Fisica, Via Archirafi 36, Palermo, Italy.
Dr D Petitgrand	Service d'Electronique Physique, CEN de Saclay, 91190 Gif-sur-Yvette, France.
Dr W A Phillips	Cavendish Laboratory, Madingley Road, Cambridge, U K.
Dr L Piche	Centres de Recherches sur les Basses Temperatures, F-38042 Grenoble Cedex, France.
Dr D W Pohl	IBM Forschungslabor, CH-8803 Ruschlikon, Switzerland.
Professor R O Pohl	Physics Department, Clark Hall, Cornell University, Ithaca, New York 14850, U S A.
Dr V W Rampton	Physics Department, University of Nottingham, University Park, Nottingham NG7 2RD, U K.
Dr G Raunio	Physics Department, Linkoping University, S-58183 Linkoping, Sweden.
Mr K Rawlings	School of Physical & Molecular Sciences University College of North Wales, Bangor, Gwynedd, U K.
Professor M Le Ray	Laboratoire d'Hydrodynamique Superfluide, 59326, Centre Universitaire de Valenciennes, France.
Professor J A Rayne	Physics Department, Carnegie-Mellon University, Pittsburgh, Pennsylvania 15213, U S A.

Professor K Renk	Fachbereich Physik, Universitat Regensburg, D-84 Regensburg, Universitatsstr 31, West Germany.
Mr W Reupert	Physikalisches Institut, Teilinstitut 1, Universitat Stuttgart, 7 Stuttgart 80, Pfaffenwaldring 57, West Germany.
Dr J Rivallin	16 Rue de l'Avignon, 78370 Plaisir, France.
Dr S J Rogers	Physics Laboratory, The University, Canterbury, Kent, U K.
Dr H M Rosenberg	Clarendon Laboratory, Parks Road, Oxford OX1 3PU, U K.
Mr M Rueff	Institute of Theoretical Physics, University of Stuttgart, 7000 Stuttgart 80, Pfaffenwaldring 57, West Germany.
Mr B Salce	Centre d'Etudes Nucleaires, Service Basses Temperatures, BP 85, Centre de Tri, 38041 Grenoble Cedex, France.
Dr T Satoh	Physics Department, Faculty of Science, Tohoku University, Sendai 980, Japan.
Mr H Schad	Physikalisches Institut, Teilinstitut 1, D-7 Stuttgart 80, Pfaffenwaldring 57, West Germany.
Mr W Schneider	Institute for Theoretical Physics, University of Stuttgart, 7 Stuttgart 80, Pfaffenwaldring 57, West Germany.
Dr F W Sheard	Physics Department, University of Nottingham, University Park, Nottingham NG7 2RD, U K.
Mr I J Shellard	Physics Department, University of Nottingham, University Park, Nottingham NG7 2RD, U K.
Dr N S Shiren	IBM Watson Research Center, P O Box 218, Yorktown Heights, New York 10598, U S A.
Mr K A Siddiqui	Physics Department, University of Nottingham, University Park, Nottingham NG7 2RD, U K.
Dr J C A van der Sluijs	School of Physical & Molecular Sciences University College of North Wales, Bangor, Gwynedd, LL52 2UW, U K.

Mr B S Smith	Physics Department, University of Lancaster, Lancaster LA1 4YB, U K.
Dr G P Srivastava	Applied Physics & Electronics Department, UWIST, Cathays Park, Cardiff CF1 3NU, U K.
Dr P Steggles	Physics Department, University of Nottingham, University Park, Nottingham NG7 2RD, U K.
Professor K W H Stevens	Physics Department, University of Nottingham, University Park, Nottingham NG7 2RD, U K.
Dr H Suzuki	Physics Department, Faculty of Science, Tohoku University, Sendai 980, Japan.
Dr K Suzuki	Department of Electrical Engineering, Waseda University, 4-170 Nishiokubo, Shinjuku, Tokyo 160, Japan.
Dr T J B Swanenburg	Philips Research Laboratories, Eindhoven, Netherlands.
Professor R L Thomas	Physics Department, Wayne State University, Detroit, Michigan 48202, U S A.
Dr M D Tiwari	Physics Department, Birla Postgraduate College, Srinagar, Garhwal (U P) India.
Dr G A Toombs	Physics Department, University of Nottingham, University Park, Nottingham NG7 2RD, U K.
Dr J W Tucker	Physics Department, Sheffield University, Sheffield S3 7RH, U K.
Mr R Twele	Universitaet Stuttgart, Physikalisches Institut, 1 Teilinstitut, D7000 Stuttgart 80, Pfaffenwaldring 57, West Germany.
Dr R Vacher	Laboratoire de Spectrometre Rayleigh-Brillouin, Universite des Sciences et Techniques du Languedoc, 34060 Montpellier Cedex, France.
Dr M M Vandorpe	CRTBT, CNRS, Chemin des Martyrs, Grenoble 38, France.
Professor G S Verma	Banaras Hindu University, Physics Department, Faculty of Science, Varanasi 5, India.

Mr J P Vigneron	Universite de Liege, Institut de Physique, 4000 Sart Tilman Par Liege 1, Belgium.
Dr D T Vigren	Institut fur Theoretische Physik, Freie Universitat Berlin, Arnim Allee 3, 1 Berlin 33, West Germany.
Mr M Villedieu	Centre d'Etudes Nucleaires, Service Basses Temperatures, BP 85, Centre de Tri, 38041 Grenoble Cedex, France.
Mr J de Vos	Solid State Department, Fysisch Laboratory, Sorbonnelaan 4, Utrecht, Netherlands.
Dr D Waldmann	Physikalisches Institut, Teil 1, D-7 Stuttgart 80, Pfaffenwaldring 57, West Germany.
Professor R Wallis	Physics Department, University of California, Irvine, California 92664, U S A.
Dr G Walmsley	Physics Department, New University of Ulster, Coleraine, Northern Ireland.
Professor H Weinstock	Physics Department, Illinois Institute of Technology, Chicago, Illinois 60616 U S A.
Professor O Weis	Institute fur Angewandte Physik, Albert Ueberle Str 3-5, 69 Heidelberg, West Germany.
Dr K Weiss	Philips Research Wy-3, Eindhoven, Netherlands.
Dr J K Wigmore	Physics Department, University of Lancaster, Lancaster LA1 4YB, U K.
Professor H W de Wijn	Fysisch Laboratorium, Rijksuniversiteit, Sorbonnelaan 4, Utrecht, Netherlands.
Mr D L Williams	Physics Department, University of Nottingham, University Park, Nottingham NG7 2RD, U K.
Dr R Windheim	KFA - Julich, Institut fur Festkorperforschung, D517 Julich, Postfach 365, West Germany.
Dr G Winterling	Max-Planck Institut fur Festkorperforschung, Postfach 1099, D7 Stuttgart, West Germany.

Dr J Wolter

Philips Research Laboratories, Eind-
hoven, Netherlands.

Dr A F G Wyatt

Physics Department, University of Nottingham, University Park, Nottingham
NG7 2RD, U K.

AUTHOR INDEX